CREATIVE NEWSPAPER DESIGN

Titles in the series

Series editor: F. W. Hodgson

Broadcast Journalism Andrew Boyd
The Computerized Newspaper Paul Williams
Creative Newspaper Design Vic Giles and F. W. Hodgson
The Freelance Journalist Christopher Dobson
A Journalist's Guide to Sources David Spark
Journalism Workbook Brendan Hennessy and F. W. Hodgson
Law and the Media Tom Crone
Magazine Journalism Today Anthony Davis
Modern Newspaper Practice F. W. Hodgson
Newspaper Language Nicholas Bagnall
Practical Newspaper Reporting Geoffrey Harris and David Spark
Practical Photojournalism Martin Keene
Printing: A Guide to Systems and Their Uses W. R. Durrant
Subediting F. W. Hodgson
Writing Feature Articles Brendan Hennessy

CREATIVE NEWSPAPER DESIGN

SECOND EDITION

VIC GILES

and

F. W. HODGSON

Focal Press

Focal Press
An imprint of Butterworth-Heinemann
Linacre House, Jordan Hill, Oxford OX2 8DP
A division of Reed Educational and Professional Publishing Ltd

A member of the Reed Elsevier plc group

OXFORD BOSTON JOHANNESBURG
MELBOURNE NEW DELHI SINGAPORE

First published 1990
Second edition 1996

British Library Cataloguing in Publication Data
Giles, Vic
Creative newspaper design – 2nd ed.
1. Newspaper layout and typography
I. Title II. Hodgson, F. W.
686.2'252

ISBN 0 240 51442 4

Composition by Scribe Design, Gillingham, Kent, UK
Printed and bound in Great Britain by The Bath Press

Contents

Acknowledgements

We are grateful to the many publishers and editors in Britain and in abroad whose newspaper pages form a vital part of this second edition of *Creative Newspaper Design*.

In particular we are indebted to David Montgomery, Chief Executive of Mirror Group Newspapers, for allowing us to illustrate with print-outs the creating of a *Daily Mirror* page on screen at their plant at Canary Wharf, and for generous on-site facilities; and to the editor of the *News of the World* and News International plc for permitting photographs of a page during a production run at Wapping. The editors of *The Sun* and *Today* (the last newspaper, alas, no longer with us) both allowed us to reproduce sequences comparing a first rough sketch of a page with the finished product.

Our chapter on design styles owes much to examples of work we were permitted to print from *The Times, The Sunday Times, The Observer, Financial Times, The Independent, Daily Express, Daily Mail, The Guardian, The Scotsman, The Herald (Glasgow), Northern Echo, Liverpool Echo, Manchester Evening News, Wolverhampton Express and Star, Evening Standard, Daily Record, Birmingham Evening Mail, Irish Independent* and *Kent Messenger*. We are likewise indebted to *USA Today* and to the editors of those overseas newspapers of which examples are used towards the end of the book.

On the technical side we were helped by being allowed by Autologic, of California, to use illustrations on the technology of type from their *Digital Type Collection*. Computer Corporation UK Ltd and Quark Systems Ltd allowed us to reproduce diagrams from their manuals to illustrate processes in page design. Peter Henderson, operations manager of Southern Newspapers Ltd, guided us through their state-of-the-art editorial and printing facility during its development at Redbridge, Southampton; Man Roland Geoman provided the picture of the revolutionary new press they have installed there.

Rolf F. Rehe's *Typography and Design for Newspapers* (IFRA, Darmstadt, 1985) was a valuable book to have by us, and we were glad to have permission to reproduce two of his designs from it. We are indebted in retrospect to our predecessors in the field: Alan Hutt (*Newspaper Design*, Oxford University Press, 1960), Edmund C. Arnold (*Modern Newspaper Design*, Harper and Row, New York, 1969), and Harold Evans, (*Newspaper Design*, Heinemann, 1973). The chapter on mastheads derived particular interest from our being allowed by *The Independent* to use some of its pre-launch designs.

Mr Dennis Morris and Mrs Helena Hind helped by lending material from their collections of historic newspapers; Brian Thomas by setting up special pictures for us, and Mirror Group Newspapers by allowing us to use the picture of Harry Guy Bartholomew.

We hope that any other acknowledgements due for the help we have had in researching and writing this book are adequately recorded in the text and captions in the following pages.

Illustrations

Colour plate section

blurb, yet it is the bold white-on-black splash headline that grabs attention first, leaving the eye to move on to the pretty girl in colour then upwards to the package of goodies
(b) The Athens broadsheet on the left, *To Fos* (The Light), is almost surrealist in its approach with a very European saturation blurb technique. *Nice-Matin* uses a more reserved typography to project similar ideas
Plate 3 (a) A page one from *Today* in its three stages of production – first the editor's rough, which is interpreted by the art desk in the second version and, third, the finished page
(b) *France-Soir* page ones are noted for the strategic positioning of colour. *Abendpost*, left, suffers, as do all German broadsheets, with 'portmanteau' words which demand plenty of Extra Condensed type. However the page – helped by effective use of spot colour – cries out to be read
Plate 4 (a) The combination of a striking blue masthead, a dominant full colour picture and use of spot colour create a unique blend of tabloid and broadsheet techniques in *USA Today*. Below: the Snapshot feature produces a down-page eye-catcher using colour drawn from the halftones in the page. Copyright 1995 and 1989, *USA Today*. Reprinted with permission
(b) Electronic pagination at work in colour: a *News of the World* page one in its later stages on an Apple-Macintosh terminal and, below it, the edition as it came off the press. Reproduced with permission

Preface

The five years since *Creative Newspaper Design* first appeared have seen huge changes in newspaper production which the designer cannot ignore. Most important, the use of the Apple-Macintosh computer with Quark XPress and other sophisticated programs have turned the dream of creating newspaper pages totally inside the computer into a reality.

Software now available ensure that types of every sort can easily be installed into editorial systems. The elegance and sharpness of well-known and much loved fonts has, if anything, been refined. Typecutters and founders of the past such as Garamond, Caslon and Bodoni, who gave us varied and beautiful fonts and were thwarted in their day by unsharp edges, pitted metal, doubtful paper and ruinous ink would have appreciated today's digital bitmapping of their artwork. Added to this, the Apple-Mac allows the modern operator to give a creative tweak to familiar typefaces.

But there are other developments, too. Pictures and graphics can now be imaged straight into the page. Movement and adjustment by mouse and keyboard have made possible the editing of pictures and designing of pages wholly on screen, including even the adverts, bringing a paperless newspaper office within the bounds of possibility.

Of course, dangers lie. A function of design is that readers should be comfortable with a simple legible read and a page that has balance and helps the eye. There are signs that some designers, given an instrument programmed to produce brilliant definition with digital dot reproduction and enabling endless variation in type sizing and 'bending', are allowing themselves to be seduced by the software. The disciplines of accurate design built up over the years are in danger of being lost amidst the technology.

The ability to use white space correctly as well as type shows signs of disappearing. The multi-talented journalists who can exploit traditional skills and deal with today's demanding software and see exactly in their mind's eye how a printed end product should look will be difficult to find unless both writing and visualizing are found a place in training programmes.

These factors are taken into account by the authors of this book. Our aim is to teach young journalists how the sophisticated tools now available can be bent, and shaped to the purpose of design while preserving disciplines built up over the years so that we can have even better newspapers.

Vic Giles
F. W. Hodgson
1996

1

The computer and design

The revolution in printing brought about by the computer over the last three decades is almost as fundamental as the invention of printing itself in the Europe of nearly five and a half centuries ago. In the context of the time span, newspapers dating back to the mid-1600s are a relative newcomer to the world of print. Yet it was Gutenberg and Caxton, the printing giants of the fifteenth century who, by the use of movable type and controlled spacing, made possible the features of newspaper design which we take for granted.

It should be remembered by journalist-designers that with all the breathtaking advances in newspaper-making that have taken place, the computer is still only a tool to enable us to do what it has always been possible to do – though what a tool!

In spelling out the fundamentals and practice of newspaper design in today's hi-tech environment, this book examines areas of change and takes account of the aids designers can call upon in their task. It is a huge field and there are complex principles at work which the young journalist needs to understand. To help identify these, this chapter looks first at the electronic methods by which pages are created. They form the essential context to today's design processes.

Cut-and-paste

In replacing hot metal printing (see pages 71-6) the new newspaper systems that began to be used in Britain in the 1970s required text to be keyboarded by writers directly from terminals into a computer whence it could be recalled for checking and editing and eventually for setting in type. The writers' keystrokes – and indeed the advertising department's – thus provided the type which, after amendment, became the newspaper. It was given the name of direct input and it spelt the end of the centuries-old method by which newspapers and other forms of typescript had to be set again in metal type by craftsman printers.

The ultimate aim of the new systems was for pages to be made up on screen from the materials held inside the computer and identified by catchlines, but difficulties over generating the graphics needed for them proved at first hard to overcome. Consequently, make-up was at first carried out by taking the materials out of the computer and using them to make up pages by a process know as cut-and-paste. It is a method still a good deal used and it works as follows.

Edited text and headlines are outputted from the computer on to sheets of bromide paper through photosetters which use high-speed revolving type masters via a cathode ray tube to translate the computer's signals into the required type. The items are sorted by catch-line and cut up and stuck on to the appropriate page cards by operators using hot wax (Figure 1) working from layouts drawn by editorial.

The page cards are replicas of the blank layout sheets on which the pages have been drawn and are graduated in centimetres and marked off in columns in non-reproducible blue ink to guide the paste-up compositor in placing the columns of matter accurately. Where stories need to be ruled off it is done in ruled sticky tape of the required points width from rolls slung on pegs at the top of the make-up frame (Figure 1). The operation is supervised by a production journalist.

The graphics – photographs, line drawings and most of the advertisements – having been edited and retouched, are also printed out

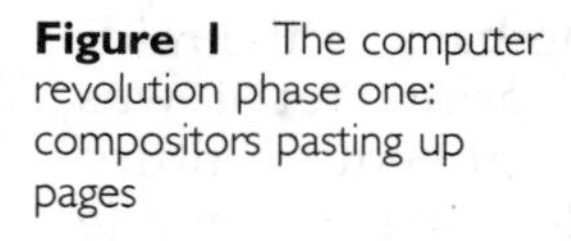

Figure 1 The computer revolution phase one: compositors pasting up pages

on bromide paper by laser printer and are likewise attached by the operator to their cards until the pages are complete. Headline and body setting are automatically spaced in the photosetter but space around text and pictures can be adjusted, if need be, by the compositor using a scalpel.

After being photocopied for passing by the editor or page executive concerned the made-up pages are photographed to produce negatives from which the printing plates are made.

From the start, cut-and-paste make-up proved both faster and more flexible than hot metal and could accept any design style. It was also cheaper in labour costs. The laser printer (usually an Autokon) used for graphics and 'blown' headlines could give special screens and tints by means of controls; the scalpel could be used to alter or improve printed-out artwork on the page; pictures and artwork could be scalpel-trimmed up to the last minute.

Moreover, the method of photocomposition, which culminated in a page negative being produced, interfaced conveniently with the plate-making requirements of web-offset printing which by the early 1980s was beginning to replace letterpress and the old rotary presses. It thus accelerated the move into the use of full colour both for editorial pages and advertising which had not been possible hitherto because of the nature of letterpress printing.

Screen make-up, or electronic pagination as it came to be known, remained the target of the new systems but it was not until 1992-3 that the problems of graphics generation began to be solved. Numbers of Apple-Macintosh desktop computers using a screen make-up program call Quark XPress, which could either be linked with the office mainframe system or be used in clusters in network, were found to give the speed and flexibility – and the huge memory required – to handle complex layouts on screen and deliver the graphics to the page.

Some problems remained, including some advertising inputs, but by 1994 most national dailies and Sundays had completed their dummy runs and moved into full screen page composition. Cut-and-paste had by no means died, however. The quality broadsheets kept a bit of photocomposition capacity handy for pages for when it was necessary to output blocks of text for pages that for various reasons could only be completed by paste-up.

According to the Newspaper Society, cut-and-paste was still being used in 1995 by more than half of Britain's provincial papers for all or part of their page production. At the same time electronic pagination was increasingly being practised alongside it in most centres, although some offices were using early Page-maker systems that lacked the flexibility of the Apple-Mac solution. In some cases the cost of re-equipping was the problem.

In offices where cut-and-paste prevailed, page production could be hastened by outputting blocks of formatted setting for things such as television programmes and race cards. In others, through

Figure 2 Computer revolution phase two: make-up moves to the screen in this print-out sequence showing the building of a *Daily Mirror* page one. For the *Mirror* as it finally looked turn to Plate 2(a) of the colour plate section. Print-outs by permission of Mirror Group Newspapers

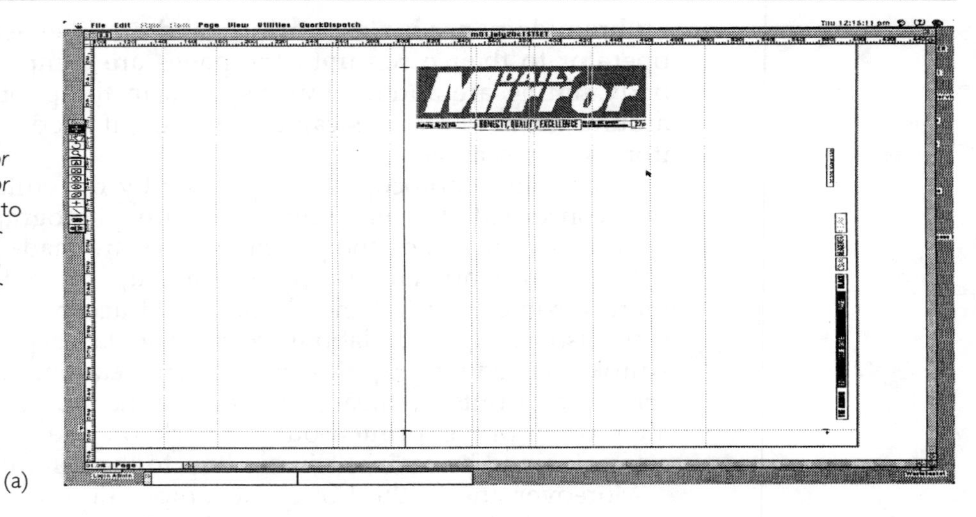

(a)

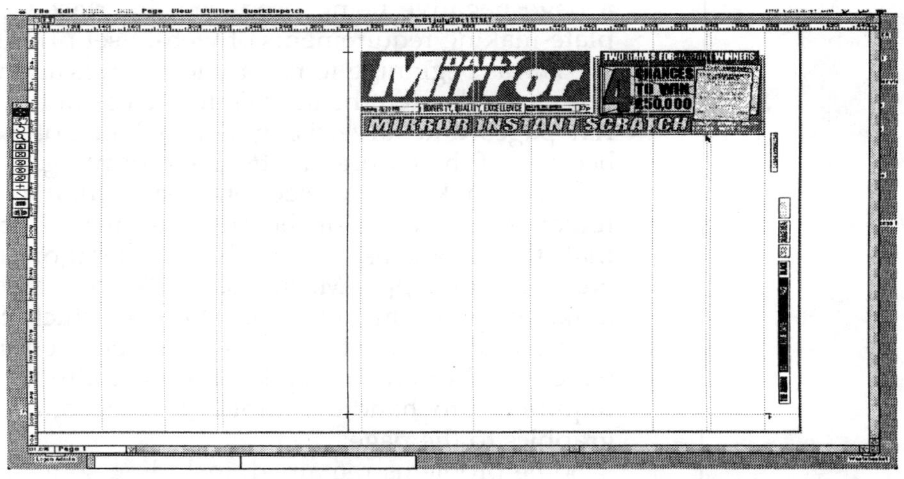

(b)

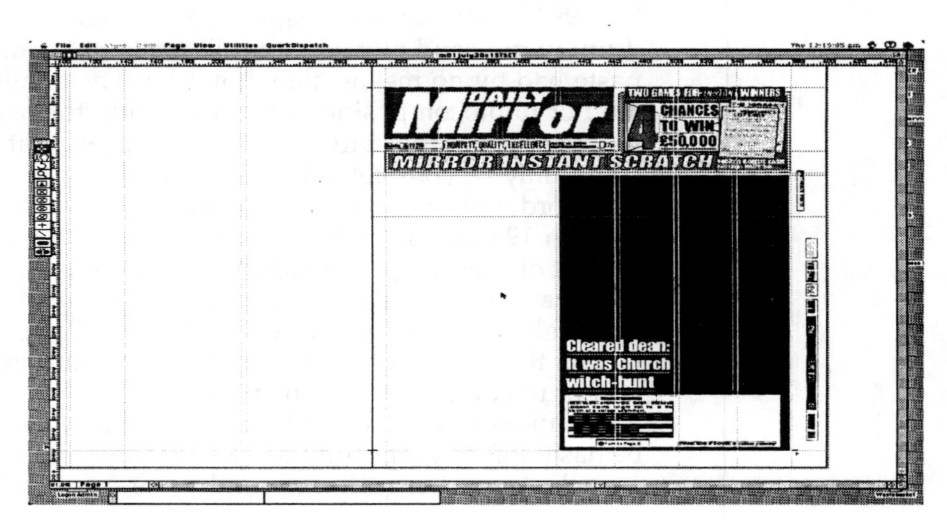

(c)

Figure 2 *Continued*

(d)

(e)

(f)

problems of graphics generation or advertising input, a page might be partly built on screen and then be printed out as a bromide with spaces to take pasted-in materials. It would then be photographed to produce the negative for plate-making. Thus a camera-driven composing room still remained the interface between the editorial department and the presses.

A typical forecast was that it would take another five years (to the year 2000) to totally replace cut-and-paste in Britain's newspaper offices.

Electronic pagination

With page make-up on Apple-Macintosh computers using Quark XPress the human advantages of fast camera and fast scalpel are exchanged for the electronic ones of mouse and keyboard control. Pages on screen are built on to layout grids the same way as with paste-up except that the materials are retrieved and placed electronically.

Using a combination of keyboard and mouse, the page elements – text, headlines, pictures and adverts – are located to the required layout by defining their space on the grid with fine-line boxes into which the items are called by their appropriate catchline. The illustrated sequence of screen print-outs of a page one from the *Daily Mirror* (Figure 2) leading to the edition as it finally appeared, Plate 2(a), shows this process at work.

It might appear that defining by boxes or windows the spaces to be filled on the page would impose a layout style of fixed modules. Not a bit of it. By 'dragging' with the mouse, the boxes can be altered endlessly in shape and size, or even tilted. Layouts can be changed on screen, headlines re-arranged to a variety of shapes at a touch, drop or stand-up letters dragged into position by the mouse with the space and kerning (track) between letters instantly adjusted. Text boxes can be filled with tint or outlined or reversed on to black or colour.

Pictures, either colour or mono, imaged electronically from the scanners into their cross-diagonal boxes, can continue to be worked on on the page. They can be enlarged in image or even adjusted in size by mouse controls if a change of layout requires it. Retouching, if needed, is carried out electronically to editorial requirements before the pictures reach the page.

If an illustration that has arrived on the page is required to be cut out, the mouse control quickly defines the outline and removes the unwanted part of the picture. If needed, at a single stroke type can be run into the irregular space that is left. Equally, type can be made to fill panels of whatever border is nominated and which can even have rounded corners.

Complicated colour blurbs – the life-blood of many a tabloid paper – suffer no restriction in concept through the use of electronic make-up, though they are sometimes prepared separately on split screen or on a slave screen to await calling up to their required space. Pictures can be fitted into the blurb with perhaps a head or foot sticking out for effect. Headlines and slogans can be tilted at any angle, the types rendered in white on black, blue on lilac or whatever colour or tint is preferred.

Effects of electronic make-up

Making up the pages electronically is the logical climax to the utilization of the computer. The main effect is to simplify and speed the editing, page make-up and printing processes once the journalist has learnt to accept the world of screens, modems and keyboards.

The method makes no demands upon reporters and feature writers that were not being made before. In fact, visual display terminals (VDUs) with their editing and correction facility are more user-friendly than typewriters; electronic files are more accessible than drawers of paper; deadlines and rejigs and updates of stories are no more difficult than before.

From the subeditor's point of view nothing has materially changed from the electronic editing already being done under paste-up. The pages still require text and headlines to be prepared and typeset to instructions that come with the page. In fact with electronic make-up a good deal of text automatically formats itself into its setting and box and requires only to be cut to fit.

Moreover, the subeditor (Figure 3) can call up the page on screen or enlarge his or her portion of it to check for fit and position.

Figure 3 Electronic editing: a subeditor using an Atex keyboard and monitor

Adjoining pages can be called up at thumbnail size to check for any clashes with headline type, illustrations or adverts before sending text and headline to the page being worked on.

For the page designer, however, there is a new situation. It is now possible to draw pages directly on to the screen.

This does not mean that paperless production has taken over. The planning of content and the juxtaposing of text, headlines and pictures in relation to the adverts in order to draw the editorial material to the reader's eye still has to be carried out. This entails discussion between page designer and executives concerned and often the production of alternative sketches on paper before the page is set up on screen. A matter of editorial policy might be in issue; a special projection on a features page might require the solving of display problems before the layout can go on-line. Only then can editing instructions be given out and the page components called up.

Sport and some special features pages might be so crowded with fine detail that only a drawn layout on paper can convey placing and type instructions to the subeditors even though the page is being assembled on screen. But broadly, an experienced layout person who is trained in the skills of design and the utilization of the computer program being used, and who understands the paper's style and market, can now design and set up the main run of pages directly on screen once the talking and the planning stage is complete.

As with pages drawn wholly on layout pads, the originating of design and the provision of artwork rests with the art bench or with the journalist designer in charge of the page. Thereafter the page is sent electronically to the chief subeditor and subeditors for 'building' on screen, checking, finishing and for passing on to the plate-makers.

Where cut-and-paste is wholly or partly in use, drawn layouts are circulated as before for the use of the paste-up compositors and the executives and subeditors concerned.

There is no doubt that the spread of electronic pagination and better computer graphics facilities is leading to more on-screen design. It is important that journalists faced with this exciting development are aware of the role of page design in the relationship between newspapers and their readers and of the principles that underpin it. It is to these principles that we now turn.

2
Theory of design

Newspapers are such familiar things that it would probably surprise the average reader to learn that there are complex design factors at work behind the pages. It might be supposed that the news of the day and the features that support it – and the advertisements for that matter – are installed in the paper as they arrive, and that the resulting pages are the arbitrary result of this process, with perhaps a little soling and heeling to get things to fit.

And yet the very familiarity of the pages of a favourite newspaper bears witness to the success of its design. The choice and use of types – were the ordinary reader to examine them more closely – stamp the newspaper unmistakably with a visual character that sets it apart from any other. The size of the headline and position of a certain story denoting it to be the most important item on the page, the readability and easily found page slot of the television programmes, the location of the Parliamentary coverage . . . all these are the clearest evidence of design.

Design thus gives a newspaper its 'feel', attracting the eye while at the same time subtly guiding the reader through the contents of the pages; blending eye comfort and familiarity with a reasonable amount of surprise. In fact, for a newspaper design to be successful, the reader should be unaware of any tricks as such, aware only that the finished product is acceptable and readable and therefore worth buying.

It can be argued that design principles must be hard to identify since there is a great deal of difference between one newspaper and another, not only in the contents but in the appearance. This is true, but it does not alter the basic concept that lies behind newspaper design. Our opening remarks can be used to summarize this concept in the form of three precise functions:

1 To attract and hold the reader's attention.

2 To indicate the relative importance and location of the contents of the pages.
3 To give a newspaper a recognizable visual character.

These functions hold good for any newspaper, national or local; morning, evening or weekly; general or special interest. The differences in appearance that arise, which can be seen from the array of titles on a news-stand, are the result of the way in which the functions of design are used to commend the ingredients of the pages to the particular market of each newspaper. These ingredients, which are the raw material out of which the page design is created, are:

- Advertisements.
- Text.
- Headlines.
- Pictures.

Principles of design

Before we examine the principles of design as expressed in the three functions we have listed, it has to be said that however significant, or entertaining, or motivated a newspaper's contents are it owes its existence to being bought and read on a regular basis. Some newspapers, even some specialist ones, rely on a percentage of casual sales – which is why page one is so important in attracting the eye on the news-stands – but it would be a hand-to-mouth existence for editors were they not able to rely on loyalty and regularity of buying habits among their readers, expressed preferably in copies ordered and delivered. This need to target a newspaper on its readers and potential readers takes us into the first function of design:

To attract and hold the reader's attention

The design, or visual pattern, of a newspaper is intended not only to grab the passing reader, but to persuade existing readers to keep coming back for more. It has been made attractive and desirable in the sort of market the editor is aiming for.

It could be argued against this that some newspapers sell on their content and that design in their case is irrelevant. This is a facile view since it presupposes that other newspapers sell on their design rather than on their content, and there is no evidence – certainly not in readership surveys – that this is so. While ideas used in page design might vary from newspaper to newspaper depending (as will be explained shortly) on the type of readership market, there is the strongest evidence that certain types of content appeal to certain readers, that there is a good deal of consistency in the pattern of newspaper buying, and that it is rooted in content. The detailed breakdown of readership of the main national and provincial newspapers, based on sex, age, education, spending habits and television viewing habits, as checked annually in the *National*

Readership Survey, published by the National Readership Surveys Ltd, points to this, and the survey is used to give each newspaper a readership profile which is greatly valued by the advertisers.

If people tend to buy the sort of newspapers whose views and content they like, where, therefore, do design techniques come in? The answer to this is that in order to attract readers in the first place a newspaper needs to advertise its existence and to establish a public knowledge of what sort of paper it is. Editors are aware that readers are being born, growing up, growing old and dying and that no readership can be taken for granted in the competitive market in which newspapers exist. The number of newspapers that have folded in recent years is evidence of readerships that have faded away or have failed to reach viable levels after launch. An editor must be constantly looking to attract new readers as well as seeking to hold on to existing readers, seeing that content is right and the standard is being maintained, and that the content is being properly drawn to the readers' attention so that they are not persuaded that their interests are better served by changing to a rival newspaper.

There is no doubt that if newspapers were the only medium of mass communication and if there were only one newspaper for a given body of potential readers, the need to project an image would not rate highly. Editors would simply be faced with the need to maximize sales in a closed readership area. They would not have to try more than a moderate amount of persuasion, and there would be no yardstick to measure how worthwhile or successful the product was. It would be like selling water to consumers to whom water is water is water.

But running a newspaper is very different from running a public utility. Since the extension of readership to all classes as a result of the 1870 Education Act and the growth of mass circulations of national daily and Sunday newspapers that followed, competition has been as much a part of newspaper publishing as it has of any other sector of the economy. If some provincial newspapers have been cocooned from this in recent decades by the collapse or absorption of rival titles, then the burgeoning of free newspapers in most readership conurbations came as a sharp reminder that the community does not owe a newspaper a living. There are no solus sites any more. Add to this the competition that comes from television, radio, teletext and news magazines and it can be seen that editors – who are appointed to succeed – have no room for complacency.

In short, behind modern newspaper production, whatever the content, is the urge to create the sort of visual image a newspaper needs in order to become known and to succeed. It is thus that cynics have likened newspaper design to the packaging of products as part of a brand image. Yet, whatever its social, political or cultural influence, a newspaper is a product in that it has to be able to sell in order to survive. It is produced commercially by a management that is answerable to shareholders; it has no subsidy. At whatever level it operates it has to depend on its readers, and its design is part of the means by which its contents are commended to these readers. Upon this rests its circulation.

To indicate the relative importance and location of the contents of the pages

Here the page design, by its use of headline type of varying sizes, maps out the items on the pages. The reader's attention is drawn to the main story on the page, then to the second most important story, and thence through intermediate stories to the smallest items.

In the case of stories or features meriting a big visual display, especially where there is a long text or a wealth of pictures, the choice and size of type is combined with pictures to project the material so that it is comfortable to the eye and inviting to read.

Regular items such as the editorial, or opinion column, the television programmes, or even the horoscopes, are located in a familiar part of the paper, and in a recognizable visual form with perhaps a specially designed logo', that helps the reader to spot them. This familiarity is an important aspect of newspaper design to which we will be returning.

To give a newspaper a recognizable visual character

No two pages of a newspaper (unless it is formularized to an extraordinary degree) ever look alike. The number of ways in which type and pictures can be used to offset each other is so varied as to be almost infinite. Indeed, not to vary page design in response to its contents would destroy the attempt to catch the reader's eye which remains the first function of design.

Yet a reader buying, say, *The Daily Telegraph* and the *Daily Mail* would recognize the papers at a glance and would be in no danger of confusing the two. Since one of them is a broadsheet (or full-size) and the other a tabloid (or half-size) it could be argued that there is little chance of confusion. Yet the visual point being made holds good if the reader were choosing, say, between *The Times* and *The Guardian*, both broadsheet papers; or between two tabloid papers such as the *Daily Express* and *The Sun*, or, in fact, between the *New York Post* and the *New York Daily News*. This is because a newspaper chooses a particular range of types and type sizes in which to present its headlines and text and broadly sticks to them. It has a recognizable type 'dress'.

All newspaper publishing houses, whatever their typesetting systems, carry a varied range of types, both serif and sans serif (see Chapter 3), some decorative, some plain, some kept for special occasions, some used on contract work for other titles. These are listed in a house type book, or called up on screen. It might seem tempting for the person drawing the page to go through the list and try everything, or to vary types at mood or whim, yet – and there are newspapers who seem to try this occasionally – the result would be a hotchpotch, with balance almost impossible to achieve and any hope of visual character destroyed. One of the accepted and universal rules of newspaper typography is that you go for a deliberate type style, using a limited range of types for headlines – probably one main typeface and a couple of subsidiary ones for contrast – and achieve variety and emphasis by using differing sizes of a type, and perhaps light or bold versions, or by a combination of capital letters and lower case letters of the same type range. This avoids visual clashes that readers would find harsh and unattractive.

The body text, or reading matter, is likewise standardized to a chosen typeface and size, varied perhaps with the occasional use of a bold version of the letters and with a bigger size of setting for the first or second paragraphs of important stories.

It is the selection and consistent use of chosen types and typesizes, whatever varieties of shape the page patterns might have and whatever the size of the pictures, that give a newspaper a distinctive visual appearance. To some extent the average length of items in a newspaper – *The Times* as against the *Daily Mirror*, for example contributes to this visual appearance, yet the image is still being created by the way in which headline type is used to break up the varying mass of reading type.

Ingredients of design

The shape or pattern that is imposed on a newspaper page is created, as we have said, from four ingredients: advertisements, text, headlines and pictures. It is useful, in examining these ingredients, to take advertisements first since, of the four, it is the one over which the designer has least control, and it constricts by the space it occupies on the page (which space is paid for) what can be done to project the editorial contents.

Advertisements

These are designed to please the client and not to fit into any particular editorial pattern. The job of the designer or creator of the advertisement ends once the space has been bought in the page. Advertisements in daily or weekly papers, local or national, can, in fact, be sold up to a month before the paper appears. In the case of some advertising campaigns the space might be booked several months before. Some companies plan a year's advertising in specific papers in advance as part of their budget spending. The content, typography and illustration used in an advertisement are chosen by the client or, on the client's behalf, by an advertising agency. The size of an advertisement is paid for at a set card rate and cannot be varied by the editor. Nor, in many cases, can its position on a particular page for which a premium might have been charged.

The income from advertising is the means by which newspapers reduce their unit costs sufficiently to have a low cover charge. High circulations and ingrained newspaper-reading habits in the main Western democracies are the product of relatively cheap papers. Effective design techniques which make the editorial content attractive and readable are the means by which editors have learned to live with advertising in exchange for being able to keep down the cost of their product.

The space occupied by advertisements can be a considerable part of the whole. All newspapers have a percentage target of advertising which ranges, depending on the paper and its type of circulation, from about 30 per cent to 47 per cent, and up to 60 per cent and higher in free publications. Within each newspaper there is an agreed pattern of placement by which certain pages – page one and sometimes the back page for instance – are kept clear. Other pages, such as the page containing the editorial opinion, are kept 'light' to allow space for regular content, while a portion of the advertising sold consists of whole pages, either a display item or collections of small classified items such as jobs, holidays and postal bargains. The remaining individual advertisements, or sometimes small blocks of like advertisements, are spread around the rest of the paper, including some pages on which specific spaces and positions have been bought. Some of these are solus positions which means that, for an extra payment, no other advertisement will be allowed on that page. The effect of this is to concentrate the non-solus advertising in greater volume on other pages.

It will be seen from this that advertising has already given the paper a certain shape before the editor begins to think what to do with the editorial contents. More worrying for the editor is that to ensure sufficient advertising revenue per copy, the number of pages (pagination) of each edition is determined normally not by editorial requirements but by the total volume of advertising sold. A rush of late space-selling can thus increase the number of pages, while a shortfall of expected advertising can diminish the number. Except on occasions of exceptional importance the editor is expected to accommodate to this.

In fact the pattern of selling by which certain pages are kept light, heavy or clear of advertising does allow the editor to stick to a general plan for the paper so that a consistent contents formula can be followed. Even so, on most of the pages advertising shapes and volume (as well as their content) can vary a great deal. 'Trading' of space and positions can be carried out on some papers (when fixed positions have not been sold) to enable editorial matter on certain pages to be given adequate space and display, especially on those that are front half of paper or right hand, where reader traffic is known (from readership surveys) to be highest. The advertising department is urged by the editor to concentrate its material on left-hand pages or at least to keep premium top right-hand positions clear. Some advertisers, aware of this tendency, stipulate (and pay for) forward right-hand positions.

It is selling practice on some papers that pages of classified advertising such as postal bargains or holidays go to the back of the paper. Sports pages are not popular with general advertisers, and are usually light on advertising.

In advert-conscious US papers page shapes can be chaotic. A few years ago a campaign to sell 'flexi-ads' left whole pages at the mercy of whatever bizarre shapes an agency could sell to its clients, so that adverts might appear in the form of a huge letter 'H' or a monstrous Christmas tree, with the editorial items fitted in the remaining space as best as possible. In British practice advert 'rate cards' stipulate available shapes and sizes and there is a general policy of selling and placing adverts to a pattern that broadly suits editorial requirements.

Another point the page designer must keep in mind with pre-sold space is to see there is no clash between the type and illustrations of editorial material and that used in the adjoining adverts. A cut-out picture of a woman's head in a story about an actress would lose its effect against a cut-out woman's head advertising face cream. A heavy sans headline on a page lead would be robbed of its dominance if a nearby advert had type of the same sort and size. A story in black panel rules would be pointless against an advert panelled with the same rules.

The way to avoid these troubles is for designers to see early proofs or copies of any display adverts on the page rather than have to alter the page at a late stage. Since advertising space is sold first this should not normally be a problem.

Text

Having taken account of the shape and contents of the adverts, the page designer's task is to locate the text of the stories on the page in accordance with the editor's estimate of its importance. On a news page the biggest story is called the *lead*, and the second biggest the *half lead*. Other stories of importance are referred to as *tops*, which derives from the practice in early newspapers of starting all the main stories at the top of the page and running them alongside each other. It is common practice today to give 'strength below the fold' by running some tops across in 'legs' down page. At the ends of columns come the one or two paragraph items referred to as *fillers*, though even the smallest filler is schemed in advance in a well-planned news page.

The text of all these stories is set in a standard size of reading type (usually 7-point or 8-point size), with a bigger size of type to indicate the start of more important stories. Newspapers depart from standard reading text only where special emphasis is called for, in which case bigger or perhaps bolder reading type is used.

Headlines

The importance on the page of the stories comprising the text is indicated, as we have seen, by the size of type chosen for the headline. Headlines, in fact, serve two functions: they draw attention to the contents of the stories, and they form part of the visual pattern of the page by creating highlights (Figure 4)). More specifically, they break up and separate the mass of reading matter to make for greater eye comfort. Varying the size also spares the reader

12 FOREIGN NEWS   Daily Post, Wednesday

Tibet rioters go on the rampage

by Daily Post Reporter

HUNDREDS of Tibetan rioters rampaged through Lhasa hours before the imposition of martial law, looting shops, starting fires, and attacking police and bystanders, the New China News Agency said.

"Despite police attempts to cool down the situation, hundreds of rioters continued to storm through the streets and lanes of the Tibetan capital," the agency said in a report monitored by the BBC.

It also gave a detailed account of rioting on Sunday and Monday and said armed Tibetans shot dead one policeman.

It said Lhasa was quiet

12 killed in new China crackdown

when martial law took effect at midnight on Tuesday.

China ordered the clampdown after three days of nationalist riots that Peking authorities say killed at least 12 people and injured more than 100. Tibetans say they believe the death toll is more than 60.

The riots have posed the biggest challenge to Peking rule since the anti-Chinese revolt of 1959 which killed thousands of people. A Reuter correspondent in Lhasa quoted Tibetans as saying they planned to press on with their drive to gain independence.

In a report from Lhasa a few hours before martial law went into effect, China's official news agency said: "Soon after 11 a.m. hundreds of rioters regrouped in front of the Bargor Street police station shouting slogans and threatening to storm the police post."

It said they later marched into another part of the street "smashing and burning goods that had survived the violence of earlier rioting. The street was shrouded in dense smoke.

"Masked rioters in eastern Beijing Road intercepted and attacked passers-by and piled up stolen bicycles and burned them," the agency added.

It said one of its correspondents saw a businessman chased and beaten by rioters.

"At 2 p.m. an estimated 200 rioters gathered at a market in Bargor Street. One rioter standing on top of a stall was shouting at the top of his voice 'Tibet is independent.'"

The agency said the trouble began shortly after midday on Sunday when priests, nuns and youths shouted "independence for Tibet" outside the Johkang Temple.

New satellites go into orbit

European launcher Ariane 4 lifts off from its base at Kourou, French Guiana

29th Euro-rocket

THE European Space Agency has launched the 29th Ariane rocket, putting Japanese and European satellites into orbit 22,000 miles up.

It was the second flight this year.

Ariane put into orbit JCSAT-1, Japan's first commercial communication satellite, and MOP-1, a European meteorological satellite which will send weather pictures every half an hour to stations on earth.

The first Ariane rocket was launched a decade ago. There have been four failed flights, the last in 1986.

A geomagnetic storm warning

A MASSIVE flare, the most intense since 1984, has erupted on the eastern edge of the sun, sending a burst of radiation that interfered with a navigation system and other communications, federal officials in the United States said.

In Boulder, Colorado, forecasters at the National Oceanic and Atmospheric Administration's Space Environment Services Centre said the flare was detected at 9:05 am (1405 GMT) by satellite instruments.

A flare occurs when energy stored by the sun's magnetic fields suddenly becomes unstable and is converted to heat and radiation within a few minutes, said Gary Heckman, director of the centre.

He said solar telescopes recorded a flare of particles rising some 70,000 miles from the surface of the sun.

He said such a burst could black out shortwave and high frequency communications that travel through the high latitudes on Earth. He said the burst could interfere with communications between aircraft flying in the far north and south, and that some television signals from satellites could be affected.

Another effect, called a geomagnetic storm, may arrive at Earth about 1400 GMT today, said Heckman.

A geomagnetic storm is a wave of magnetically charged particles smashing into the Earth's magnetic field, disrupting natural currents and triggering the aurora borealis, or northern lights.

DIGEST

N-snub

FOUR Russian ports have refused entry to one of their own ships because of fears over nuclear power. The huge nuclear-powered container ship has been turned away after public protests from Vladivostok, Nakhodka and Magadan in the Soviet Far East, and a workers' protest in Vostochny made officials ban the vessel.

Suite music

FURNITURE which belonged to Beethoven until the composer's death in 1827 was sold for £4,800 at a Paris auction yesterday.

Reminder

THE West German city of Nuremberg plans to turn the vast coliseum in which Hitler held his Nazi rallies into a museum which will chronicle aspects of the era.

Two die

A 'PLANE carrying civilian passengers was destroyed on the runway by a rebel missile in the Afghanistan city of Jalalabad yesterday. Two women passengers were killed.

Gun crazy

A GUNMAN burst into a house in a black suburb of the city of Pietermaritzburg and killed four people, one of them a 12-year-old girl, police said yesterday.

Russian-U.S. talks open in good mood

THE new U.S. administration and the Kremlin made their first high-level contact yesterday and declared it a good beginning.

Soviet Foreign Minister Mr Eduard Shevardnadze and U.S. Secretary of State Mr James Baker spent two hours together in talks in Vienna that highlighted differences on human rights, arms control, the Middle East and Nicaragua.

But Mr Shevardnadze told reporters before leaving for Moscow: "I have the impression that Mr Baker and I can work together successfully."

And Mr Baker said before leaving Vienna: "Mr Shevardnadze said he thought this was a good beginning and I agreed with that."

Differences

They met during the launch of new disarmament talks between the 23 Nato and Warsaw Pact states, and their wide-ranging discussions pointed up some major policy differences.

Mr Shevardnadze said Moscow refused to accept that aircraft and troop cuts should be relegated to second place at the Conventional Forces in Europe talks.

The U.S. and its 15 Atlantic allies want to give priority to discussing cuts in tanks, artillery and armoured troop carriers, which they say could be part of a surprise attack.

The two men agreed to meet again in Moscow in May to discuss prospects for a first summit between Presidents Bush and Gorbachev.

Mr Shevardnadze said he expected their Moscow talks could also produce a date for resuming the U.S.-Soviet START talks in Geneva to halve strategic nuclear stockpiles.

The two men looked solemn as they emerged from their talks but they smiled and chatted together as they posed in brilliant sunshine for photographers.

Mr Baker criticised Moscow's failure to condemn publicly Ayatollah Khomeini's call for British author Salman Rushdie to be killed for his novel The Satanic Verses.

Eduard Shevardnadze

French jail Briton

AN ENGLISHMAN was jailed for 12 years by a French court yesterday after being found guilty of attempting to murder a young French woman in 1986.

Stephen Coates, 25, from Hemel Hempstead, Herts, tried to strangle Nadine Dauber after meeting her on a train between Paris and Marseille.

First he tried to steal her handbag. When she screamed, he strangled her until she stopped moving.

"I panicked," Coates told the French jury during his trial in the town of Carpentras.

"Thinking that she was dead I opened the window of the compartment and threw her out."

Mrs Dauber survived, suffering only cuts and bruises.

15 hurt at clinic

ISRAELI soldiers stormed a UN medical clinic in the occupied Gaza Strip yesterday.

Palestinians were demonstrating in the building. Staff said 15 Arabs were wounded.

Reuter joournalist Paul Taylor saw three women taken away on stretchers and said that the soldiers' commander told his men: "Good job, I congratulate you all."

Staff at the clinic said that an eight-year-old girl was among the injured, also two ambulance drivers and a local journalist.

PROPERTY POST Homemaker

EVERY SATURDAY!

. . . starting this week

DAILY POST

Three survive desert ordeal

A PREGNANT woman, her husband and their two-year-old son survived a 13-day ordeal lost in the searing heat of the Kalahari Desert, partly by sucking moisture off plants during the night, the Botswana Press Association said yesterday.

Jonathan Hayman, his wife Laura and their young son were driving across the Central Kalahari Game Reserve when their truck broke down.

After six days they ran out of food and water. They had not seen any other people, so they decided to begin walking the 120 miles back the town of Serowe.

The mother and child were too weak to continue after three days, but Mr Hayman carried on. After four more days farm workers found him on Saturday and took him to Serowe.

A Botswana army helicopter found his wife and son, the news agency said.

"I had given up hope," said 23-year-old Mrs Hayman. "I thought we were both dead."

She and the child survived by drawing water from plants during the night.

Daytime temperatures in the Kalahari Desert often reach 104 degrees Fahrenheit at this time of year.

Figure 4 How a newspaper page strikes the reader: the visual highlights in type and pictures aim to attract the eye and guide it round the page

the confusion that would occur if all headlines were to shout equally for attention.

Pictures

The remaining ingredient of the page design is the pictures (or any other form of illustration). These, like headlines, have two functions, in that they may form part of a particular story which they have been chosen to illustrate, while at the same time they create visual highlights in the page pattern (Figure 4). The placing and size chosen for the main picture can be the key element in the page design and the contents of the pages are usually allocated so that no page is left without pictures. Thus the design element as well as their illustrative role in a news story or feature can be a factor in the choice of pictures.

Influences on design

The job of the journalist designer is to display the editorial contents of the pages to the best possible advantage within the design formula adopted by the paper, and in the space allocated for this purpose. There are a number of influences to take into account here and the success of a page design can be measured by how far it copes with and responds to them. Advertising, as can be seen from the above discussion, is one of them. Yet it must not be overstated. By far the most important factor is the nature of the actual editorial material – the text and pictures – from which the page is to be composed.

Contents

The planning of the paper and the main decisions about the stories and pictures to be used are taken at the editor's daily conference. Decisions on specific stories might be taken by the editor or the editor's delegated executive – who may or may not be the person designing the page. Thus not only must the subject or nature of the contents and length of stories be taken into account, but also any decisions about the space or prominence to be given to particular items and the accommodating of any headlines that have been put up at the planning stage. The aim of these decisions, and of the ideas that go into the page design – called the *layout* or *scheme* – is to root the design in the materials of which the page is composed. A good picture, as we have seen, can be the key to a bold layout, while the wording of a headline can sell a page. Any design which ignores these points or subserves them to some abstract whim of a shape, or to a rigid formula, is failing in its purpose.

The length and number of stories to be carried will decide whether it is to be a 'busy' page or one in which the design is going to have to cope with presenting a wordy text in a readable way. The purpose of later chapters in this book is to show how design copes with the different kinds of content.

Style

In drawing the pages the journalist designer will need to take into account the type 'dress' of the paper – the types and typesizes normally used – and the attitudes to picture size and presentation and story length so that the pages are not visually out of character or clash one with the other. Thus, different papers and different sorts of papers become known in terms of their design style.

Principal optical area

In placing the elements on to the page the designer is influenced by the known psychology of reading habits – the fact, based on readership research, that the top right-hand part of the page, and especially the right-hand page, is where the reader's first attention will go. This has been defined by Edmund Arnold, the American writer on newspaper design, as the principal optical area (POA).

Production methods

The sophistication of the editing and type-originating system and the availability of colour and high-speed printing and advanced camera room facilities can affect design techniques by limiting or expanding what is possible in the time available. The number of columns in the column format used is an important governing factor, although the modern tendency is more and more to standardize tabloids into a seven-column format and broadsheet papers into eight-column.

Editor/proprietor whim

Lastly, and of importance on some papers, is the influence that can be brought to bear by editors or proprietors to present certain stories and certain pages in a particular way as a result of some personal interest or involvement. It is as difficult for the journalist designer to resist an editor's whim as it is for an editor to resist a proprietor's whim, even though there might be some reasoned grounds for doing so. Some of the oddities of picture choice and story presentation that go against normal style and practice in a newspaper can be traced back to such pressures.

3
Typography

Typography is the basic stuff of newspaper design. The transformation brought about by the computer has not altered the importance of understanding and knowing how to use type in creating newspaper pages. In fact, it has made these qualities more necessary now that machines are carrying out work at one time the preserve of craftsman printers trained in the history and uses of 'movable letters'.

Typefaces commonly available on computer software today have their roots in developments that go back to the invention of printing in Europe in the fifteenth century and earlier. The word text, which is commonly used for the body type of newspapers and books, is derived from textur, which was the name given in Germany to the script letter form imposed on his empire by Charlemagne in the ninth century. The tiny condensed lower case letters, referred to as the Caroline (Colour plate 1), were a breakthrough from the then flowery style used in the scriptoriums and enabled monastic scribes to produce manuscripts at great speed. So tight did this script become that it appeared at a distance like woven material with a texture of stitches – hence *textus*. Decoration and relief for the eye was achieved by drawing illuminated capital letters at the start of sections, which practice survives today in the initial drop or stand-up capital letter round which the first few lines of a paragraph are set, especially in features display and in magazines.

As used at speed in the many Vatican and religious texts of the Middle Ages, the characteristic letters of the Caroline acquired a slope, to which was given the name *italic*.

In reaction against the plain but readable Caroline, scribes in various regions of Europe moved gradually from the rounded lower-case letters into condensed sharper versions. Out of this grew the black letter with its spikes and serifs formed by calligraphic diamond shapes as the quill reached the head and feet of the letters.

This was referred to in England as the Old English, while in Germany it took on the name *Fractur* because of its broken look. Fractur, as later mechanically founded and refined by designers, remained the national type of Germany until 1940.

Serif types

The invention of printing from movable type by Gutenberg in Germany in 1450, which spread rapidly to France, England and the Low Countries, produced the first great age of type design, from which the early Bembo (1495) and the designs of Christopher Plantin (1570) are still popular. Claude Garamond (1499-1561), Europe's first commercial typefounder, produced under the patronage of Francis I of France the first mathematically designed type in the 1540s, naming it Grecs du Rois. His Roman was to become the standard, or upright, letter for all printing. Garamond went on to produce a family of roman types based on the letters of the old Latin inscriptions, characterized by flat serifs, or tails, on the end of the letter strokes that were to influence generations of type design. Derivatives of Garamond's early faces are still in use.

Sir William Caslon, the first great English typefounder, issued his type specimen sheet in 1734. Caslon, a refinement of Garamond (Figure 5), was more comfortable to read, with the serifs cleanly bracketed, and it remains popular in British and many American newspapers. Official copies of the United States Declaration of Independence were printed in Caslon, the face being introduced into the American colonies by Benjamin Franklin. Computer digitizing has further cleaned up the well-cut characters and given them a new lease of life.

A more graceful serif type, though less favoured in newspapers, was Baskerville, introduced by John Baskerville, a teacher of calligraphy, twenty-five years after Caslon's first sheets. His faces, too, remain available in modern computer systems, though usually under changed names. It was an eighteenth-century Italian designer, Giambattista Bodoni (1740–1813), however, who was to produce type designs that would be taken up by every country using the European alphabet. Bodoni bold and italic, with their fine serifs and elegant lowercase letters being still today the most used stock headline type in newspapers. Bodoni, a meticulous artist, was reputed to have spent three years designing the letter 'a' for the alphabet of his revolutionary Ultra Bodoni in which he refined his faces down to the finest of thin lines in the up strokes to a point where they almost disappeared, with the down strokes exaggerated to an enormous thickness (Figure 5).

Among much used modern seriffed faces are those of Stanley Morison, with his Times New Roman, designed in 1931 for the relaunch of *The Times*, a classic series used in different variations by many broadsheet newspapers; and the widely used Century bold, produced by the prolific L. B. Benton for American Typefounders Inc. This was first cut in 1890 for *Century magazine*, which wanted a heavier seriffed face with the hairlines given greater thickness than those then available.

abcdefghijklmnopqrstuvwxyz
(a) ABCDEFGHIJKLMNOPQRSTUVWXYZ

abcdefghijklmnopqrstuvwxyz
(b) ABCDEFGHIJKLMNOPQRSTUVWXYZ

abcdefghijklmnopqrstuvwxyz
(c) ABCDEFGHIJKLMNOPQRSTUVWXYZ

abcdefghijklmnopqrstuvwxyz
(d) ABCDEFGHIJKLMNOPQRSTUVWXYZ

ABCDEFGHIJKLMNOPQ RSTUVWXYZ
(e) abcdefghijklmnopqrstuvwxyz

abcdefghijklmnopqrstuvwxyz
(f) ABCDEFGHIJKLMNOPQRSTUVWXYZ

abcdefghijklmnopqrstuvwxyz
(g) ABCDEFGHIJKLMNOPQRSTUVWXYZ

Figure 5 Serif faces that reflect 400 years of type design: (a) Garamond; (b) Caslon Regular; (c) Baskerville Old Face; (d) Bodoni Bold; (e) Ultra Bodoni Italic; (f) Century Schoolbook; (g) Times Bold

abcdefghijklmnopqrstuvwxyz
(a) ABCDEFGHIJKLMNOPQRSTUVWXYZ

abcdefghijklmnopqrstuvwxyz
(b) ABCDEFGHIJKLMNOPQRSTUVWXYZ

ABCDEFGHIJKLMN rstuvwxyz
(c) OPQRSTUVWXYZ abcdefghijklmnopqrstu

Figure 6 Slab serif types. Note the heavy squared serifs: (a) Egyptian Bold Condensed; (b) Rockwell Bold; (c) Playbill

Sans serif types

By the late nineteenth century American newspapers on the new frontiers were inventing different type formats nearly every day, one of the favourite faces being the Egyptian family of slab-seriffed types (Figure 6). Yet it was the British entrepreneur and newspaper owner Lord Northcliffe who started the move into sans serif for tabloid papers with the *New York World*, which he launched in America four years after founding the *Daily Mail* in 1896. Alfred Harmsworth, as he then was, favoured the new Grotesques then coming into use which, in fact, derived from the primitive sans serif faces that had first appeared in the type specimen books of Figgins and Thorowgood in 1832. They called the faces Sans Serif, the first coining of this term. As used in the *New York World*, and soon to be copied in British papers (Figure 7), these early crude sans faces have heavy solid strokes of equal width and are not attractive.

The Grotesques, or Grots as they became known, were to be the basis of all sans serif designs since, culminating in the elegant faces of typography and sculptor Eric Gill (1882-1940) with his Gill Sans, Perpetua, Joanna and Pilgrim (Figure 8), and the German designs of the Bauhaus period of the 1920s, leading to Futura, sometimes called Future in computer systems, cut by Paul Renner in 1927. The Germans, in fact, despite their dalliance with the mediaeval look, had a major impact on modern sans serif, reforming, expanding and condensing to achieve a wide range of designs to which the digital computer designers of the late twentieth century have added their refinements.

A popular American sans development in the 1890s was Medium Condensed Gothic – at one time referred to as News Gothic because of its versatility – along with its companion Extra Condensed Gothic. Both these types, and later the adaptable Tempo bold and heavy range, were used in the *Daily Mirror*'s development of the tabloid style under Harry Guy Bartholomew in the 1930s and 1940s, and later under Hugh Cudlipp, the *Daily Mirror* being responsible for the successful exploitation of condensed sans faces by the British popular tabloids.

Univers, however, could be described as the modern typographer's most favoured face. Adrian Frutiger put this elegant and wide-ranging font together for Deberny and Peignot in 1957 and it appears in nearly all systems, a body type version of it being Helvetica, available in bold and roman (Figure 9). The beauty of this sans face is its clinical presentation and legibility in all sizes and weights.

Body types

Many of the body types in today's computerized systems (Figure 9) have an old pedigree, being derived in some cases from faces that appeared in early nineteenth-century type sheets. Ideal and Ionic, the latter still used, were shown in a Stephenson Blake specimen sheet as early as 1830, the long popularity of Ionic being due to its legibility under the stressful conditions of high-speed printing. The

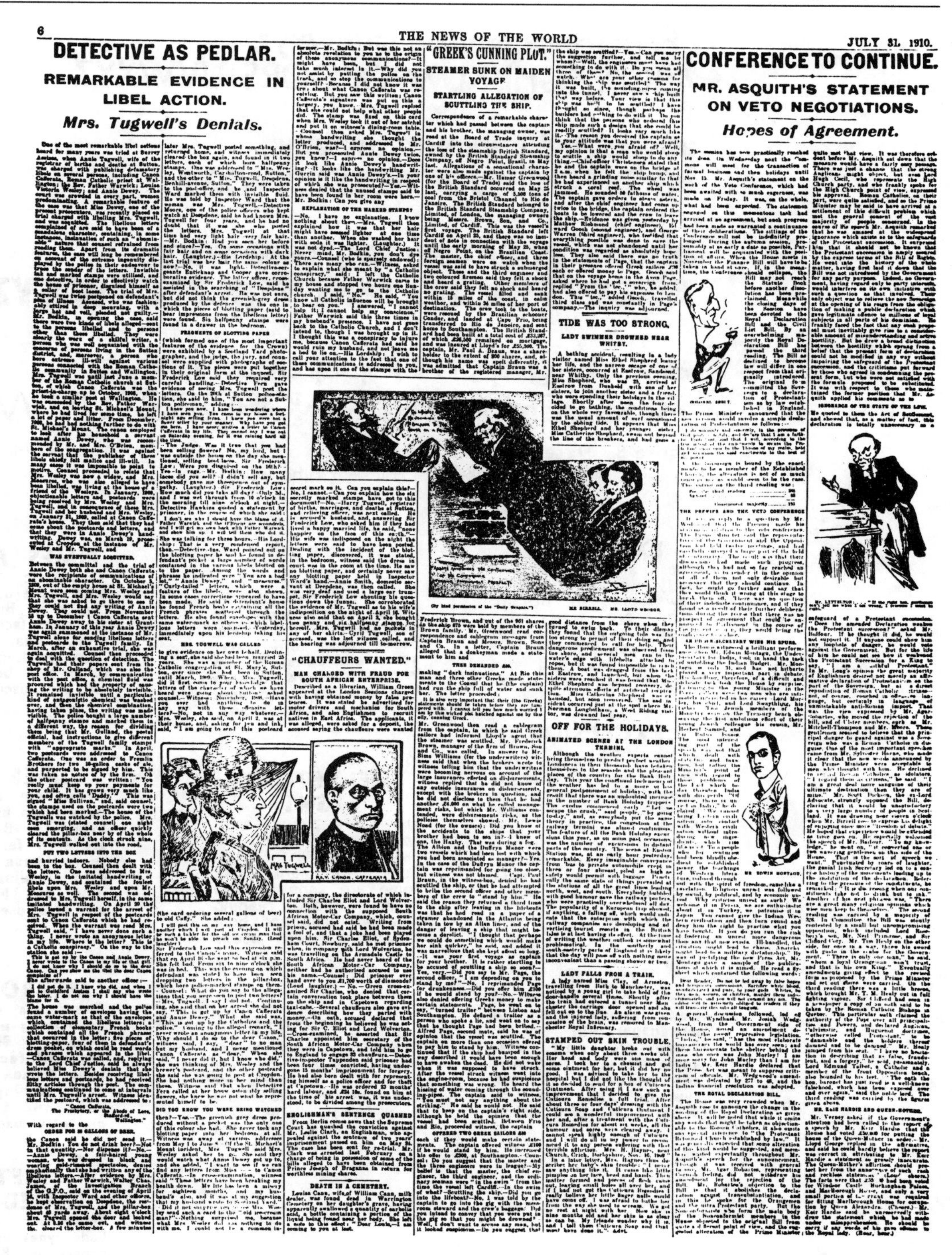

6 THE NEWS OF THE WORLD JULY 31, 1910.

DETECTIVE AS PEDLAR.

REMARKABLE EVIDENCE IN LIBEL ACTION.

Mrs. Tugwell's Denials.

"GREEK'S CUNNING PLOT."

STEAMER SUNK ON MAIDEN VOYAGE

STARTLING ALLEGATION OF SCUTTLING THE SHIP.

TIDE WAS TOO STRONG.

LADY SWIMMER DROWNED NEAR WHITBY.

CONFERENCE TO CONTINUE.

MR. ASQUITH'S STATEMENT ON VETO NEGOTIATIONS.

Hopes of Agreement.

"CHAUFFEURS WANTED."

MAN CHARGED WITH FRAUD FOR SOUTH AFRICAN ENTERPRISE.

OFF FOR THE HOLIDAYS.

ANIMATED SCENES AT THE LONDON TERMINI.

LADY FALLS FROM A TRAIN.

STAMPED OUT SKIN TROUBLE.

THE ROYAL DECLARATION BILL.

ENGLISHMAN'S SENTENCE QUASHED.

DEATH IN A CEMETERY.

Figure 7 This 1910 page from the *News of the World* shows the shift into a primitive form of sans type for page lead headlines alongside the paper's ancient Cheltenham Condensed

abcdefghijklmnopqrstuvwxyz
(a) ABCDEFGHIJKLMNOPQQRSTUVWXYZ

abcdefghijklmnopqrstuvwxyz
(b) ABCDEFGHIJKLMNOPQRSTUVWXYZ

abcdefghijklmnopqrstuvwxyz
(c) ABCDEFGHIJKLMNOPQRSTUVWXYZ

abcdefghijklmnopqrstuvwxyz
(d) ABCDEFGHIJKLMNOPQRSTUVWXYZ

Figure 8 These sans faces typify the best in modern type design: (a) Futura Medium; (b) Gill Sans Bold Condensed; (c) Franklin Gothic Bold; (d) Univers

EXCELSIOR
These are examples of body types used in newspapers in Britain. Some are serif face, some are sans. Roman is the standard reading type but italic and bold variants are also used.

EXCELSIOR ITALIC
These are examples of body types used in newspapers in Britain. Some are serif face, some are sans. Roman is the standard reading type but italic and bold variants are also used.

DORIC
These are examples of body types used in newspapers in Britain. Some are serif face, some are sans. Roman is the standard reading type but italic and bold variants are also used.

HELVETICA BOLD
These are examples of body types used in newspapers in Britain. Some are serif face, some are sans. Roman is the standard reading type but italic and bold variants are also used.

HELVETICA LIGHT
These are examples of body types used in newspapers in Britain. Some are serif face, some are sans. Roman is the standard reading type but italic and bold variants are also used.

BELL GOTHIC
These are examples of body types used in newspapers in Britain. Some are serif face, some are sans. Roman is the standard reading type but italic and bold variants are also used.

TIMES ROMAN
These are examples of body types used in newspapers in Britain. Some are serif face, some are sans. Roman is the standard reading type but italic and bold variants are also used.

TIMES BOLD
These are examples of body types used in newspapers in Britain. Some are serif face, some are sans. Roman is the standard reading type but italic and bold variants are also used.

METRO BOLD
These are examples of body types used in newspapers in Britain. Some are serif face, some are sans. Roman is the standard reading type but italic and bold variants are also used.

METRO MEDIUM
These are examples of body types used in newspapers in Britain. Some are serif face, some are sans. Roman is the standard reading type but italic and bold variants are also used.

Figure 9 Some examples of 8 point newspaper body type set across 8½ picas

shortness of descenders and ascenders also allowed for economy of space. The modern version generally used was cut by Linotype for the *New York Herald Tribune* in 1926, with Intertype providing a version for the *New York Times* a few years later.

William Addison Dwiggins, a student of the designer Goudy, designed all the versions of the sans body face Metro in 1929-1930. Linotype produced this font with the names Metrolight, Metroblack. Metromedium and Metrothin. The square characters of the black became popular throughout Britain for wide intros for splash stories, and for captions and panels, being favoured over the older stock bold face, Doric, a Figgins and Thorowgood type dating back to 1854. The Metros sometimes appear in computer systems as Chelsea.

Another bold face, noted for its economy in depth, had a curious history. In 1938 the Bell Telephone Company of America needed a more economic type for the huge and ever-growing New York State telephone directory. A Bell typographer, who never merited a byline, joined forces with the Linotype Corporation to invent Bell Gothic (Figure 9), a clean and readable sans serif of uniform weight of line which offered them the visual boldness of types one or two sizes larger.

Imperial, cut by Edwin W. Shaar for Intertype, was another popular body type in hot metal cases, the bold version being noted for its blackness when used in captions or panels.

Faces and families

Every alphabet is in some sense unique. Some are dramatically different, others are adaptations of a previous face created to suit a particular purpose. Each is called a *typeface* and is given a name often, as can be seen, that of the designer. In the days of metal matrices, each letter was mathematically created on paper, transferred to metal, cut and countersunk. Outstanding type cutters such as Nicholas Jenson and John Baskerville would have been amazed and delighted at today's computer digitation of their type characters in which refinement and authenticity are produced by the computer breaking down the shape of the letters and re-creating them by bitmap editing to produce a result better than the original artist could possibly have hoped for. Every size is created in its own right with nothing being scaled down for fear of losing a characteristic.

Each typeface has its *family* which is made up of fonts of characters which are differentiated from other fonts by *weight* and *interpretation*.

The word font derives from the hot metal tradition and was the name given to a cabinet full of all the characters and sizes of a given type including punctuation, ligatures, cyphers, figures, fractions and asterisks, and which could range from 81 up to 160 characters depending on usage. The weight of a piece of type is determined by the visual heaviness or lightness of the strokes that make up the characters, giving bold or light letters. By interpretation the face designer meant the condensing or expanding of the characters to produce narrower and wider variants for particular purposes. The

Figure 10 Weight and interpretation in type design: 18 point Helvetica lower case in its various presentations

Helvetica Helvetica
Helvetica Helvetica
Helvetica Helvetica
Helvetica Helvetica
Helvetica Helvetica
Helvetica Helvetica
Helvetica Helvetica

weight and interpretation are signalled in the compounded names of types – Metroblack, Metrolight, Helvetica Bold, Helvetica Medium Italic, Extra Condensed Gothic and so on (Figure 10). Today, the horizontal hold button on a computer can produce a percentage squeeze or expansion to the designer's choice.

Categories of type

To make the type map intelligible to the user it has become a convention to list typefaces, a selection of which we have now introduced to the reader, in five main categories:

Old English

Old English (Figure 11) is referred to as Fraktur in Europe. These are the alphabets derived from the hand-produced quill pen letters of the scriptoriums and sometimes referred to under their mediaeval name of Black Letter. Fonts can be either in-lined or solid and vary in the degree of decoration applied to them. They survive mainly in newspaper mast-heads and in ornamental lettering.

Serif

This is an important range of typefaces, old and new (Figure 5) used in all manner of printing, which is characterized by letters that do not finish square on their strokes but carry a horizontal line or tail, sometimes at a slight angle (that is, the *serif*) which has the effect of adding dignity or authority to the letters. At the bottom of the letter the serif may be attached by a curve or bracket, to the main body of the character, or sometimes an angled attachment (Figure 12). The serif, too, traces back to the flourishes given by the quill pen, as can be seen in the characteristic serifs in the range of *old style* faces such as Bembo and Garamond, which are not horizontal but sloping and rounded. By comparison the range known as *modern* (as in Bodoni) has thin unbracketed serifs, except in the capitals M and N, with a

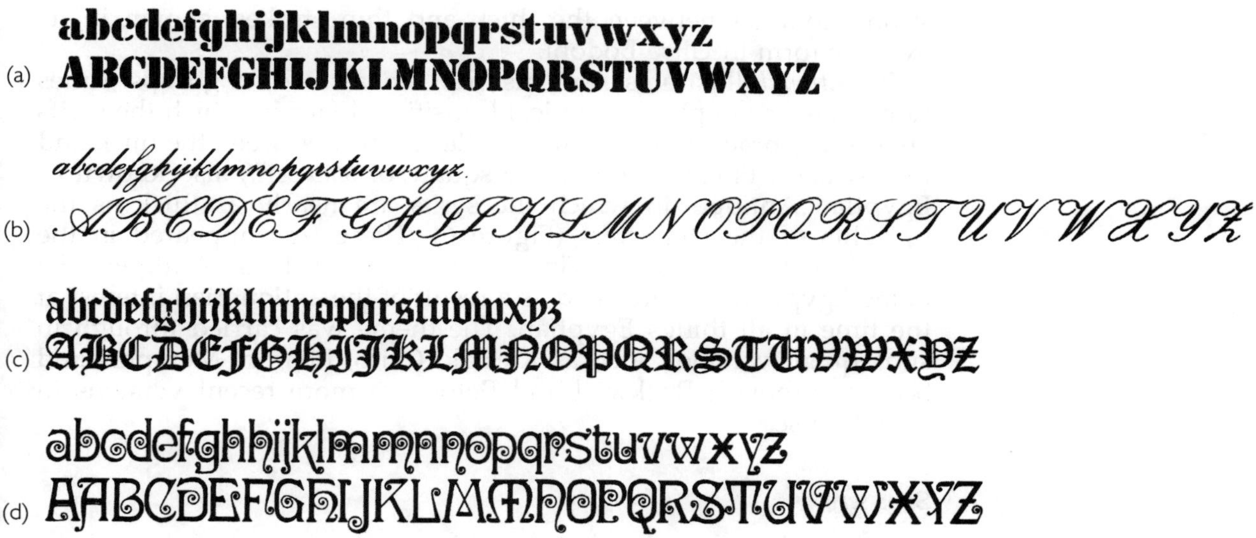

abcdefghijklmnopqrstuvwxyz
(a) ABCDEFGHIJKLMNOPQRSTUVWXYZ

abcdefghijklmnopqrstuvwxyz
(b) ABCDEFGHIJKLMNOPQRSTUVWXYZ

abcdefghijklmnopqrstuvwxyz
(c) ABCDEFGHIJKLMNOPQRSTUVWXYZ

abcdefghhijklmmnnopqrstuvwxyz
(d) AABCDEFGHIJKLMMNOPQRSTUVWXYZ

Figure 11 Novelty and other little used faces, common in advertising, might find a place on feature pages or in a special projection that looks to type to set the mood: (a) Stencil Bold; (b) Palace Script; (c) Old English; (d) Ringlet

Figure 12 The terminology of type: this illustration from Autologic's *Digital Type Collection* shows the traditionally named parts of letters and the significance of the x-height

strong contrast between the thick and thin strokes, shown in its extreme form in Ultra Bodoni.

Transitional illustrates elements from the last two subgroups. Times Roman, for example, is a typical transitional face in which the serifs are slightly bracketed with just a little contrast between the thick and thin strokes. Finally, there is the square or *slab serif*, also known as Egyptian Antique, in which the serifs are nearly as thick as the uprights of the characters (Figure 6). These first appeared in the American type books of Vincent Figgins in 1815 as Antique. The name Egyptian was attached as a result of the antiquarian interest at the time in all things Egyptian. The theme was carried through in subsequent versions such as Karnak, Cairo, Pharon, Memphis and Scarab, although Rockwell and Beton are more recent versions in popular use.

Sans serif

Sans serif describes alphabets in which the letter strokes are without serifs (Figure 8). The characters have generally a uniform thickness and weight throughout. The name Gothic has come to be attached to some of these designs – a misnomer since its origins go back to the Black Letter script of the monasteries. Helvetica, Univers, Grotesque, Franklin, Tempo, Futura and Metro are examples.

Scripts

Script and cursive – not for the news pages – are faces based on brush or pen strokes. Brush stroke letters can induce an air of urgency, while script type, much used on invitation and visiting cards, produces an effect of officialdom and authority. Occasionally, such typefaces will have some application on a features page but the subject needs to be exactly right (Figure 11).

Novelty faces

These are the odd ones out. Modern designers encouraged by the transfer typographic sheets from Letraset and Mecanorma have produced alphabets constructed from sunbursts, brickwork and other elements. They can work on advertising posters and on television but have little application in newspapers (Figure 11). The versatile computer makes these variants easier to achieve.

Measurement

Despite the onset of metrication in page design, typesizes even in the most modern of computerized systems are still measured in a 250-year-old system of points, of which there are, as a guide, approximately 72 to the inch and 28 to the centimetre. Strictly there are two points systems, the American, used throughout the English-speaking world, and the European, used in most of Europe, but the difference

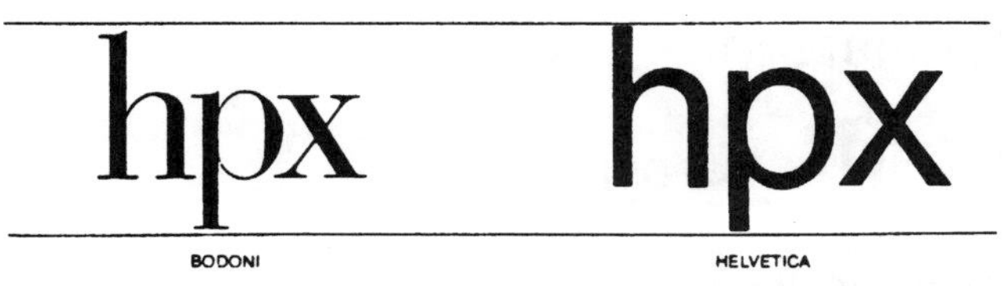

Figure 13 How variations in x-height can create an optional illusion. These letters in Bodoni and Helvetica are the same size but have different x-heights

between the two is so small that in the ordinary run of typesizes it would pass unnoticed. Here we are concerned with the American system. We relate points to inches purely for convenience of learning since the two are different entities. If we are to be exact there are 72 points (0.996264) to an inch.

Type size is measured by its height in the page, but it should be remembered that sizes are derived from the metal base upon which the characters stood in the old hot metal system. This had to allow space for ascenders and descenders in the lowercase letters f, g, h, j, k, l, p, q, t and y, and for the differences in height between capital and lowercase letters. Thus, to say that a 36 point letter is half an inch deep is not strictly true, since without ascenders and descenders the actual size on the page is less than half an inch. The problem is complicated further by the fact that the proportion of space taken up by the ascender and descender varies in different type ranges so that the *x-height* – the mean height of a letter without the ascender or descender – can vary from one type to another (Figure 13). Thus when a type is described as being 'big on its body' it means that its x-height is greater in relation to its descenders and ascenders than with other types. Such a type consequently looks bigger in any given size. An oddity occurs with certain types called *titling*, used for big headlines or posters, which exist only in capitals, since here the characters take up the full height available on their base. Consequently a headline in 72 point titling really is an inch deep.

Typesizes, originally for convenience of manufacture in hot metal days, are designated in a strict series of sizes. In body type they start at 4½ points, ranging through 5, 6, 7, 8, 9 to 10 points, then jumping in series to 12 and 14 points. Headline series run: 14, 18, 24, 30, 36, 42, 48, 60, 72, 84, 96, 108, 120 and 144 points (Figure 14). Sizes can go beyond 144 points in computer software, and systems typographers, in response to American demand, have introduced 16, 20 and 22 point sizes to prevent distortion of existing sizes by having to reduce or enlarge type to get headlines to fit.

Computer systems are programmed at the outset with traditional series sizes for the very good reason that they enable page designers and production journalists to achieve a deliberate type balance at the visualizing stage which is essential if a newspaper's character is to be preserved. While the sizing can be overridden by keyboard

Century bold 18 POINT

Century bold 24 POINT

Century bold 30 POINT

Century bold 36 POINT

Century bold 42 POINT

Century bold 48 POINT

Century bold 60 POINT

Century bold 72 POIN

Figure 14 Century Bold lowercase showing stock sizes from 18 point to 72 point

controls during design and editing, and types bastardized to within fractions of points of each other, the facility should be used sparingly if design is to be taken seriously.

Setting widths have always been in picas and points, a pica being a unit of 12 points (or one sixth of an inch) and this remains so with computerized systems. Standard column widths range from 8½ picas to 10 picas or more depending on the number of columns. The old English term for a pica, an em (or mutton), which was based on the width of the standard 12 point roman letter 'm', has now fallen largely out of use, and with it the en, or nut, based on the standard 12 point roman letter 'n'.

In subediting practice, except where setting format codes are used, setting is called up by keying in such terms as '12rom, 14p6', which means 12 point roman body type across 14 picas and 6 points, or, for example, '10MM, 8p4', which is the usual way of saying 10 point Metromedium across 8 picas and 4 points. Standard single or double column setting usually has a simpler one-stroke letter code to save keyboarding time. A system can be formatted with whatever sorts of standard width body setting are required to give maximum speed and simplicity in keystroking.

Letter spacing

Letter spacing is a main area of faults in newly installed systems. If newspapers want to avoid interrupting the reading habits of their buyers, spacing is too important to be left to the systems manufacturers. This means that as far as possible it should imitate the hot metal system that went before. It must therefore be programmed in at the start in body setting no less than in headline type. Body setting should tend towards being tight rather than slack, since open setting lacks impact, although justifying the lines on the keyboard will inevitably produce variations, especially in narrow measures. To achieve a good effect, the units of space should relate to the typesize rather than standard spacing being given across the board. A good test to detect faults in spacing is to prop a newspaper page upside down against a wall. The eye, untrammelled by having to sort out the sense of the words, can thus concentrate on the pattern of space between letters and words and faults quickly become apparent.

Kerning

For ideal reading at all sizes, spacing between some letters needs to vary. While manufacturers might supply a system in which the letter spacing is satisfactory overall, it is down to the user to modify this by using preference menus in the software kern programme. Readability can then be improved by deleting or inserting units of space between certain letters, and also between words in which particular end letters occur. This is called kerning (Figure 15).

Characters that will always require visual balancing, especially in headlines, are the obviously eccentric shapes such as capital A, W, V and Y, where they fall together. Rounded letters suffer in the same way. Hot metal systems were clumsy in this respect in that the satisfactory lodging together of letters such as the Y, O and T relied on the eye of the operator of the casting machine. With hand-set headline type this often entailed permanently defacing metal characters by cutting with a saw into the waste or beard of a letter in order to draw it closer to its neighbour.

Computer setting allows for a much more precise attitude towards spacing since the system can produce units of space related to the widths of given characters and a judicious improvement can be carried out by keyboard commands to track or kern. Using a one hundredth part of the width of a letter in larger sizes as a unit, an instruction can be given to the computer to remove space from

Figure 15 These letter combinations show the effect of kerning, or tracking: from normal spacing in the top line down to 4 units kerned in the bottom

AV Te Yo Y.

AV Te Yo Y.

AV Te Yo Y.

between the letters by deleting the number of units required in proportion to allow a line a given density. Entire alphabets of every font in the house can be kerned in a uniform fashion in this way to the editor's taste. It can be extended and standardized into formats across the setting patterns so that greater readability is achieved.

The text

The act of reading is performed by the eye moving across lines of type in short jumps. These jumps take place at regular intervals with brief pauses for about a quarter of a second. The perception of the words covered takes place during the fixation period which is called the eye span and which, according to experts in these matters, covers in average newspaper reading text just 2 centimetres, or 5 picas. The eye then moves on to the next pause.

However, the eye also has to cope with the descent from line to line. To ease this function and to avoid lines becoming excessively long in characters, and the number of jumps required, it has been found helpful to set book texts in 10, 11, or even 12 point type. Newspapers, with their wider sheet and need to pack the space with content, have historically opted for columnar formats, which enable a smaller body face to be used, thus maintaining lines of acceptable length while easing the eye's descent through the text.

As a result of having narrower lines than books, newspapers find that they can set comfortably for the eye in a body face of 7 or 8 point, with even 6 point being used in tabulated information. Where prominence needs to be given to the beginnings of stories, however,

In a mass media system, the typographer is usually a member of the encoding team. He stands between the original source of the message and the channel (here: the printed page), which will carry the message to an audience. He has considerable control over the coding process by selecting type faces, type sizes, and by

9 point solid (no leading).

In a mass media system, the typographer is usually a member of the encoding team. He stands between the original source of the message and the channel (here: the printed page), which will carry the message to an audience. He has considerable control over the coding process by selecting type faces, type sizes, and by

9 point with ½ point leading.

In a mass media system, the typographer is usually a member of the encoding team. He stands between the original source of the message and the channel (here: the printed page), which will carry the message to an audience. He has considerable control over the coding process by selecting type faces, type sizes, and by

9 point with 1 point leading.

Figure 16 Lines of 9 point Crown set across 14 picas demonstrate the effect on body type of line spacing or leading (ledding)

especially important stories, a bigger typeface is used, often of wider measure. It follows, therefore, that subeditors must increase the body type size when setting to wider measure.

Where required for special setting, or to fill, line spacing equal to 1 point or 1½ points can be inserted by keyboard commands (Figure 16).

Intro

A page lead introduction or *intro* across two or more columns might be established in importance for the reader by being in 12 point, or even 14 point, perhaps in a bold version of the standard face, and still be comfortable to read. A single column story might carry an intro in 10 point or even 8 point, on a body size of 7 point. A caution here: practice in intro setting varies from paper to paper, popular tabloids going for bigger and bolder intro sizes, while the qualities and more traditional papers settle for one or two sizes up in the stock body face. *The Times*, for example, does not deviate from a fixed one size up from its 8 on a 9 point body, relying on the effect of fine line spacing thus produced to help the eye through the longish texts.

Also a matter for house style is whether the first paragraph starts with the first word in capitals, or not (as in most of the qualities) or whether drop letters are used (Figure 17). More universal is the practice of starting the first paragraph full out, while giving all the others a standard indent on the first line of 6 points or sometimes 12.

Drop letters

Drop letters are large size initial capital letters of the first word of the intro of a story set in a special house type style, from three up to six or seven times the size of the body, and round which the first few lines of the paragraph are set. The size indicates how many lines of shorter setting are needed – that is, a two-line, three-line drop, or in the case of the more magaziney layouts, up to a ten-line drop, a device much used on *The Times* features pages.

Drop letters, as we have seen, are descended from the illuminated capitals of monkish manuscripts and their job is unashamedly to decorate the text and attract the eye. For this reason they are used nowadays more on features pages where longer texts and deliberate projection give more call for them (Figure 17), though two- and three-line drops still appear on the main stories on some newspapers. The drop letter also exists as a *stand-up drop*, in which it is aligned with the base of the first line and stands above the paragraph. Magazines and some up-market newspapers use these paragraph markers at up to 120 point as stand-ups giving a huge amount of display space above the first line of the paragraph. Body type can be set easily round the contours of drop letters by the use of keyed formats giving a snug fit and useful relief to the shape of the page.

To be effective, the size of a drop letter should be in proportion to the width of the setting, and thought must be given as to whether to use serif or sans serif letters. If a serif face is used then the body type of the paragraph should be larger than normal since the serifs on the letter will push the rest of the line further away from the initial. A sans letter will allow the rest of the word to be tightly set alongside it, thus increasing its readability. In either case the rest of the first word should be in caps of the body.

A more adventurous way in magazine or colour supplement layout is to draw a squared piece of artwork as a drop, more on the lines of the illuminated letter. A ploy here is to suspend it in the white of the gutter in front of the column. In this case it is best used squared up to the top of the first line of the body type it accompanies. Check that the gutter space available, however.

A warning: the drop or stand-up capital is for decorative use only and should not be peppered around the page indiscriminately. Avoid

16 *Lifestyle* THE SCOTSMAN Wednesday, March 8, 1989

Skirts for the flirts

When it comes to sheer sex appeal the Italians lead the world. In Milan this week some are going to great lengths with skirts which sweep the floor but others are taking hemlines up, up and almost away. **Ann Chubb** reports

AT a time of recession, should fashion quietly evolve wearable good taste classics or titillate our appetites with clothes that are covetable and exciting? The Italian designers, first in the international fashion circuit to unveil their wares for autumn are doing both.

Georgio Armani, whose younger, sportier Emporio line kicked off the shows on Sunday morning, made a strong statement in favour of understatement with softly tailored sportswear that you could buy to wear for ever. Gianni Versace, on the other hand, stuck his neck out for the shorter (and even shorter) skirt in a show that revelled in pure, unadulterated sex — something the Italians are rather good at. And at a time when too many designers offer too many options, we have to admit admiration for such positive thinking.

Armani, whose main line collection will be unveiled tonight, is something of an international force these days. Not only with an awesome programme of international expansion — his Emporio superstore opened in London last month, ten more are planned for around the country soon and a new Armani shop opens in London this weekend.

The man whose tailoring is the most covetable in the world has softened his line considerably. Shoulders are now dropped and slightly rounded, jackets shorter and curvier, trousers often track-suit-style or else flattering jodhpurs of grey jersey. The cape, a strong Milanese theme, appeared in every shape from voluminous and flowering to bolero brief. And while the Milanese women happily saunter the streets in the currently sweltering spring weather wearing opulent minks and sables, Armani comes out strongly in favour of the frankly fake with Fifties-style stoles and shrugs.

Versace, on the other hand, came on strong with a collection that majored in sex. His skirts were universally cut short, and not only were they cut to mid thigh or higher but they were skin tight and often side slit to boot. Sometimes they were simply tiny fans of Fortuny-style diagonally pleated fabric. And the short skirt was worn with everything from swirling ankle length great coats to long tailored jackets, a mere one inch shorter than the skirt or with brief bejewelled bustier tops.

On less than perfect models (Versace imported three of the world's best at vast expense to show exclusively for him) it was the look that could have been tarty — in fact it looked terrific. These are the sort of clothes that men will positively urge women to buy.

ELSEWHERE in Milan, the longer hemline proliferates, often soft and voluminous and ankle length, with trousers the only other alternative. Dolce e Gabana actually manage to make the long skirt look sexy by cutting it tight and curvy in jersey and teaming it with corset-style tops or fresh white cotton shirts edged with lace or translating it into wispy black chiffon. The essential accessory to complete the outfit is a Corsican-style crocheted shawl with deep fringed edging. It helped, too, to have actress Charlotte Lewis as star model looking just like a young Claudia Cardinalle. Byblos on the other hand offered pleated trousers so wide that they looked almost like ankle length skirts — very pretty for evening in black teamed with simple black waistcoat and crisp white shirt.

Colours are definitely brightening up here in Milan. Byblos showed Harris tweed jackets in delectable shades of yellow or mint green or apricot for colourful hacking jackets over tapered cord pants. And they gave tweeds a new twist with neat suits in cheerful checks of bright colour enlivened with Paisley print lapels and cuffs.

Huge Paisley patterns were also used for jolly sweaters in bright colours teamed with suede walking shorts and bright hooded coats. The newest winter coat, which appears in almost every Italian collection, is wide and swingy, with dropped shoulder lines, cape or shawl collar and is often made of unlined blanket-like wool.

There was a Scottish air about the Missoni collection which had wonderfully practical quilted cire coats lined with tartan or tapestry and bright tartan-inspired knits for knickerbockers partnered with striped multi-coloured sweaters and jauanty Tam O'Shanter hats. There were also big-collared tartan-style knit coats mixing bright colours together, while those famous Missoni multi-coloured zigzag knits looked young and lively used for long lean cardigans teamed with short flirty skirts.

A Chinese influence brought kimono-style jackets in tree bark effect knit and trouser suits in quilted print teamed with coolie hats of black or brown velvet.

As always the Italians excel at throwaway luxury looks of sumptuous quality. Good examples this week have included Mario Valentino's sheepskin coats in delectable shades of pale blue, pink or mimosa yellow, their wide shawl collars delicately fretworked to imitate lace; Laura Biagotti's softest suede jogging suits of palest petal pink and her gently tiered cashmere smock dresses; Genny's bitter chocolate leathers, glossy and paper-fine, used for wonderful three-quarter length coats, beaver-collared, and widely belted over short ocelot print skirts.

Maybe we're all taking fashion too seriously though. Certainly designer Franco Moschino who doesn't, is currently a huge commercial success with his wild and wicked designs that frequently contain some sort of joke — he has no less than 23 licensees making everything he designs from jeans to lingerie, most of which are now available in Great Britain.

He thinks that clothes (including his own) are vastly overpriced so writes "very expensive jacket" across the back of a plastic hacking jacket, or embellishes black shirts with their actual price written in gold embroidery. And last night he closed the whole of chic Via Sant Andrea for one mammoth street party to launch what he calls "his first and last shop".

LEFT: Gianni Versace is taking things to extremes with this short daytime suit. The chic jacket and skirt are designed to be worn without a blouse — in winter.

BELOW: Keep out the chill with a heavy sweater, leather culottes and a wide, soft coat from Byblos. Similar coats appear in many of the Italian collections this year.

A lesson on the catwalk

A SPARTAN split-level catwalk dominated by a curious human silhouette functioning as a surrealistic clock created a suitably avant garde backdrop for this year's Glasgow School of Art Fashion Show. The audience might have anticipated the performance from the set: aggressive, up-tempo, and out for thrills, although there were regrettably less than usual of the latter.

Two girls skipping in soft, ballet-inspired costumes, and then suddenly adopting the stance of body builders supplied the sort of visual aid a disappointing collection required to give it impact. But, in the end, the medium was more pleasing than the message.

Ethnic influences were writ large, if belatedly; there were sarongs, saris, and an abundance of gold filigree as the young designers opted for the frankly opulent, led boldly (to where no man has gone before) by the exquisite printed gents' shawls designed by Alistair McCauley.

Glasgow has long been celebrated for its printed textiles, and *The Glasgow Print* was very positively tattooed on the forearm of this show. The two models who strode ferociously with their pink, turquoise and orange umbrellas spearheading their matching, animal print dresses epitomised the atmosphere of captive energy.

The triumph of colour confidence was probably Susan Telford's red, pink and green dress and jacket, delineated with sculptured flowers, which, unlike most of the evening's offerings owed no obvious acknowledgements to last season's couture collections.

GAULTIER'S recent venture into external boning was responded to at regular intervals, most notably in a curious black and green sculpted wedding dress. Jasper Conran and John Galliano both breathed their beady influences into the rich embroidery and chiffon of Donna Lewis's blouses, breeches and pants, but there was an occasional showstopper nonetheless. Pauline Collins should be a name to watch, as her grey ruched blouson top, skirt and leggings stretched the limits of the fashion designer towards the three-dimensional vision of the sculptor.

Other intimations of (potential) immortality came from Gillian Orr, with an olive green silk gents' suit and Claire Hendry whose men's chiffon flares added avant-garde bite.

All in all, the Class of '89 had a lot to be proud of, though their deliberately moody and mannered show displayed a lot that could be improved on, too. If such obviously talented young designers would trust more to their own instincts, (as Geraldine Mitchell did with her glorious printed hats, swimwear and duffle bags), then the shows of the future could be more fun, more inspired and altogether more salesworthy.

Iain McCallum

Jane Torday examines how a fatal attraction has become a drag

Light at the end of the haze

THE theatre is packed. On stage, faint ribbons of smoke coil like chiffon around Maria Aitken, as she paces restlessly about the cream drawing room set. The smoke drifts over the audience and it smells very pleasant, not any old vulgar Virginian, and yet not entirely French or Turkish, but something rather herbal which will not offend the non-smokers in the audience.

Continuous smoking is required of the actors in the current production of Noel Coward's *The Vortex*, focusing as it does on a fading society beauty who is about to lose her 1920s toyboy and gain a much sadder young man, her own son of the same age who returns home from Paris as a cocaine addict.

In Coward's day, smoking was seen as an entirely civilised pleasure in sophisticated circles. At the other end of the scale, it was one of the few forms of pleasure available to workers on low incomes with little leisure, though some still insisted it was a decadent and disgusting habit.

Cigarette smoking really took off as a glamourous activity for the masses under the influence of Hollywood. Marlene Dietrich and other screen idols used cigarettes a to accentuate their smouldering looks — and the public followed suit.

Smouldering star: Marlene Dietrich set the trend

Victorian country houses had smoking rooms which were exclusively male preserves. Before that time, prejudice against smoking in polite society was often as pronounced as it has become again today. The Duke of Wellington made his guests smoke, if smoke they must, in the servants hall. The difference today is that smokers and non-smokers alike are united in agreement that smoking is an unhealthy habit.

Smoking has become unacceptable for valid reasons, and even a small child can tell you what those are: smoking can endanger your health and can cause lung cancer and heart disease. Cigarettes are seen as an implement of death and have become an unwelcome intrusion into the lives of non-smokers because of the threat of passive smoking.

But it has now got to the stage that when moderate smokers (who smoke a few cigarettes a day, or less than a packet a week) light up at a social function, they are increasingly regarded as murder suspects instead of mere suicidal simpletons. Non-smoking hosts often can't find ashtrays for their guests, having either burned them or given them away to jumble sales.

National No Smoking Day is thus a glorious opportunity for non-smokers and reformed smokers to point the finger at their fallen friends and relations who still indulge. Whether or not this year's No Smoking Day will provide a genuine opportunity for smoking sinners to renounce the evil of their ways and tread henceforth on the path of righteousness, remains to be seen.

Some interesting advice is offered in the leaflet for No Smoking Day, compiled by the admirable "Europe against Cancer" group: "Instead of smoking, breathe deeply, move around, do something else, think about the harm that cigarettes will do you, clean your teeth, take sips of water."

Today, could prove to be almost as funny as Red Nose Day on March 10, if wherever we go we can be entertained by haunted figures with anxious expressions, breathing deeply and moving around aimlessly in between teeth-cleaning sessions and sips of water. Tea and coffee will not be a comfort because as the leaflet says: "Tea and coffee can trigger off the smoking urge." The leaflet also advises us: "Go easy on the alcohol, it weakens resolve."

National Days dedicated to the promotion of tree-planting, adult education or breast feeding are all providing a form of social service. So are days dedicated to the reduction of the use of aerosol sprays, lead petrol and cigarettes. But perhaps another National Day could be added to this list. National Tolerance Day, when for 24 hours, we could all unashamedly abandon our usual prejudices against beliefs, lifestyles and habits which do not correspond with our own. The experience could prove to be so beneficial, so relaxing, so liberating, that smokers like myself might not feel the need for a cigarette the whole day long.

Figure 17 Typographical variations in a features page from *The Scotsman*. Drop letters and white space are used as eye-breakers in place of crossheads. The body setting is standard width, but with the main feature indented to show columnar white, and the column seven story set ragged right, or unjustified. The 'standfirst' under the headline is set bastard measure, also ragged right, to fill the space between the intro and the cut-out picture. Reproduced by permission

mixing it with cross-heads and other eye-catching ploys (see Chapter 12). Drop letters should be elegantly spaced out on a page projection so that they do not line up or cover each other. Rivers of white space can be caused by badly positioned drops or stand-ups.

Justified setting

The practice in newspapers, as with books, is to use justified setting that is setting that is squared off to an even margin down both sides of the column. It helps the eye by providing lines of the same width for it to scan. The machine provides artificially a setting system that breaks and hyphenates words where necessary in order to fill the lines to meet the spacing parameters. This process is known as hyphenating and justifying, or to H and J. The only relief in solid setting is the normal indent on the left of the column at the start of each paragraph.

In the case of computerized systems a pattern of hyphenation is programmed in at the outset to ensure that words are correctly broken syllabically, avoiding breaks in one-syllable words or proper names thus following the style practised under hot metal.

Unjustified setting

Justified setting has remained unchallenged from the days of the *Gutenberg Bible* – even from the period of quill pen manuscripts – to the present day. Now advertising and magazine design has brought unjustified setting into fashion in an attempt to appear different or give attractive variation (Figure 17).

A cardinal rule here is never to unjustify at the front end. The eye requires a constant reference line on the left and is not equipped for the shock of having to look for the start of the next line. On the occasions when unjustified right was used under the hot-metal system the Linotype operator would be told to set the text 'ragged right', allowing the machine to run to the extremes of a set width. Problems would occur by massively unregulated indents on one-word lines, dropping a full word to the next line. In this respect the electronic method is more accurate and better regulated. There can be commands even for 'soft unjustification' or 'hard unjustification' depending on how the designer wants the space to be used. Contrary to expectation, the text does not usually take up more space than with justified setting.

Unjustified setting is little called for on news pages, though it can give effective variation and a feeling of lightness on features, provided measures are not wide or the text long. Used in parts of the paper on a regular basis it can key the reader into a specific feature such as the editorial comment – but beware of overkill! Blanket use of such setting can weary the eye and remind the reader of the familiar comfort of normal justified lines with their regular traverse for the eye. Beware also the temptation of using ragged right on too narrow a measure. At 8 ems and under this could result in one-word turns and be unreadable.

Setting variations

Variations in body setting are used for two reasons: to give emphasis to certain parts of a text, or to certain stories; and to ease the eye in its perusal of the pages.

Most body setting has italic and bold variants, or is used along with a bold intro setting of a compatible type. Special stress can be given to the odd paragraph by setting in bold or italic, while particular stories on a page can likewise be differentiated by being set in bold or (less commonly) in italic.

Paragraphs

While H. W. Fowler (of *Modern English Usage*) describes a paragraph as 'essentially a unit of thought, not of length' he also says: 'The purpose of a paragraph is to give the reader a rest.' In the case of newspapers the latter function takes precedence. With type in columns long paragraphs are tiring to the eye and make the text look solid and unattractive and so, depending on the pace of the writing, newspaper paragraphs tend towards shortness. This gives the eye the relief of extra white space resulting from the pica or 6 point paragraph indent. Some quality papers also relieve their longer texts by putting a 2 point space between paragraphs, as well as setting stock type on a larger body to give fine linear white.

Crossheads

A much used method to provide eye breaks with runs of text of more than 4 or 5 inches is to drop in crossheads, which are usually single lines of display type in a regular stock face of the house taking one key word from the part of the story that follows. Sometimes, for uniformity, the type chosen is a smaller version of the headline type of the story. (See development of this in Chapter 12.) Some newspapers use pieces of decorative rule in place of crossheads. Features pages sometimes rely simply on drop letters with a pica of space above, or run the text with heavy leading (ledding) between paragraphs making crossheads unnecessary (as on page 35).

Indention

Another way to vary setting is to indent it either 6 points (nut) front or 6 points each side (n.e.s.), which gives vivid columnar spacing (Figure 17) to set a story off against its fellows. A variation of this is the *hanging* or *reverse* indent, which has the first line of a paragraph standing proud full out and the rest indented usually 6 points. This works best with bold setting and with single column type. It is also known as nothing and one, or 0 & 1, and has periods of vogue on various newspapers, often being rediscovered by new editors.

Bastard measure is any type setting of non-standard width (Figure 17) and can occur against a non-standard width picture. Sometimes a story will start, for design purposes, in a bastard measure intro, and then turn into single column. Type is also set bastard measure to fit into panelled shapes to suit a page design, in which certain texts are enclosed in a frame of print rules or borders. As with other type and setting variants the designer should avoid using too many of the devices in the same page or design format.

Headlines

Headlines are a dominant element in any page design. At its simplest, headline typography consists in the use of a few chosen types and type sizes to produce page designs that conform generally to a house style and give the pages their particular visual character. There are factors in the constructing and placing of headlines that are common to all newspapers. For instance, as we have seen, all pages have a lead story with a dominant headline. This can either take the form of a *banner* across the page, or a number of lines stacked on top of each other. The use of a *strap-line,* in smaller type, above the main headline is also common. Likewise the descending order of importance of stories is indicated by items of double or single column in reducing order of type size as shown in the tabbed page (Figure 18). And yet within these parameters there are plenty of variations of approach.

In many broadsheets and the more traditional of tabloid papers the practice is to design pages to a modular or formula style with headlines of predictable shape and size, and even position (see page 146). With the exception, perhaps, of the lead story the headline shapes come first and the words are generally made by the subeditors to fit. Headlines are located in regular sizes and widths, sometimes centred and sometimes set left according to a paper's style, occasionally with a wordy single column kicker to give variety to the middle of the page.

Features pages in most newspapers tend towards free-thinking in headline display (see pages 160 and 162). It is important, with this approach, that the shape and size of the headline on the lead story should reflect the mood of the headline words and the projection decided upon, even if some of the supporting stories are of more conventional shape.

Here, the working relationship between subeditor and page designer (the same person on many provincial newspapers) is vitally important as the page begins to take shape on the layout pad or screen. This can entail perhaps a main headline consisting of one or two words, exaggeratedly big, with perhaps a wordy, discursive second deck. It can have a mood headline linked to a dominant picture (see page 162). It can involve special type, or type used in an innovative way to match the mood or content of the text. A sense of fun or a feeling of tragedy can be induced by the choice of words or type. It can entail layout in which pictures rather than headlines become the dominant element. It is an approach to design that offers a more challenging and imaginative world for the designer, though the context of a type style remains important. (See development of this in Chapters 9, 10 and 11.)

Type choice

Fashion can dictate type use and the prevailing fashion is for wholly lowercase headlines, especially in those newspapers using a serif type dress. Lowercase type, because of its varied contours, is considered to be generally more readable, yet the danger, in unsophisticated hands lies in producing pages of a lulling uniformity – a fault that occurs even in such a design-conscious paper as *The Guardian*. For those papers using a free-style approach or seeking impact, especially on features pages, an embargo on headlines in caps can be constricting, the page one splash headline being a particular victim of this fad. However, the careful juxtaposing of light and bold lowercase faces and a balanced variation in sizes can produce pages of great elegance, especially where there is good picture content, as with *The Times* or *The Independent*.

The essential thing in type choice (examples on pages 159 and 177) is not to introduce too many permitted variants but to utilize type weights and sizes of a given range so that the page has a sense of pattern. Of the main variations *italic* is one that has been discontinued on many papers. Its use now is often as a form of labelling, sometimes reversed as white on black, to indicate to the reader where a particular subject or section is regularly located.

In more traditional layout styles, special emphasis within a type range can be given by the occasional multiple line heading of smaller than usual size with extra line spacing, while a predominantly sans format can give emphasis to a soft feminine subject by using a fine sensitive serif face as an alternative.

The main difference between the broadly traditional pages and the free-style tabloid approach lies in the type sizes used. While 72 point is often the largest type on a traditional broadsheet page, it is customary, in the search for impact, to go up to 120 point or even 144 point and bigger, on a poster-style tabloid front page or Sunday supplement feature spreads.

Arrangement

A *strap-line* can run either above the main headline, or be bunched together alongside it. It should be about four sizes down from the main head, in order to fulfil its explanatory role and not to appear

ANATOMY OF A MIRROR FRONT

MASTHEAD, LOGO or SEAL

DATE LINE

WHITE ON BLACK (WOB)

BOLD - SINGLE COLUMN

BY LINE

INDENTED COLOUR PICTURE

SOLID BLOBS

42 POINT SUB-HEAD

DAILY Mirro

Friday, April 28, 1989 National Sale: 4,025,418 Incorporating the Daily Rec

Life for boy who wanted to wipe out the world

By FRANK CORLESS

A BRILLIANT science student experimented with deadly chemicals in a plot to destroy the human race.

He planned to invent a deadly plague and use mutant flies to spread it.

A judge, ordering evil Matthew Williams to be locked up for life yesterday, said he was clearly highly intelligent . . . and VERY dangerous.

Williams, 20, whose father once stood as a National Front candidate, conducted a terror campaign, leaving bombs in phone boxes and launching arson at-

PLAN: Williams

tacks. But it wasn't until he fired a crossbow at neighbours because their radio was too loud that police caught him.

They then discovered the secret laboratory in his quiet Merseyside home — and his diaries.

The diaries told how in his lab, under a large picture of Hitler, he planned the destruction of the human race.

In his diaries, Williams — who had 10 O-Levels and 4 A-Levels — wrote: "I hate people. The majority of people I come into contact with are filthy, ignorant, agressive scum who should not exist.

"I intend to destroy them all by whatever means I can."

● Full story — Page 9

FIN EVI PAI

By SYLVIA JONES and RAMSAY S

AN evil couple are being hunted by police investigating the baby food blackmail plot.

The pair hatched their twisted scheme to spike tots' food FIVE YEARS ago, it was revealed last night.

● £18,000 taken from joint account

● Now they ask for another £1m

HILLSBOROUGH APPEAL £302,850

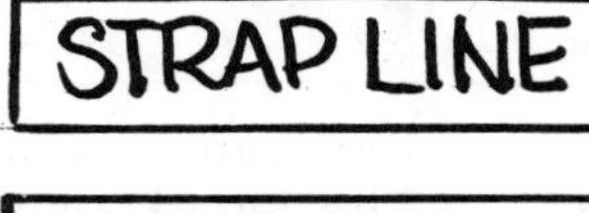

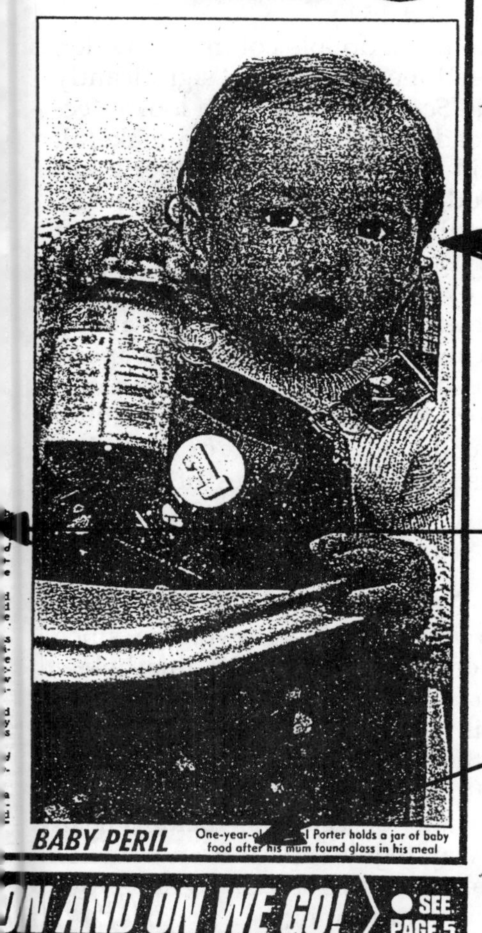
REVEALED: THE BLACKMAILERS' FIVE-YEAR PLOT

THIS

BABY PERIL One-year-ol l Porter holds a jar of baby food after his mum found glass in his meal

ON AND ON WE GO! ● SEE PAGE 5

Figure 18 The anatomy of a page – the designer's terminology demonstrated in this tabbed explanation of a *Daily Mirror* page one

too dominant, and should be lowercase on top of caps or vice versa. The length of the lines is a matter of fitting the page design. It is sometimes an advantage, depending on the design style of the paper, to reverse the strap as a white on black (WOB) or white on 50 per cent tint (WOT), with the dramatic alternative of reversed colour panels. Another variation is to run the strap in black type on a lighter (25 per cent) tint. It should be well separated by space from the main head, a reversed strap requiring more space. Underscored straplines can look effective, provided underscores (usually in 2 point or 4 point) are not overused on the page. Contrast can be enhanced with a very light typeface on top of a thicker black underscore of even up to a pica in the case of bold tabloid pages. The thicker the underscore the more massive should be the space below it. A useful guide is that a 72-point headline and accompanying strap with, say, a 4-point underscore, needs to be separated from the intro below by 18 points of space, with 24 points of space if the main headline or underscore is bigger. (See examples on pages 157 and 159.)

Centred headlines, whether single column, double column or wider, are now in style, although no headline should be significantly narrower than the space allotted to it. *Set-left headlines* can look effective but only as a variation to a broadly centred style. To make all headlines set left, as some papers did at one time, has neither point or advantage to it, and often left splodges of white all over the page. Headlines on the right-hand edge of the page should not be set left on account of the raggedness it leaves on the outside column. *Set-right headlines* baffle the reading eye and should not be used. *Staggered heads* with the first line set left and the last set right are an American fashion which has never caught on in British newspapers.

Two-deck headlines are commonly used and enable two angles of a story to be explored. The second deck (in fact, a separate headline) should be a smaller version of the main type. *Three-deck* or *multi-deck* heads are seldom found these days outside the pages of the *Wall Street Journal*.

Line spacing

Capital letter headlines are easier to line-space than lowercase, since there is not the problem of ascenders rising out of the line below against descenders hanging down from the top line. Whatever system has been formatted, it is still visual experience that counts here. To make the space acceptable the format should be overridden at the keyboard with a base-line shift. With cut-and-paste the scalpel can be used to 'overlap' ascenders and descenders, tucking each neatly in to the white space. In this way regular space from x-line to x-line can be maintained. With small sizes this is not possible, but then the space inconsistencies are not so affronting.

The general rule is to maintain a consistent standard of line spacing using computer formats as far as possible. Excessive line space makes a headline look weak, and is also a waste. As far as possible a line should fill laterally the space it is given, unless indented white is used for effect. Space, remember, is there to aid the eye. (For general letter spacing in typesetting see pages 31–2.)

Effect of the computer

The digital conversion of typefaces from original matrices for the purposes of computer setting has enhanced the work of the old type designers.

A leading method in this digitizing has been the Ikarus Concept, developed by the German firm URW, of Hamburg. One of the typesetter manufacturers using it is Autologic, who describe in their *Digital Type Collection Book* how the work is done: 'The production process begins with the drawing of each character based on the historical development of the face. Along with original drawings come preliminary space specifications. Since Ikarus deals with contours rather than with bit-mapped information, the necessary outline points are defined and fed into the system for processing. . . . After conversion from Ikarus to digital format, each character is studied carefully on a bit-mapped screen and edited as needed (Figure 19). The kind of refinement accomplished in this way is exemplified by Autologic's rendition of Athena, originally the Linotype Corporation's Optima face.'

The change of name of familiar faces which is common in the new digitized type masters is a tacit acknowledgement that what is being offered to clients is an edited refinement, or improvement, of a known typeface in the same way that in the past, tried and popular faces were constantly being improved upon and brought out under variant names.

In the process of refinement subtle bows and cuppings are retained even down to the 10 point master through the patterning techniques now being used. There could have been no such advanced refining by even the most modern metal type-cutting experts. Master matrices would not have had the efficiency of reproduction that applies with computer outputs. Herein lies the breakthrough in modern computerized typesetting.

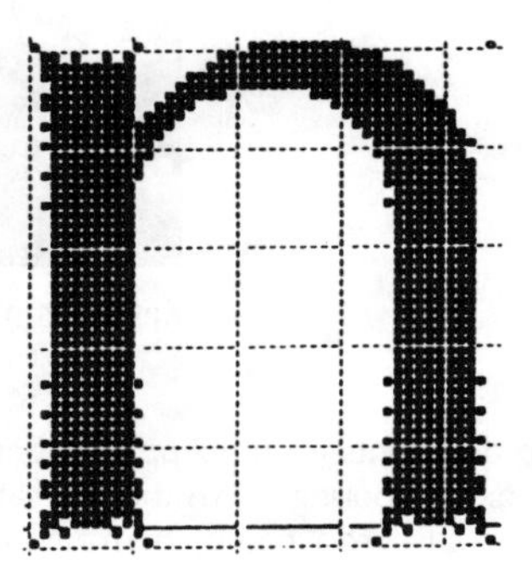

Figure 19 The Ikarus concept: this enlargement of a 10 point Athena 'n' is produced from the bit structure on the right which is based on a recreation by Ikarus of the original letter

Using Key Caps

1. **Choose Key Caps from the Apple menu.**

   The menu bar changes and this window appears:

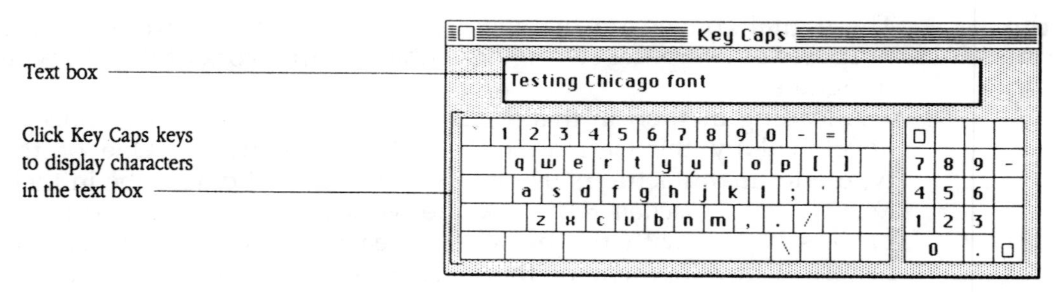

Key Caps

The window shows a keyboard template, and the font displayed first is the system font, Chicago, used extensively by the Macintosh. But you can change the template to show any font installed in the startup System file.

2. **To see different characters, pull down the Key Caps menu and choose another font from the list displayed.**

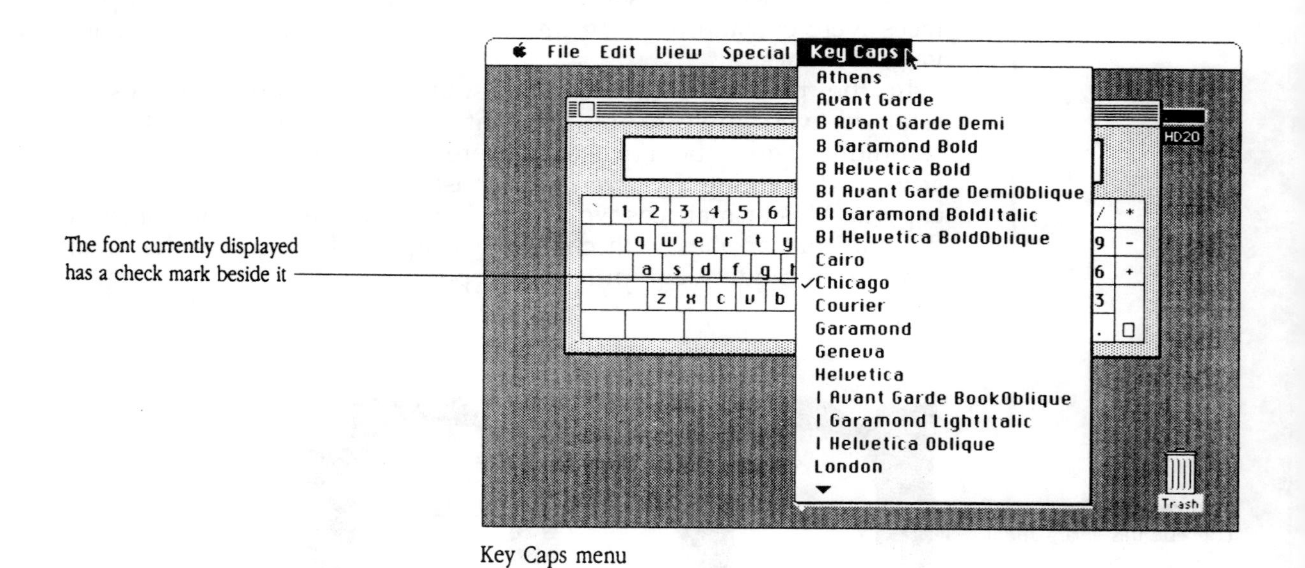

Key Caps menu

Figure 20 Calling up type fonts on the Apple-Macintosh: (a) choosing key caps from the Apple menu displays the characters in the text box; (b) pulling down the key caps menu lists the fonts available, with a check mark showing against the one currently displayed. Reproduced by permission

The notes in the Autologic manual are instructive to systems users in that they point up the subtlety of individual sizes and counsel caution in the use of compensatory kerning and other house methods to vary letter spacing. The manual continues: 'Proofs showing all relevant character combinations are generated for each master size and put through a final spacing check. If necessary, adjustments are made and proofs submitted to quality control for approval. For smaller masters, to retain the appearance of continuity with the large masters, it is often required to modify individual character structures. The ability to detail fonts in this manner is one of the advantages of a multimaster database arrangement. A 9 point typeset image does not and should not emanate from the same data as a 36 point image of the same character.'

An oddity of type design is that round letters, in addition to needing to be closer to each other than square characters, also need to be marginally deeper in body in order to convey visual balance. It is for this reason that the practice of the original type designers, who made round letters – o, e, c, etc. – in a given size fractionally deeper than their fellows, has been followed in the digitizing process. The justifiability of this difference in actual height between round letters and square did not stop some old-time printers from trimming off the tops and bottoms while tidying up a page!

Unless careful checks are made and formatting carried out on a house-style basis at the fitting-out stage in a computerized system, there will be dissatisfaction and messy changes to make later, and the type character of the newspaper may be damaged. A fully trained systems person and an editorial expert should combine to examine the countless permutations and possibilities offered and to see that sizes and spacing requirements are formatted exactly as wanted, while at the same time making discreet adjustments that might be overdue. A useful function, for example, is to have paragraph spacing on a command key. Even drop letters and big quotes can be so programmed. But above all, in a system where types are being presented under new names, an editor or proprietor must be able to ensure that a successful newspaper is not landed with unwelcome changes in visual appearance. To cover negligence at the fitting-out stage by suggesting to the readers that a newspaper is undergoing a vibrant and exacting re-design can presage disaster.

A variation of the saw-happy old warrior who trimmed off metal letters is the philistine subeditor who instructs the machine to 'squeeze it' to prevent a wanted headline from 'busting'. This is the way to destroy the careful original programming that was carried out to retain type balance in the pages. Such a headline will instantly look wrong. There are occasions when, as a deliberate plan, the versatility of the computer can be utilized – when a headline stands alone on a page of pictures, for example.

In the early days of newspaper design, one of the artists whose job it was to enliven the features pages would be asked to read the copy and come up with a drawn headline which reflected the mood of the piece. These exponents of incidental type design are redundant in the face of the achievements of systems and systems people whose creative key-stroking produces undreamed of typographical effects.

4
Historical influences

Newspaper design is an organic thing; it is not a few notions thought up on the spur of the moment aided by a quick glance through a type book, important though type is. The concepts behind page layout have been refined through more than 200 years of evolution in the use of typography and response to readership needs. An examination of early papers and their readers will disclose a world far removed from the sophisticated products we are familiar with, and yet embedded in them are the roots of today's practices.

Early newspapers

The first newspapers, carrying mainly foreign news because of the censor, began circulating in Britain in the 1620s though few copies have survived. With the outbreak of the Civil War, newspapers were harnessed by each side along with pamphlets in the propaganda war. *A Perfect Diurnall of the Proceedings* of Parliament, of 1645, of which copies can still be found (Figure 21) shows vividly how these partisan early journalists and editors set about wooing their readers on their front pages during the war of words.

Through the Commonwealth period and the restoration of the Stuart monarchy from 1660 the censor clamped down again and newspapers, hampered as they were in the coverage of national and political news (Parliament was not open to the press until the 1780s), concentrated on performing a service for the sort of people who could afford to buy them. The heavy *Stamp Duty* imposed in 1711, and which was to last, at sums that varied to suit government policy, until 1855, ensured this, though there were some short-lived political sheets that circulated illegally untaxed.

As a result of the cost, readers comprised mainly farmers, landowners, brewers, merchants, traders, the clergy, the aristocracy and academics. As newspaper publishing took off through the eighteenth century it was inevitable that editors (who were frequently the printers) went strongly for shipping news, grain and

Figure 21 A *Perfect Diurnall*, December 15 1645

(691) Numb.124

A Perfect Diurnall

OF SOME

PASSAGES

IN

PARLIAMENT,

And from other parts of this Kingdome.

From Munday the 8. of *Decem.* till Munday the 15 of *Decem.* 1645.

Collected for the satisfaction of such as desire to be truly informed

Printed for *Francis Coles*, and *Laurence Blaikelock*, and are to be sold at their Shops in the *Old-Baily*, and at *Temple-Barr*.

Beginning Munday the 8. of *December.*

The House of Commons spent much time this day in debate of the businesse of *Ireland*, and how the Government of that Kingdome for the future should be setled; whether in one hand; or more; and after much debate thereupon it was voted. *That the Government of the said Kingdome should be put into one hand, and to continue for the space of one yeare but not longer, and so a new Governour to be chosen every yeare, and the further debate of this businesse was put off till Munday next.*

A Letter was read in the Commons House from a Gentleman of worth and quality in Sir *Thomas Fairfax* Army setting forth the great necessities of the Souldiers of that Army for Cloathes for this winter season and the House thereupon entring into debate, ordered that 6000. suits of Apparell should be speedily provided and sent to them, which we hope will bee accordingly dispatched without Intermission as the occasion requires and calls out for.

The City of *London* being formerly desired to advance the summe of 30000. l. for the Scots Army in this Kingdom upon their advance to beseige *Newark* and upon the further renewing of that ... The City having expressed much willingnesse to advance the said summe the House of Commons taking speciall notice

Iiiiii

The further proceedings upon the Propositions for Peace. The result upon the Letter from the King to send Propositions to the Parliament. As also a new Order of Parliament giving longer time to Delinquents to come in to the Parliament, and compound for their Delinquency with the penalty set upon such as shall refuse to come in according to this last Order. The last Intelligence from Chester and from Sir Thomas Fairfax Army, as also the proceedings of the besiegers against Newarke and Belvoyre Castle, An accompt of some late Occurrences from Oxford, and from P. Rupert and Maurice, now againe with the King at Oxford, with severall passages of newes from divers parts.

stock prices, the weather, royal news, law reports, social gossip and the advertising of goods and services appropriate to their patrons.

This modest content was presented usually in four of what we would call tabloid-sized pages, though the broadsheet size was becoming more common by the end of the century. Newspaper circulation was concentrated in London and the home counties, by far the biggest population centre, though the first provincial weekly paper, the *Norwich Post*, had appeared as early as 1701, in Britain's then second biggest city.

A four-column format was usual, with simple labels or even just place names above each item. *The Public Ledger* for September, 1760, for instance (Figure 22) carries on page one the very detailed running order for the Coronation of George III under the heading of PLAN OF THE PROCESSION AT THE CORONATION, a poem on the occasion entitled THE PATRIOT CORONATION, a letter TO THE PRINTER OF THE PUBLIC LEDGER, an item headed STATE LOTTERY 1761 and a collection of shipping movements datelined from various ports, under the heading of PORT NEWS, turning at the foot of column four without warning to the next page (turn lines had yet to be invented).

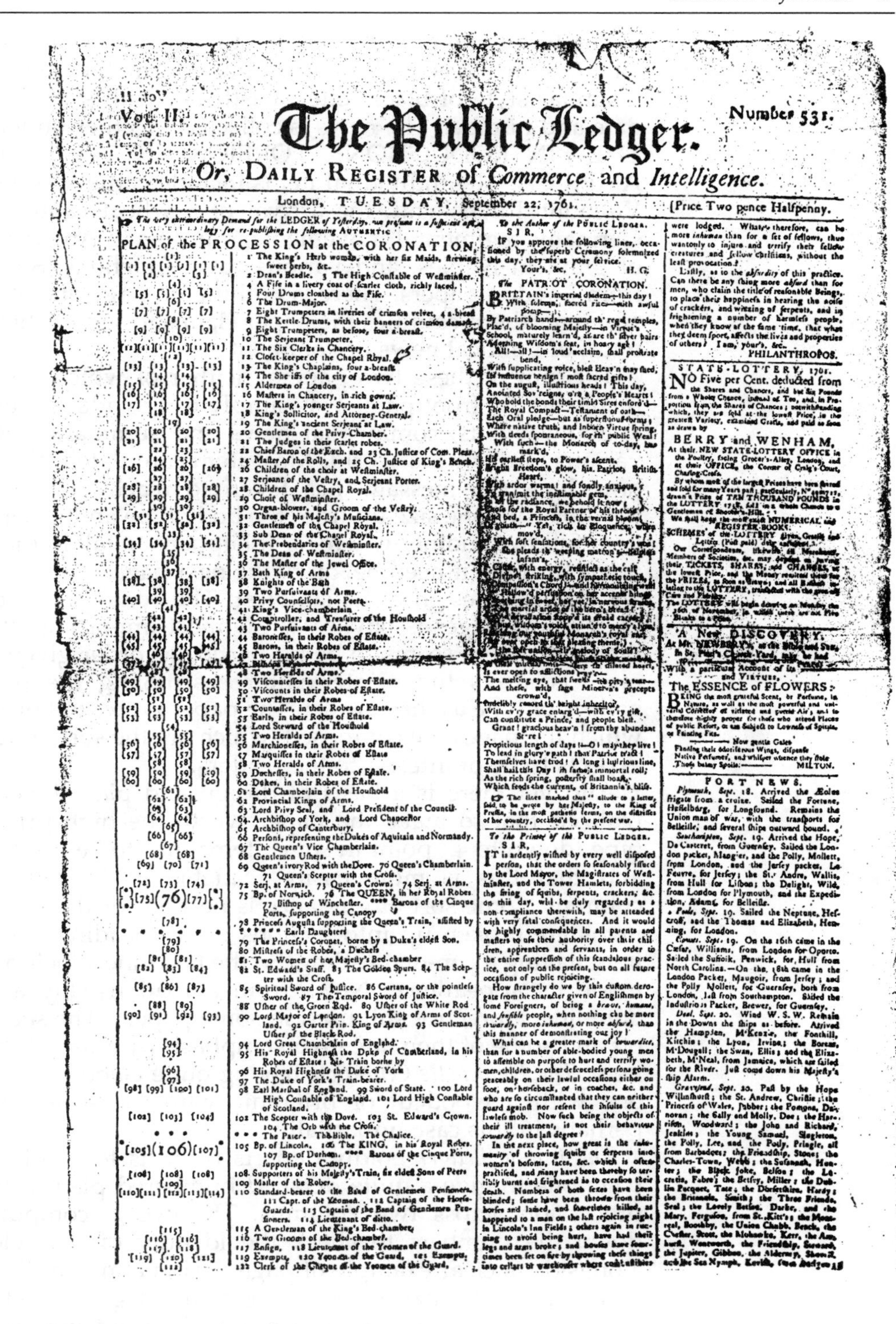

Vol. II.

The Public Ledger.

Number 531.

Or, Daily Register of Commerce and Intelligence.

London, TUESDAY, September 22, 1761. [Price Two pence Halfpenny.

The very extraordinary Demand for the LEDGER *of Yesterday, we presume is a sufficient apology for re-publishing the following* AUTHENTIC

PLAN of the PROCESSION at the CORONATION.

1 The King's Herb woman, with her six Maids, strewing sweet herbs, &c.
2 Dean's Beadle. 3 The High Conſtable of Weſtminſter.
4 A Fife in a livery coat of ſcarlet cloth, richly laced.
5 Four Drums cloathed as the Fife.
6 The Drum-Major.
7 Eight Trumpeters in liveries of crimſon velvet, 4 a-breaſt.
8 The Kettle-Drums, with their banners of crimſon damaſk.
9 Eight Trumpeters, as before, four a-breaſt.
10 The Serjeant Trumpeter.
11 The Six Clerks in Chancery.
12 Cloſet-keeper of the Chapel Royal.
13 The King's Chaplains, four a-breaſt.
14 The Sheriffs of the city of London.
15 Aldermen of London.
16 Maſters in Chancery, in rich gowns.
17 The King's younger Serjeants at Law.
18 King's Sollicitor, and Attorney-General.
19 The King's ancient Serjeant at Law.
20 Gentlemen of the Privy-Chamber.
21 The Judges in their ſcarlet robes.
22 Chief Baron of the Exch. and 23 Ch. Juſtice of Com. Pleas.
24 Maſter of the Rolls, and 25 Ch. Juſtice of King's Bench.
26 Children of the choir at Weſtminſter.
27 Serjeant of the Veſtry, and Serjeant Porter.
28 Children of the Chapel Royal.
29 Choir of Weſtminſter.
30 Organ-blower, and Groom of the Veſtry.
31 Three of his Majeſty's Muſicians.
32 Gentlemen of the Chapel Royal.
33 Sub Dean of the Chapel Royal.
34 The Prebendaries of Weſtminſter.
35 The Dean of Weſtminſter.
36 The Maſter of the Jewel Office.
37 Bath King of Arms
38 Knights of the Bath
39 Two Purſuivants of Arms.
40 Privy Counſellors, not Peers.
41 King's Vice-chamberlain.
42 Comptroller, and Treaſurer of the Houſhold
43 Two Purſuivants of Arms.
44 Baroneſſes, in their Robes of Eſtate.
45 Barons, in their Robes of Eſtate.
46 Two Heralds of Arms.
47 Biſhops [illegible]
48 Two Heralds of Arms.
49 Viſcounteſſes in their Robes of Eſtate.
50 Viſcounts in their Robes of Eſtate.
51 Two Heralds of Arms
52 Counteſſes, in their Robes of Eſtate.
53 Earls, in their Robes of Eſtate.
54 Lord Steward of the Houſhold
55 Two Heralds of Arms.
56 Marchioneſſes, in their Robes of Eſtate.
57 Marquiſſes in their Robes of Eſtate
58 Two Heralds of Arms.
59 Ducheſſes, in their Robes of Eſtate.
60 Dukes, in their Robes of Eſtate.
61 Lord Chamberlain of the Houſhold
62 Provincial Kings of Arms.
63 Lord Privy Seal, and Lord Preſident of the Council.
64 Archbiſhop of York, and Lord Chancellor
65 Archbiſhop of Canterbury.
66 Perſons, repreſenting the Dukes of Aquitain and Normandy.
67 The Queen's Vice Chamberlain.
68 Gentlemen Uſhers.
69 Queen's ivory Rod with the Dove. 70 Queen's Chamberlain. 71 Queen's Scepter with the Croſs.
72 Serj. at Arms, 73 Queen's Crown. 74 Serj. at Arms.
75 Bp. of Norwich. 76 The QUEEN, in her Royal Robes 77 Biſhop of Wincheſter. **** Barons of the Cinque Ports, ſupporting the Canopy
78 Princeſs Auguſta ſupporting the Queen's Train, aſſiſted by ****** Earls Daughters
79 The Princeſs's Coronet, borne by a Duke's eldeſt Son.
80 Miſtreſs of the Robes, a Ducheſs
81 Two Women of her Majeſty's Bed-chamber
82 St. Edward's Staff. 83 The Golden Spurs. 84 The Scepter with the Croſs.
85 Spiritual Sword of Juſtice. 86 Curtana, or the pointleſs Sword. 87 The Temporal Sword of Juſtice.
88 Uſher of the Green Rod. 89 Uſher of the White Rod
90 Lord Mayor of London. 91 Lyon King of Arms of Scotland. 92 Garter Prin. King of Arms. 93 Gentleman Uſher of the Black Rod.
94 Lord Great Chamberlain of England.
95 His Royal Highneſs the Duke of Cumberland, in his Robes of Eſtate: his Train borne by
96 His Royal Highneſs the Duke of York
97 The Duke of York's Train-bearer.
98 Earl Marſhal of England. 99 Sword of State. 100 Lord High Conſtable of England. 101 Lord High Conſtable of Scotland.
102 The Scepter with the Dove. 103 St. Edward's Crown. 104 The Orb with the Croſs.
* * * The Pater. The Bible. The Chalice.
105 Bp. of Lincoln. 106 The KING, in his Royal Robes. 107 Bp. of Durham. **** Barons of the Cinque Ports, ſupporting the Canopy.
108. Supporters of his Majeſty's Train, ſix eldeſt Sons of Peers
109 Maſter of the Robes.
110 Standard-bearer to the Band of Gentlemen Penſioners. 111 Capt. of the Yeomen. 112 Captain of the Horſe-Guards. 113 Captain of the Band of Gentlemen Penſioners. 114 Lieutenant of ditto.
115 A Gentleman of the King's Bed-chamber.
116 Two Grooms of the Bed-chamber.
117 Enſign, 118 Lieutenant of the Yeomen of the Guard.
119 Exempts, 120 Yeomen of the Guard, 121 Exempts,
122 Clerk of the Cheque of the Yeomen of the Guard,

To the Author of the PUBLIC LEDGER.
SIR,

IF you approve the following lines, occaſioned by the ſuperb Ceremony ſolemnized this day, they are at your ſervice.

Yours, &c. H. G.

The PATRIOT CORONATION.

BRITAIN'S imperial diadem—this day!
With ſolemn, ſacred rites—with awful pomp—
By Patriarch hands—around th' regal temples,
Plac'd, of blooming Majeſty—in Virtue's School, maturely learn'd, as are th' ſilver hairs
Of ripening Wiſdom's feat, in hoary age!
All!—all!—in loud acclaim, ſhall proſtrate bend,
With ſupplicating voice, bleſt Heav'n may ſhed,
Its influence benign! moſt ſacred gifts!
On the auguſt, illuſtrious heads! This day,
Anointed Sov'reigns, o'er a People's Hearts!
Who hold the bonds their timid Sires enforc'd—
The Royal Compact—Teſtament of oath—
Each Oral pledge—but as ſuperfluous forms;
Where native truth, and inborn Virtue ſpring,
With deeds ſpontaneous, for th' public Weal!
With ſuch—the Monarch of to-day, has mark'd,
His earlieſt ſteps, to Power's aſcent.
Bright Freedom's glow, his Patriot, Britiſh Heart,
[illegible]

[illegible]
Is ever open to afflictions pray'r—
The melting eye, that ſwells—in pity's tear—
And theſe, with ſage Minerva's precepts crown'd,
Indelibly recorded th' bright inheritor,
With ev'ry grace enlarg'd—with ev'ry gift,
Can conſtitute a Prince, and people bleſt.
Grant! gracious heav'n! from thy abundant Store!
Propitious length of days!—O! may they live!
To lead in glory's path! that Patriot track!
Themſelves have trod! A long illuſtrious line,
Shall hail this Day! in fame's immortal roll;
As the rich ſpring, poſterity ſhall boaſt,
Which feeds the current, of Britannia's bliſs.

☞ The lines marked thus " allude to a letter, ſaid to be wrote by her Majeſty, to the King of Pruſſia, in the moſt pathetic terms, on the diſtreſſes of her country, occaſion'd by the preſent war.

To the Printer of the PUBLIC LEDGER.
SIR,

IT is ardently wiſhed by every well diſpoſed perſon, that the orders ſo ſeaſonably iſſued by the Lord Mayor, the Magiſtrates of Weſtminſter, and the Tower Hamlets, forbidding the firing of ſquibs, ſerpents, crackers, &c. on this day, will be duly regarded; as a non-compliance therewith, may be attended with very fatal conſequences. And it would be highly commendable in all parents and maſters to uſe their authority over their children, apprentices and ſervants, in order to the entire ſuppreſſion of this ſcandalous practice, not only on the preſent, but on all future occaſions of public rejoicing.

How ſtrangely do we by this cuſtom derogate from the character given of Engliſhmen by ſome Foreigners, of being a *brave, humane*, and *ſenſible* people, when nothing can be more *cowardly*, more *inhuman*, or more *abſurd*, than this manner of demonſtrating our joy!

What can be a greater mark of *cowardice*, than for a number of able-bodied young men to aſſemble on purpoſe to hurt and terrify women, children, or other defenceleſs perſons going peaceably on their lawful occaſions either on foot, on horſeback, or in coaches, &c. and who are ſo circumſtanced that they can neither guard againſt nor reſent the inſults of this lawleſs mob. Now ſuch being the objects of their ill treatment, is not their behaviour *cowardly* to the laſt degree?

In the next place, how great is the *inhumanity* of throwing ſquibs or ſerpents into women's boſoms, faces, &c. which is often practiſed, and many have been thereby ſo terribly burnt and frightened as to occaſion their death. Numbers of both ſexes have been blinded; ſome have been thrown from their horſes and lamed, and ſometimes killed, as happened to a man on the laſt rejoicing night in Lincoln's Inn Fields; others again in running to avoid being hurt, have had their legs and arms broke; and houſes have ſometimes been ſet on fire by throwing theſe things into cellars or warehouſes where combuſtibles

were lodged. Whatever therefore, can be more *inhuman* than for a ſet of fellows, thus wantonly to injure and terrify their fellow creatures and fellow chriſtians, without the leaſt provocation?

Laſtly, as to the *abſurdity* of this practice. Can there be any thing more *abſurd* than for men, who claim the title of reaſonable Beings, to place their happineſs in hearing the noiſe of crackers, and whizzing of ſerpents, and in frightening a number of harmleſs people, when they know of the ſame time, that what they deem ſport, affects the lives and properties of others? I am, your's, &c.

PHILANTHROPOS.

STATE-LOTTERY, 1761.

NO Five per Cent. deducted from the Shares and Chances, and but Six Pounds from a Whole Chance, inſtead of Ten, and, in Proportion [illegible], the Shares of Chances; notwithſtanding which, they are ſold at the loweſt Price, in the greateſt Variety, examined Gratis, and paid as ſoon as drawn by

BERRY and WENHAM,

At their NEW STATE-LOTTERY OFFICE in the Poultry, facing Grocer's-Alley, London, and at their OFFICE, the Corner of Craig's Court, Charing-Croſs.

[illegible]

We ſhall keep the moſt exact NUMERICAL and REGISTER BOOKS.

SCHEMES of the LOTTERY given Gratis [illegible]

[illegible]

the PRIZES, as ſoon as drawn; and all Buſineſs relating to the LOTTERY, tranſacted with the greateſt Care and Fidelity.

The LOTTERY will begin drawing on Monday the 16th of November, in which there are not Five Blanks to a Prize.

A New DISCOVERY.

At Mr. NEWBERY's, at the Bible and Sun, in St. Paul's Church-Yard, may be had [illegible]

With a particular Account of its [illegible] and VIRTUES,

The ESSENCE of FLOWERS:

BEING the moſt grateful Scent, or Perfume, in Nature, as well as the moſt powerful and univerſal Corrector of vitiated and putrid Air; and is therefore highly proper for thoſe who attend Places of public Reſort, or are ſubject to Lowneſs of Spirits, or Fainting Fits.

——— Now gentle Gales
Fanning their odoriferous Wings, diſpenſe
Native Perfumes, and whiſper whence they ſtole
Thoſe balmy Spoils. ——— MILTON.

PORT NEWS.

Plymouth, Sept. 18. Arrived the Æolus frigate from a cruize. Sailed the Fortune, Heſſelburg, for Longſound. Remains the Union man of war, with the tranſports for Belleiſle, and ſeveral ſhips outward bound.

Southampton, Sept. 19. Arrived the Hope, De Carteret, from Guernſey. Sailed the London packet, Maugier, and the Polly, Mollett, from London, and the Jerſey packet, Le Feuvre, for Jerſey; the St. Andre, Wallis, from Hull for Liſbon; the Delight, Wild, from London for Plymouth, and the Expedition, Adams, for Belleiſle.

Poole, Sept. 19. Sailed the Neptune, Heſcroft, and the Thomas and Elizabeth, Henning, for London.

Cowes, Sept. 19. On the 16th came in the Cæſar, Williams, from London for Oporto. Sailed the Suffolk, Penwick, for Hull from North Carolina.—On the 18th came in the London Packet, Maugoir, from Jerſey; and the Polly Mollett, for Guernſey, both from London, laſt from Southampton. Sailed the Induſtrious Packet, Brewer, for Guernſey.

Deal, Sept. 20. Wind W. S. W. Remain in the Downs the ſhips as before. Arrived the Hampden, M'Kenzie, the Fonthill, Kitchin; the Lyon, Irvine; the Boreas, M'Dougall; the Swan, Ellis; and the Elizabeth, M'Neal, from Jamaica, which are ſailed for the River. Juſt come down his Majeſty's ſhip Alarm.

Graveſend, Sept. 20. Paſt by the Hope, Willinſhurſt; the St. Andrew, Chriſtie; the Princeſs of Wales, Jabber; the Pomona, Dunovan; the Sally and Molly, Dee; the Harriſon, Woodward; the John and Richard, Jenkins; the Young Samuel, Singleton, the Polly, Lee, and the Polly, Pringle, all from Barbadoes; the Friendſhip, Stone; the Charles-Town, Webb; the Suſannah, Hunter; the Black Joke, Belſon; the Lucretia, Fabre; the Betſey, Miller; the Dublin Pacquet, Tate; the Dorſetſhire, Hardy; the Britannia, Smith; the Three Friends, Seal; the Lovely Betſey, Darke, and the Mary, Ferguſon, from St. Kitt's; the Montreal, Boothby, the Union Chabb, Beach, the Cheſter, Scott, the Mohawke, Kerr, the Amherſt, Wentworth, the Friendſhip, Steward, the Jupiter, Gibbon, the Alderney, Shone, and the Sea Nymph, Kevill, from Antigua [illegible]

Figure 22 *The Public Ledger*, September 22, 1761

The first issue of *The Newcastle Chronicle* of March 24, three years later (Figure 23) is of the same size and format, with page one having a similar style of black letter masthead, with subtitle and ruled-off date and number. The news is collected together under the headings THE INTRODUCTION, FOREIGN NEWS, LONDON and AMERICA. The individual items include cattle plague, the death of Viscount Townshend, the arrival of a wheat cargo, beef prices in Germany, regimental movements, murders in America, the shortage of grain in Naples, and various royal functions. Some are quoted from London papers. There is nothing on page one of interest to the seven northern counties the paper claims to serve. There is also not a whiff of political news or comment.

The aim of these two papers, neither of which carries advertisements on the front page, was simply to provide readers with public announcements, commercial news and summaries of events from various parts of the world. Most items are dated, the date showing how long the message has taken to reach the paper. The label headings are one size up from the body type and in capitals, and one item follows another wherever it happens to fall in the column, with items turning, if need be, on to the next page.

It is page design at its simplest. The content of the first item in each of the pages is obviously the most important on the page but thereafter there seems to be no particular order given to things and one visualizes the printer-editor assembling the pages from the galleys of type as they are set. The paper grows out of the contents of which the pages are composed, which are the sort of items the editor expects to be of interest to the readers.

And yet there is a basis of design and plan there. Already it is realized that to make the paper worth the high price, fairly small type (down to 4½ point in some items) has to be used to get a lot in, and that this is more easily readable if divided into narrower columns. Some sort of guidance is needed for the reader and so the items are broken up into sections under headings, sometimes each item having a run-on heading. So the headline, albeit primitively, has made its appearance. News from abroad usually appears with a date-line, a practice still used in some papers. The sections are ruled off and some of them are given emphasis by a drop letter on the first item; *The Newcastle Chronicle* having quite elaborate drops including one with a woodcut of Britannia with the scales of justice – or perhaps in this case impartiality.

The Public Ledger boldly departs from its four column format on page one to give a quite detailed numbered plan of the coronation procession across two columns. The eye is compulsively drawn down the numbers to check the identity of the celebrities listed alongside, a clear example of page design aimed at the reader's eye. One could visualize it as a subject for a graphic chart. There is, however, no illustration, and although woodcuts and copperplate engravings were already common in book production, and had

The Newcastle Chronicle.

Or, GENERAL WEEKLY ADVERTISER.

Vol. I. SATURDAY, March 24, 1764. Numb. 1.

This Paper is circulated by the most speedy Conveyance thro' the Counties of *Northumberland, Durham, Cumberland, Westmorland*, and *Lancaster*, and the *West* and *North Ridings* of the County of *York*; also thro' *Annandale, Nithsdale, Gallsway, Tiviotdale, Liddesdale, Tweedale*, and the *Mers*, or *Berwickshire*, in *Scotland*.

The INTRODUCTION.

THE Public generally expect something or other by way of prelude, or introduction, to publications of this sort; but as we have already given our friends their bill of fare in our proposals, which we hope they have perused, it only remains necessary to enlarge upon and explain the general heads thereof.—And,

1st, By way of apology, we may possibly be arraigned by several in points of prudence and policy, for an undertaking of this kind at this time; and by others, for introducing a new paper where they may imagine the country is already sufficiently supplied; but we hope by substituting ingenious and social entertainment in place of the horrid details of war and bloodshed, to recommend our paper to all the humane and generous part of mankind, and even to convince every one that our scheme is not chimerical, but founded on a plan full as extensive, and as well calculated for public utility, as that of any paper extant: For

"*As long as chequ'd variety shall please,*
"*And labour'd science wear the dress of ease;*
"*Whilst morals, bus'ness, novelty unite,*
"*Be ours the task to profit and delight.*"

We do not hereby mean to insinuate any better abilities towards conducting a paper than our neighbours, (tho' we will exert our utmost) or that the public should receive our attempts to serve them as a favour, when we candidly confess, that we were in the course of trade unwillingly necessitated to it; and tho' the term *necessitated* may seem to want explanation to the public in general, yet it must be very plainly understood and assented to by our senior news-writers here, who consequently must least of all disapprove our measures. Our friends, however, who please to favour us with their advertisements, may be assured that we have a double and indispensible interest in having this paper to succeed (and even to exceed) both as to numbers and extent of circulation, to which no pains or expence will be spared; and if thereby it answers our own purposes in point of advertising, &c. it must in consequence answer the same to the public in general.

We would willingly have obliged our friends in and about town with a Wednesday's paper, as was most generally wished for and expected, but the expence of circulation would have been intolerable, and not sufficiently expeditious on any but a market and post day.

2dly, We have proposed, that in this paper will be exhibited news foreign and domestic, the prices of grain, stocks, &c. and to render it a magazine of entertainment as well as a paper of intelligence, there will frequently be inserted characters of, and extracts from, new books, with other select and original literary articles in prose and verse:—As for the *select*, we have no great doubts about having fund enough to collect from; but for *originals*, should our literary correspondents which we have engaged, and rely on for that purpose, fail us, we are alas! unwarily bound by virtue of our proposals (like other ingenious publishers) to *write to ourselves*.

3dly, We have proposed the strictest impartiality in all points:—But as different people often tack different definitions to one and the same word, we beg leave to explain our own ideas in this point, *i.e.* What we would express by the word *impartiality*, is not to avoid all disputes whatsoever, offensive and defensive, since disputes will naturally arise among individuals, and often for the good and edification of the whole, such as criticisms, different opinions, &c. but to publish indiscriminately for all, so consequently to please and displease all parties by turns, which bids the fairest for impartiality in our sense of the word; for as few persons or parties are right in every particular, "that man who goes blindly into all "the measures of his party, and vindi-"cates all their proceedings, cannot justly "vindicate himself."

4thly, And consequently in matters of state, what we mean by not infringing the liberty of the press, or running into licentiousness, is, That after rendering to our superiours all deference and due respect, we shall not be biassed by, or particularly attach'd either to those who are Out, or those that are In, but occasionally insert interesting particulars for and against both: We shall always be particularly attached to our gracious sovereign, and to the interest of all his worthy, loyal subjects, who (tho' out of power) are ready upon all occasions to exert themselves in the support and defence of their king and country in their just rights and liberties, and who, as the sinews of state, better merit the freedom hitherto enjoy'd, than those greater personages (if any there are) who, from the pride and luxuriancy of private or public wealth, &c. and for sinister or ambitious motives, wantonly provoke and maintain party-spirited faction, to the prejudice and abuse of public tranquility, and constitutional welfare.

5thly, Despising the base, despicable spirit of calumny and detraction, we shall never officiously meddle with private characters, either by insinuation or otherwise, but when we can do it with justice, and to their advantage; and tho' we will not admit of scurrility on any occasion, or an aggressing affront or offence, to oblige any person or party, yet the injured and aggrieved shall never want an asylum in our paper, or even a weapon of defence to pelt down licentious insolence and tyrannic oppression, *i.e.* Our title of General Advertiser, shall continually remind us, that we will not, from sinister or selfish motives, refuse any fair and warrantable advertisements (the subject and language thereof being decent and delicate) to the diminution of his majesty's revenue, the prejudice of commerce, or to the disadvantage of any individual in his own legal property.

We thought it necessary to explain ourselves on the fore-mentioned heads, that we might not hereafter be mistakenly taxed with deviating from our own proposals; consistently with which, as above explained, the public shall be most zealously served, by

Their very humble servants,
The PUBLISHERS.

Saturday and Sunday's Posts.

FOREIGN NEWS.

Rome, *Feb.* 18.

WE hear of nothing but clamours and insurrections on the confines of the kingdom of Naples, where the scarcity of grain is extraordinary and general. This has engaged our magistrates of the office of provision to send commissaries into divers parts to provide sufficiently for the ecclesiastical state, for fear a scarcity with us should be productive of the same tumults and disorders.

Paris, March 2. The author of *An Appeal to Reason*, (a work condemned to the flames by the chatelet) is condemned by a sentence of the same court to perpetual banishment, after being exposed to the carcan (an iron collar wherewith a malefactor is tied to a post); and his printer, the Sieur Grange, is banished the kingdom for five years. It is known, that this author is the Abbe de Caveirac, the same who some years ago had the wickedness and temerity to publish an apology for the massacre of St. Barthelemi.

By an arret of the 29th ult. four other pieces are ordered to be burnt.

LONDON, March 15.

They write from Brandenburgh, and other parts of Germany, that the distemper among the horned cattle was very much abated, which has been effected by hanging four or five onions about the beasts necks directly after they are taken ill, and will not eat. Those onions draw the infection out, and look the next day as if they had been boiled. This remedy is to be repeated several times, and the onions which have been used, are to be buried in a deep hole. And in a few days after the cattle are taken with a running at the nose, which carries off the distemper; it is also proper at that time to hang up some onions in the distempered cattles stables.

Yesterday was held a general court of the Russia company, at their court room over the Royal Exchange, on some affairs of importance; when, after debate, the question was put, whether an agent was necessary or not? which, on a division, passed in the affirmative by a majority of six. After which, an elegant entertainment was provided at the King's Arms tavern, in Cornhill, at which were present the earls of Sandwich, Egmont, and Halifax, with several other persons of distinction, as also M. De Grosse, ambassador extraordinary from the empress of Russia.

On Tuesday night an express arrived at the Rt Hon. Lady Townshend's in Whitehall, which brought an account of the death of her husband, the Lord Viscount Townshend, who died on the road coming from Bath on Monday evening. By his death 3000 l. per annum devolves to her ladyship. His lordship is succeeded in title and estate by his eldest son, the Right Hon. George, now Viscount Townshend, of Rainham, in the country of Norfolk, lieutenant-general of the ordnance, colonel of the 28th regiment of foot, as also of the western battalion of militia of the county of Norfolk, a major-general, and knight of the shire for the said county of Norfolk.

It is said orders are given for all the officers in Great-Britain and Ireland, upon absence of leave, who belong to the forces in the East-Indies, to hold themselves in readiness to embark on board the first ships that shall sail for those parts.

We hear his majesty will certainly visit his German dominions, soon after the rising of the parliament, and that he will be accompanied in his journey by his brother Prince Henry.

A grant of 650,000 l. is ordered immediately to pay off part of the navy debt.

Yesterday the money brought from Lisbon in the Hampden packet, Hill, on the merchants account, was lodged in the Bank.

On Tuesday Capt. Griffin, of the Lapwing, took leave of the directors of the East-India company, going express to Bengal.

Tuesday and yesterday several hundred men and boys, many of them disbanded soldiers and sailors, were sent down the river, in order to be shipped off for the East Indies, to complete the broken regiments there.

It is said that press-warrants will shortly be issued out, in order to man the fleets going to the East and West Indies.

On Tuesday 2200 quarters of wheat were shipped for the Straits; and a considerable quantity of gunpowder for Virginia.

The proceedings of the supreme Council at Martinico against the jesuits, dated in Dec. last, are just published at Paris. It appears that the jesuits there are to be treated exactly in the same manner as in the jurisdiction of the parliament of Paris.

By a letter received on Monday from Hamburgh we are informed, that beef, mutton, and pork, are sold there for sixteen-pence a pound; and the other necessaries of life equally as dear.

A shock of an earthquake was felt the 26th of last month at Mantua, but it did no damage.

They write from Dublin, that on a late calculation, the number of Roman Catholics in Ireland are found to exceed his Majesty's Protestant subjects in that kingdom by ninety-five thousand souls, which is thought to have greatly augmented the late opposition with regard to granting the former the proposed security on lands.

Monday and Tuesday's Posts.

From the LONDON GAZETTE.

Turin, *March* 3.

HIS Royal Highness the Duke of York is in perfect health, and greatly satisfied with the manner of his reception here. His Royal Highness has employed his time in visiting some of the strong places in the neighbourhood of Turin, and whatever else is worthy his attention; and has been often at court in his private character, where he is constantly received by the whole royal family with every possible mark of esteem and regard.

The Duke of Modena has sent hither the Count Belgiojoso, a person of the first distinction of that country, to compliment his Royal Highness the Duke of York, and to invite him to honour Milan with a visit; offering him an apartment in the Ducal palace; or, upon his refusal of that, a hotel in the town. His Royal Highness made a suitable answer to these civilities, and proposes spending the three last days of the carnival at that place. The Duke of Parma has also followed this example, and has sent the Marquis Calcagnini to compliment his Royal Highness, and to invite him to his capital, which invitation it seems to be his Royal Highness's intention to accept, upon his return from Rome and Naples.

Ratisbon, March 4. The college of princes, having received an imperial rescript, brought hither by Prince John of Lichtenstein, notifying to them the resolution taken the 20th past by the electoral ambassador at Frankfort to proceed to the election of a king of the Romans on the 3d day of this month, and having presented this evening the conclusion, notifying their assent thereto, by Count Saurau the Saltzbourg minister, to the prince of Tour and Taxis, the chief imperial commissary to the diet, came immediately to an agreement to separate till the 1st of April.

Brussels, March 6. The scarcity of provisions is so great in the ports adjacent to Francfort, and near all the great rivers, occasioned by the late vast inundations, that a very large quantity of provisions is dispatched daily from hence for that city, where the consumption is immense, thro' the unusual concourse of people assembled there upon account of the approaching election and coronation of the king of the Romans. [Gaz.

FOREIGN NEWS.

Lisbon, Feb. 14. The court is at Salvatiera. Recruits continue to be raised; and being allowed to bear arms, they have committed great disorders in the city, and rob and assassinate with impunity. The inhabitants are afraid to leave their houses after dark. These excesses are the more inexcusable, as the troops are paid with the greatest exactness.

Naples, Feb. 18. Since the 11th inst. eleven ships are arrived here with corn, and more are impatiently expected. These, however, have had a happy effect in calming the populace, whom necessity had made outrageous.

AMERICA.

Charles-Town, Jan. 11. Several expresses have arrived here from Augusta, &c. with accounts of the murders committed at Long Canes by the Indians. The news threw the people, in and about Augusta, into great consternation; all the settlers on Savannah river, above that place, fled thither with their families for shelter, even some of the inhabitants of Augusta have sent their families farther down the country for safety.

Jan. 18. The Indians who murdered the settlers at Long Canes, are now undeniably known to be Creeks, and it likewise appears that they are not of the Mortar's gang, so that the number of our enemies among them is greater than was imagined; and it is supposed they will soon commit other hostilities.

LONDON, March 17.

The Hereditary Princess of Brunswick has received a letter from the King of Prussia, written by himself, congratulating her on her safe arrival at Brunswick, accompanied with an invitation for their Serene Highnesses to pass some time this summer at Berlin; at which place (it is said) they will be met by his Royal Highness the Duke of York.

A ship lately arrived from Newfoundland, has brought over a few chaldrons of coals, the produce of that island; which for pitchiness and clear burning, are esteemed equal to the best Long Benton.

We hear that the fleet destined for the East-Indies will consist of three seventy-four, two sixty, and three new fifty-gun ships, with two frigates; each ship to be manned with their full complement of marines, as in time of war; which squadron will be under the command of commodore Arbuthnote.

Lord Robert Manners's regiment is ordered to march from Chatham barracks to Portsmouth, in order to embark for Jamaica.

Figure 23 *The Newcastle Chronicle*, March 24, 1764

appeared in newspaper advertisements as early as 1652 (in an advert for goldsmith's work in *The Faithfull Scout*), they were not thought necessary on editorial pages, which simply listed what editors thought the readers wanted to know.

With circulations remaining modest – only a few thousands an issue for the most successful – despite their growing numbers, and their readership still tied to the educated and better-off classes, newspapers did not have a great deal of incentive over the next few decades to improve their appearance. A more important influence on their development was the political freedom and extension of voting franchise that came with the nineteenth century. These brought a rapid growth in political news and comment as the press spearheaded the development of democracy. The result was a great expansion in words, political analysis and endless punditry which make Victorian newspapers such heavy fare when one turns back their tightly packed pages.

A peep at the first issue of the broadsheet *Sunday Times* in 1822 (Figure 24), ten years before the first Reform Act, shows a modest four-page sheet of five 16-pica columns of mainly 6 point type, in which page one is divided between advertising and a lashing attack on the activities of the government. The two page one headings, ORIGINAL STRICTURES and ASPECT OF PUBLIC AFFAIRS are in capitals but still only a size up from the body setting. The title-piece is in the familiar black letter type, but ruled off below it, and significant of the new ethos of the press is a quotation in 9 point and running across the page: 'Let it be impressed upon your minds, let it be instilled into your children, that the LIBERTY OF THE PRESS is the PALLADIUM of all the Civil, political and RELIGIOUS RIGHTS of an ENGLISHMAN.'

The inside pages still have the news divided up into sections, but the content is broader and the items appear under the headings: POETRY, AGRICULTURE, SPORTS, MISCELLANEOUS, FOREIGN PAPERS, POLITICS OF THE TIMES, THE STAGE, IRELAND and LAW, with the headings set in an expanded seriffed face of about 10 point caps. Under each head are run-on headings in small capitals of the body face denoting the items: MARRIAGE PREVENTED, DESTRUCTIVE FIRE, MELANCHOLY CASUALTY, HORRIBLE OUTRAGE, HORRID MURDER, ANOTHER COACH ROBBERY, WEST INDIAN PIRACIES and the like. Though still labels, these now show more attempt at reader involvement. The main breakthrough is in content. Sport introduces readers to prizefighting, a walking contest and trotting races. The Stage column reviews the opening of the new Drury Lane Theatre with *The School for Scandal*; with *The Jealous Wife* and *Venice Preserved* at Covent Garden. There is a letters column (despite being issue number one) and some animated court reports as well as the traditional market prices and a small column of stock prices.

The Times of November 14, 1854, (Figure 25) apart from a massive increase in classified advertising now, in common with other

The Sunday Times.

"Let it be impressed upon your minds, let it be instilled into your children, that the LIBERTY OF THE PRESS is the PALLADIUM of all the Civil, Political and RELIGIOUS RIGHTS of an ENGLISHMAN."

No. I. LONDON: SUNDAY, OCTOBER 20, 1822. PRICE 7d.

THE TICKLER, MONTHLY MAGAZINE, PRICE SIXPENCE.

STANDARD SCHOOL BOOKS.

PROSE TRANSLATIONS OF LATIN AUTHORS.

FOURTH EDITION.

THE COSTLY LITERATURE OF MODERN TIMES PLACED WITHIN THE REACH OF THE LOWER INCOMES! PRICE TWO-PENCE.

ALI PACHA; THE AFFAIRS OF GREECE!!!

REVERSION.

POSTING REDUCED ON THE GREAT NORTH ROAD.

WINE ROOMS, QUEEN'S ARMS, CHEAPSIDE.

ORIGINAL STRICTURES.

TO OUR READERS.

ASPECT OF PUBLIC AFFAIRS.

Figure 24 *The Sunday Times*, October 20, 1822

(a)

The Times.

No. 21,898. LONDON, TUESDAY, NOVEMBER 14, 1854. PRICE 5d.

(b)

THE DAILY GAZETTE.

LATEST NEWS.

Local and District News.

MIDDLESBROUGH POLICE COURT.

(c)

The Daily Telegraph.

LONDON, WEDNESDAY, JULY 3, 1872. [TWELVE PAGES, ONE PENNY.

(d)

ASK FOR CHRISTMAS HARMSWORTH.

Daily Mail.

ASK FOR CHRISTMAS HARMSWORTH.

NO. 1,117. LONDON, MONDAY, NOVEMBER 20, 1899. ONE HALFPENNY.

PERSONAL. SHIPPING, TOURS, &c. CONCERTS. ENTERTAINMENTS, &c. ENTERTAINMENTS, &c. HOTELS & RESTAURANTS. AUCTIONS.

Owing to the immense number of advertisements received for the "Daily Mail," we are obliged to hold over several pages of them. They will be inserted as speedily as possible in the order in which they have been received.

BUSINESS NOTICES.

PRIVATE ADVERTISERS.

LOST AND FOUND.

DUNLOP TYRES

Figure 25 (a) *The Times*, November 14, 1854
(b) *The Daily Gazette*, November 22, 1869
(c) *The Daily Telegraph*, July 3, 1872
(d) *Daily Mail*, November 20, 1899

Victorian papers, occupying the whole of the front and back pages, and an even greater length of texts, shows little change in presentation from 32 years earlier. If anything it is set more solid with not a whisker of white space impinging on the words. It contains, in this issue, the superb war reports of William Russell from the Crimea; politics and government, with Britain midway between the two great Reform Acts, have become a staple ingredient. It has now moved into a six-column format and a wider page. The type, though still small, is a clearer, more legible face, well printed by the standards of the day, as befits the advanced plant of Britain's top-selling daily, with its circulation of 40,000 an issue. Its news is still grouped under headings in capitals scarcely bigger than its body type. It has three and a half columns of editorial opinion without a headline, law reports, wide foreign coverage, court circular, city news, market prices, weather, shipping . . . but, unlike *The Sunday Times*, no sport for its daily readers.

Its price is still 5 pence, including *Stamp Duty*, and its readership, despite its campaigning record, remains the elite of the day.

The new readership

If, by 1850, newspapers showed only modest advances in presentation and circulation, the second half of the century was to change that. And the key to both was to be readership.

In 1855 Stamp Duty came off for good and for the first time a penny newspaper became a possibility. By now, the take-off of the Industrial Revolution had begun to alter the face of Britain's cities, bringing regular work, higher wages and, despite the social problems that went with it, an improving standard of living. More significantly it began to spread wealth down the social scale by producing, for the first time, a salaried middle class of managers and technocrats and a new upwardly mobile class of skilled workers. More people could afford more papers, and had more reasons for buying them. At the same time the new rail network had made national distribution on the day of publication a reality. Whether editors realized it or not the scene was set for expansion.

The vitality of *The Times* as the top-selling daily from 1820 under its great editors, Thomas Barnes (1785–1841) who built up the world's first network of news correspondents, and John Thaddeus Delane (1817–1879), ran out on the threshold of the new challenge, and it was *The Daily Telegraph* (Figure 25) that took up the lead. Founded as the first penny national paper the year Stamp Duty ended, the *Telegraph* quickly moved from a radical standpoint to a broadly Right-wing stance that suited the new middle class readers who found its wider news content, heavy crime coverage and sport and theatre reviews preferable to the elitist content of *The Times* and the old Victorian dailies.

The *Telegraph* was also the first paper to use stunts and promotions to push its sales some three decades before that arch publicist, Lord Northcliffe. In 1873 it discovered the 'Loch Ness Monster', in 1877 it gave a Jubilee party for 30,000 children in Hyde Park which was reviewed by the Queen. It went for young readers, crusaded against capital and corporal punishment, and claimed to be the first paper 'to treat prostitution boldly and plainly'. It passed *The Times* circulation of 60,000 in 1860, went to reach 200,000 before the end of the century and remained the top-selling daily for forty years until 1900.

Cheaper papers and the arrival of reading literacy with the Education Act of 1870 led to a rash of new weekly and evening papers appearing all over the provinces, including many of the titles still being published today. An inside page from an early copy of *The Daily Gazette* (Figure 25), which went on sale in Middlesbrough in 1869 at a halfpenny (and is still appearing) is typical of local papers of the period. It has a column of LATEST NEWS from London 'by telegraph' quoting items from *The Times*, the *Morning Post* and *Daily News*, and four well-filled columns of local news and readers' letters.

The prestigious London dailies could now aim at a national circulation, with the expanding rail network providing the means of reaching breakfast tables from Plymouth to John O'Groats. *The Daily Telegraph* was the first paper to cater in this way for the new readership produced by the Industrial Revolution, and yet it made disappointingly little attempt to change its presentation. Opening the edition of July 3, 1872 (Figure 25), we find twelve pages of six-column format, seven of them, including the front and back, crammed with classified advertising. The news, varied though it is, is set tightly into the remaining pages, with headlines little bigger than fifty years earlier. There are five columns of editorial opinion without headlines, though on the main stories, such as the sensational:

DR LIVINGSTONE'S SAFETY

OUTLINE OF HIS DISCOVERIES

THE NILE SECRET SOLVED

the three-decker headline has arrived, even if still only in 14 point.

A look at the densely packed pages of small classified advertisements in these Victorian papers carries the clue to the conservativeness of their presentation. Until there was a truly mass market to be exploited in terms of spending power and expectations, the need to display a newspaper's contents visually, either editorial or advertising, did not arise. The competitive consumer environment of mass circulation newspapers and mass produced food and goods in which communicators exist today had simply not dawned in the 1870s and 1880s; and while such a market might be just round the corner, editors, who live notoriously day to day, were unlikely to be its prophets. In the conservative world of newspapers the *Telegraph* had already taken a giant leap ahead of its fellows in recognizing that newspaper readership was changing.

Lord Northcliffe

The catalyst of change was to be Lord Northcliffe (Figure 26). Born Alfred Harmsworth, the son of an impoverished barrister in Dublin, and largely self-educated, he discovered as a young reporter from his contributions to the magazine *Titbits* (founded in 1881 by George Newnes) the new world of 'snippet' journalism. *Titbits* thrived on stories of adventure, gossip about the famous, readers' letters, advice columns, competitions and potted paragraphs giving unusual facts. It was aimed perceptively at a largely family and young working class market that had not yet developed a newspaper buying habit. By 1890 *Titbits* had reached the unheard of circulation of 200,000 an issue.

Figure 26 Lord Northcliffe (1865–1922), 'father' of the popular press

Envious of its success, Harmsworth launched his own magazine *Answers* at the same market in 1888. After several successful magazine launches in the new popular field he bought the failing *London Evening News* in 1894, took it down market and made it profitable. In 1896 he realized his greatest ambition when he used his now considerable fortune to launch his own daily, the *Daily Mail*, as the first national paper to reach, in its content, the working class reader who had hitherto been ignored.

Northcliffe's new daily (he was given his barony by Edward VII in 1905) was scorned as a 'newspaper for office boys' but its rising circulation soon proved that a paper strong on readers' letters, tales of battle, adventure and human interest was what the new readership wanted and its sales, in 1900, became the first to pass the million mark. The popular press of the twentieth century had been born.

Northcliffe sensed that there was spending power to be tapped in the new wage earners, and he cleverly married the new mass production of goods to the growing consumer market to create in the *Daily Mail* a truly national advertising medium. The income from this enabled him to price the paper at a halfpenny by offsetting its production costs against advertising revenue. Like all good ideas, Northcliffe's recipe for success was blindingly simple. The cheaper the paper the more people bought it; the more the circulation grew the more attractive it became to the advertiser; the greater the advertising revenue the more Northcliffe spent on improving the paper and its sales by hiring the best journalists, getting the best stories and financing stunts to get the paper talked about – and the richer became Britain's first newspaper tycoon. The press had become big business.

The effect of Northcliffe's ideas on the newspapers of the period was two-fold: it spawned a new growth of down-market dailies – the *Daily Express* (1900), the *Daily Mirror* (1903), and *Daily Herald* (1912). It drove the *Daily News* and *Daily Chronicle* to widen their market towards their eventual merger as the *News Chronicle*, and it spelt a decline in fortunes of the more conservative Victorian dailies – the *Morning Post* and *The Standard* – causing even *The Times* to totter and sell out to Northcliffe in order to survive. In the big conurbations of Manchester, Birmingham, Newcastle and Bristol the competition between titles likewise intensified in the race to capture the new markets. The battle for readers had begun, a battle in which some old and influential titles were to go to the wall.

It was inevitable that with a new and very different body of readers, and in an age in which spending power was expanding across the board, newspapers would have to change in image as well as content. What the critics of the press described as its commercialization had revealed newspapers to be a product with a capacity for making money as well as having influence. Yet influence on its own was not enough. There can be nothing less influential than a paper that is failing to attract readers in an expanding market.

When it came to design, a vital part today in the image-making process, it has to be said that Northcliffe was no innovator. To him ideas and words were the market weapons. He left advertising to occupy the front and back pages of the *Daily Mail* (Figure 25) as in the old Victorian papers, although display adverts with line drawings on such big sellers as Yorkshire Relish sauce (with a cookery book offered for a shilling to those reading the advert), and the many soaps, cocoas and medicines began to adorn the paper. Looking inside (Figure 27), we see that the type sizes in 1899 are little changed from 1855, though the busy *eight*-column format he instituted is easier to read, with shorter and more varied items, many with two-decker headlines, and with distinctive features pages (Woman's Realm, Daily Magazine) lightened by line drawings.

It fell to rival Sir Arthur Pearson's *Daily Express*, much influenced in content and market by the *Mail*, to be the first national daily, in 1901, to put news back on to the front page, though it was not until 1966 that *The Times* became the last British paper to give in to this trend.

Photographs

The breakthrough that made modern design techniques possible came with the perfecting of the means of reproducing photographs on newsprint by the halftone process. The first successful photograph had been taken by Henry Fox Talbot at Lacock, Wiltshire, as early as 1835. The need for a viable means of reproduction, however, delayed the use of photographs in newspapers until March, 1880, when the *New York Daily Graphic* became the first when it published a picture of the city's shanty town using the new halftone process. In this the tones which made up the details of the photograph were broken up into tiny dots on an engraved plate which, on picking up ink, reproduced by their varying density the tones of the actual picture. The method solved the problem of ink pick-up and only by looking extremely closely could the eye detect that the picture had been 'reconstructed' by means of a screen.

It was another decade before newspapers in general began installing process-engraving departments to produce halftone blocks, though even then the pictures used were mainly studio portraits, and it was not until 1904 that Northcliffe's new *Daily Mirror* became the first newspaper in the world to employ its own staff photographers to provide news photographs as a day-to-day ingredient. By this time page design was already on the move with newspapers beginning to utilize artists' line drawings of news events and personalities, as can be seen in the *Daily Mail* (Figure 27).

It was an example of Northcliffe's indifference to visual presentation that, while accepting pictures as viable news content, he did not rate them in the same class as words, and the idea of photojournalism as an independent journalistic skill would have been anathema to him. His ideas were pervasive and news pictures up to 1914 were considered to be more appropriate in picture papers such as the

DAILY MAIL, LONDON, THURSDAY, DECEMBER 28, 1899. 7

Daily Magazine

OUR VOLUNTEERS—WHAT THEY DO AND WHAT THEY MIGHT DO.

How Great Britain's Citizen Soldiery have been Prepared for Active Service.

PREVIOUS BOER SIEGES

The Way Our Soldiers Spent Christmas in the Transvaal Nineteen Years Ago.

1st V.B. DURHAM ARTILLERY MOUNTING A GUN.

For guns equal to and similar to the pattern used by the Regular artillery. Are there not other nations who buy cast-off cannon without foisting them on our Reserve, to find in the day of trial that the Auxiliary artillery is only half-trained?

And last, but not least, for a large-hearted recognition by the War Office and the public generally, that the Volunteer, be he officer or rank and file, is doing a noble work, and one which reflects the utmost credit on him. Chaff, ill-natured criticism, and official coldness must go once and for all to the limbo of obsolete things, never to be disinterred.

PREPARING DINNER FOR THE 9TH V.B. LANCS ARTILLERY.

CITIZEN SOLDIERS!

You will be delighted with to-morrow's grand Patriotic Number of the "Penny Pictorial Magazine," and especially with the special illustrated article, "The Making of a Volunteer." Make certain of getting a copy by ordering the "Penny Pictorial Magazine" this very day.

1ST V.B. MIDDLESEX ENGINEERS BRIDGE BUILDING.

MEMS. FROM THE FRONT.

ORDER TO-MORROW'S WAR NUMBER "ILLUSTRATED MAIL."

Feuilleton.

Days of Doubt.

BY ALICE MAUD MEADOWS.

Figure 27 New-style features presentation – inside the *Daily Mail* of 1899

THE DAILY GRAPHIC, JULY 26, 1909.

HOW I FLEW ACROSS THE CHANNEL: BY M. BLERIOT. (See Page 7.)

THE DAILY GRAPHIC

ONE PENNY

No. 6122.—Vol. LXXIX. LONDON: MONDAY, JULY 26, 1909. Registered as a Newspaper.

BRAVO BLERIOT!

M. BLERIOT AND HIS MONOPLANE, IN WHICH HE YESTERDAY ACCOMPLISHED THE FIRST 'CROSS CHANNEL FLIGHT BY A "HEAVIER-THAN-AIR" MACHINE.—THE FLIGHT FROM BARAQUES, NEAR CALAIS, TO DOVER OCCUPIED TWENTY-THREE MINUTES.

345

THE DAILY MIRROR, Tuesday, November 12, 1918.

ALLIES' DRASTIC ARMISTICE TERMS TO HUNS

The Daily Mirror

CERTIFIED CIRCULATION LARGER THAN THAT OF ANY OTHER DAILY PICTURE PAPER

No. 4,696. Registered at the G.P.O. as a Newspaper. TUESDAY, NOVEMBER 12, 1918 One Penny.

HOW LONDON HAILED THE END OF WAR

The King and Queen appeared on the balcony at Buckingham Palace to acknowledge the cheers of the crowd that gathered to congratulate their Majesties on the victory.

BRITISH OFFICIAL MONS CAPTURED

How news of the armistice signature came over the wire to the newspaper offices. A facsimile of it as automatically printed on the tape machine. The cheers which greeted it were the first to be raised.

There never again will be such news for the Mercuries of the streets to cry.—(Daily Mirror.)

A historic message as it came over the wire. It is dramatic that the last British war communiqué should proclaim our forces at Mons.

Goddesses in the car, accompanied by a man of war, celebrate the occasion.—(Daily Mirror.)

Nothing gave greater satisfaction to all of us than the news that the cessation of hostilities found the British armies once more in possession of Mons, where the immortal "Contemptibles" first taught the Huns what British valour and steadfastness could do. They left the town as defenders of a forlorn hope; they re-entered it conquerors indeed.

Figure 28 Early picture newspapers exploit the new photojournalism: the *Daily Graphic* of July 26, 1909, and the *Daily Mirror* of November 12, 1918

Daily Graphic and his own *Daily Mirror* (Figure 28) which he had founded in 1903 as a paper for women readers (about whom he did not entertain a very high opinion). Such papers existed side-by-side with the orthodox 'news' dailies, which continued to concentrate on the words.

The vintage years

Edwardian newspapers with their conservative up-and-down layout show the cautious onset of ideas – two and three decker headlines in spindly 14 to 30 point, with static photographs of seldom more than two columns still vying with artists' drawings. The *News of the World* for July 31, 1910 (Figures 4 and 29) was advanced in its time in using a crude sans typeface alongside its Cheltenham Condensed, and for its daring gimmick of printing an entire popular song sheet as a circulation booster.

It was the First World War, with all its visual horror and squalor, that brought home to editors the vital role of news pictures as an ingredient on a par with – and occasionally superior to – the words,

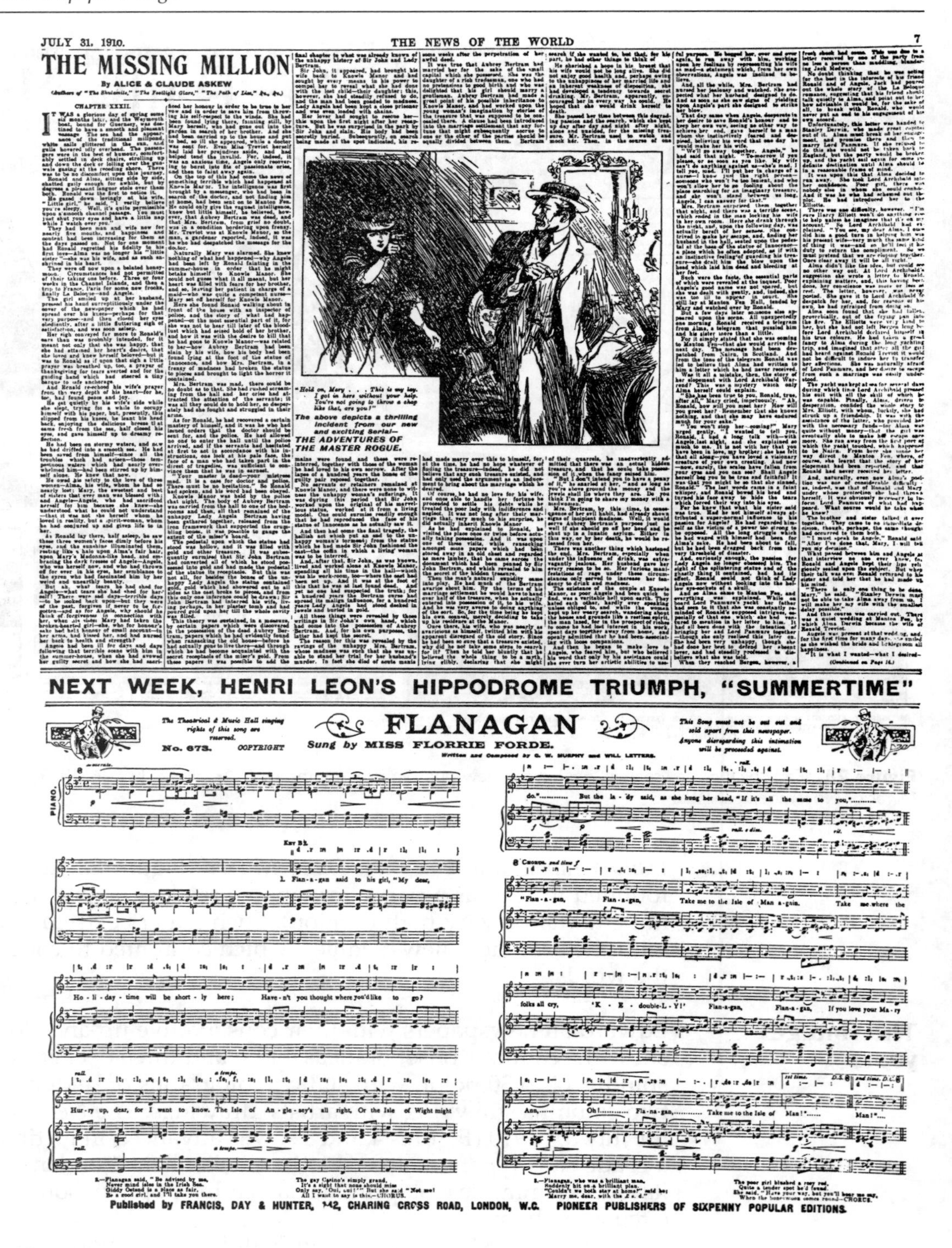

JULY 31, 1910. THE NEWS OF THE WORLD 7

THE MISSING MILLION

By ALICE & CLAUDE ASKEW

(Authors of "The Shulamite," "The Footlight Glare," "The Path of Lies," &c., &c.)

CHAPTER XXXII.

IT WAS a glorious day of spring some six months later, and the Weymouth boat, bound for Guernsey, was destined to have a smooth and pleasant passage. The sea had the appearance of the traditional millpond, white sails glittered in the sun, and gulls hovered idly overhead. The passengers were in the best of humour, comfortably settled in deck chairs, strolling up and down the deck or lolling over the gunwale gazing at the receding shore. There was to be no discomfort upon this journey.

"Hold on, Mary . . . This is my lay. I got in here without your help. You're not going to throw a chap like that, are you!"

The above depicts a thrilling incident from our new and exciting Serial—

THE ADVENTURES OF THE MASTER ROGUE.

(Continued on Page 14.)

NEXT WEEK, HENRI LEON'S HIPPODROME TRIUMPH, "SUMMERTIME"

The Theatrical & Music Hall singing rights of this song are reserved.

No. 673. COPYRIGHT

FLANAGAN

Sung by MISS FLORRIE FORDE.

Written and Composed by C. W. MURPHY and WILL LETTERS.

This Song must not be cut out and sold apart from this newspaper. Anyone disregarding this intimation will be proceeded against.

1. Flan-a-gan said to his girl, "My dear, Ho-li-day-time will be short-ly here; Have-n't you thought where you'd like to go? Hur-ry up, dear, for I want to know. The Isle of An-gle-sey's all right, Or the Isle of Wight might do."...... But the la-dy said, as she hung her head, "If it's all the same to you,"......

CHORUS. "Flan-a-gan, Flan-a-gan, Take me to the Isle of Man a-gain. Take me where the folks all cry, 'K-E-double-L-Y!' Flan-a-gan, Flan-a-gan, If you love your Ma-ry Ann,...... Oh!...... Fla-na-gan,...... Take me to the Isle of Man!"...... Man!"....

2.—Flanagan said, "Be advised by me,
Never mind isles in the Irish Sea.
Giddy Ostend is a place as fair,
Be a good girl, and I'll take you there.
The gay Casino's simply grand,
It's a sight that none should miss
Only say, 'Oui, oui!'" But she said "Not me!
All I want to say is this.—CHORUS.

3.—Flanagan, who was a brilliant man,
Suddenly hit on a brilliant plan.
"Couldn't we both stay at home?" said he;
"Marry me, dear, with the £ s. d."
The poor girl blushed a rosy red,
Quite a tender spot he'd found.
She said, "Have your way, but you'll hear me say,
When the honeymoon comes round—CHORUS.

Published by FRANCIS, DAY & HUNTER, 142, CHARING CROSS ROAD, LONDON, W.C. PIONEER PUBLISHERS OF SIXPENNY POPULAR EDITIONS.

Figure 29 Circulation pullers of their day: a *News of the World* fiction series, *The Missing Million*, circa 1910, sharing a page with one of its famous half-page song sheets of popular tunes of the time

and newspapers from the 1920s onwards show a lively awareness of this in their presentation. The page patterns of the *Daily Express* (Figure 30) demonstrate both increased picture use and more creative headline writing, though type quality is still fairly poor. The front and back pages of the *Daily Express* of October 6, 1930, with its famous coverage of the R101 airship disaster, is an example of what can be achieved by picture journalism. It also shows the advent of the new Century type, which was to be used by the *Express* and the *Mail* with such distinction during the next three decades.

But it was in the words and the headline writing, and ultimately in typography, that the newspapers of the 1930s made their greatest advances. Journalists look back on the period as the golden age of reporters when, untrammelled by rival media, newspapers combed the world for exclusive stories, fought to be the first on the street with them, regarding detailed coverage of their market as the entitlement of readers. Writers were backed up by technician editors such as Arthur Christiansen of the *Express* and Harry Guy Bartholomew (Figure 31) of the *Mirror* (and co-inventor of the first picture wiring system), who knew their markets and who harnessed typography and powerful headlines to hit their readership targets and make their newspapers the best and most talked about on the news-stands.

It was the age of 'busy' papers with multideck headlines that covered stories from every angle, pictures scattered about like an optical joyride, and endless fillers rolling up the columns, ensuring that no newsworthy items were left unexploited. Prodded by the ebullient Lord Beaverbrook, Christiansen exploited traditional forms to their limit, reducing headline sizes with the importance of the story as the eye moved down the page, polishing every line of every filler so that each word earned its place on the page. It was the something-for-everybody approach in which the aim was to dazzle the eye with a cornucopia of goodies spread around the basket.

And yet design techniques were moving forward. The text size pages might carry a lot of headlines, but editors were becoming fussy about type style. Types were being used more consistently and in a set pattern of sizes so that the *Daily* Express, the *Daily Mail* and the *Daily Herald* and the other dailies were assuming a distinctive appearance of their own. Good pictures and series of pictures were being given special spreads. The circulation war of the 1930s resulted in a greater stress on exclusiveness, on appearance, on the use of blurbs and promotions and in an increase in type sizes and more imaginative type use as page one became a weapon on the news-stands.

The tabloid revolution

The sale of the *Daily Mirror* by Lord Rothermere in 1934 to a new company, Daily Mirror Newspapers, threw into prominence Harry Guy Bartholomew, its former picture editor under Northcliffe, who became the editorial director. It was the moment Bartholomew had been waiting for to put into effect a theory he had long held that

DAILY EXPRESS, Monday, October 6, 1930.

Express

TO-DAY'S WEATHER:
Rain at Times.

NO. 9,493. MONDAY, OCTOBER 6, 1930. ONE PENNY.

The R101 Crashes And Is Destroyed: Lord Thomson, Sir Sefton Brancker And 44 Others Burned To Death.

THE TRAGIC AFTERMATH.—Dead bodies shrouded in sheets near the wrecked airship.—*Special "Daily Express" Picture.*

GREAT AIRSHIP STRIKES A HILL AFTER BATTLE WITH A STORM.

SLEEPING PASSENGERS ENVELOPED BY SWIFTLY RUSHING FLAMES.

THE giant airship R101, which left Cardington at 7 p.m. on Saturday for India, crashed on a hill near Beauvais, France, at 2.5 a.m. yesterday, and 46 of those on board—including Lord Thomson, Minister for Air, and Air Vice-Marshal Sir Sefton Brancker—were burned to death.

A quarter of an hour before the crash those on board had no warning of danger, although the weather was stormy. At 1.50 a.m. they received a message from the French Air Ministry, telling them they were one kilometre from Beauvais. They replied "Thank you"—nothing more.

The next few minutes sufficed to produce the worst disaster in air history. Low-lying clouds had prevented the R 101 rising, and the storm, which left masses of rain-water on the top of the envelope, forced her down until she struck a low hill by a wood near the village of Allonne.

MAJOR SCOTT IN THE R 101.

A few moments and the vast airship was a mass of flame. Many of the victims were killed in their sleep, so swift was the rush of the fire when the hydrogen of the R 101 was ignited.

Some of the survivors owed their escape to the fact that the water tanks broke, and the flood washed them through the flames to safety. Rare heroism was displayed by three of the survivors who, though injured, returned to the burning wreck in the hope of finding some comrade to help.

One survivor's last glimpse of the wreck was Flight-Lieutenant Irwin standing at his post quietly giving orders. He was still there when the flames enveloped him.

Lord Thomson was considered by many to be the Government's nominee as next Viceroy of India. Flight-Lieutenant Irwin, the captain of the airship, was among the victims, who include a number of distinguished officers and air experts.

Biographies of victims are on Pages Two and Ten.

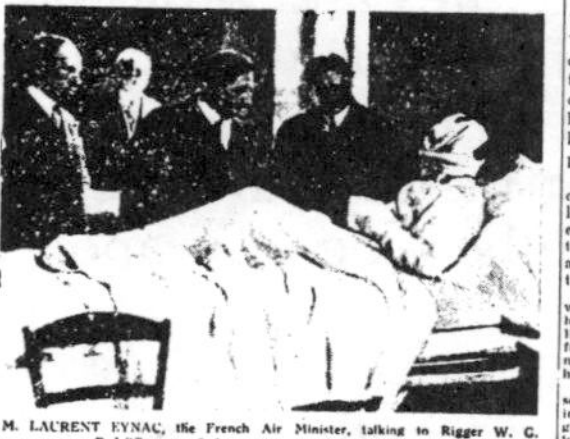

M. LAURENT EYNAC, the French Air Minister, talking to Rigger W. G. Radcliffe, one of the survivors, in Beauvais Hospital.

LATEST NEWS.

An emergency meeting of the Air Council was held at the Air Ministry last night to hear Air Chief Marshal Sir John Salmond's report on the disaster in which five chiefs of the Ministry perished.

• • •

It is announced by the Air Ministry that a public inquiry will be held in this country, subject to co-ordination with the arrangements which are being made by the French Government.

• • •

The bodies of the victims will be brought to England for burial. They will be carried across the Channel in a warship.

• • •

Engineer H. J. Leech was in a cabin with two other engineers when the crash came. "We shook hands," he said, "and swore we would not be burned to death, but that somehow we would make our way out. We had no idea how." All three escaped.

Lord Thomson, his valet, and all the officers of R 101 had taken out special accident policies for substantial amounts. It is understood that Sir Sefton Brancker had not taken any special precaution against accident, and that the crew were not covered.

• • •

In common with all other Government property, the airship itself, which cost £[illegible], was not insured.

• • •

"There can be no question of a stowaway being on board the airship," said an Air Ministry official last night, referring to French reports that more than fifty bodies had been recovered.

• • •

An expert inquiry will be opened to-day on the scene of the crash, Major Cooper, Inspector of Accidents, Air Ministry, presiding.

• • •

When the disaster happened only twelve men were on duty, navigating the airship. Every one else was asleep.

Mr. H. J. Greenwall's full story of the disaster and rescue scenes is on Page Three. Survivors' narratives are on Page Eleven.

"A DRUNKEN, REELING MONSTER."

By HAROLD PEMBERTON.
"Daily Express" Special Correspondent
(who flew from London at dawn yesterday to the scene of the disaster).

BEAUVAIS, Sunday, Oct. 5.

LET me tell the story of the most terrible disaster in the history of British flying as simply and as faithfully as possible. Let me reconstruct an accurate picture of how brave men of estate high and low met their death in the fabulous palace of the air.

It needs no gloss. No mind could conceive a drama more pitiless than the one unfolded to me by eyewitnesses in this stricken and terrified town, a few short hours after the flaming sky had lighted terror in a thousand eyes.

From my airplane flight over the wrecked R 101, and stories told to me by survivors as having something of Divine Providence about them, it is from these that I reconstruct for you a narrative of facts that must sear the heart.

Beauvais, a small French town some forty miles from Paris, was sleeping peacefully when, about 2 a.m., the great ship appeared above the house-tops.

It was a fearful shadow, for the droning engines only just cleared the roofs, and set up a terrific echo. It was a shadow with a silhouette, for the navigating gondolas were a blaze of light, and there was just a passing glimpse of grim, set faces. People here talked of an explosion, but I imagine what they heard was the noise of the engines.

According to one survivor, there had been some cause for anxiety ever since the ship had sailed into the blue from Cardington. She had never gained any flying altitude. She was a giantess of the air with a burden which could not be shaken off until the tanks became lighter through fuel consumption.

At Beauvais, where the full drama begins, the burden was added to by the gale, and floods of rain, water piling up on her mighty back, water that became like lead, beating her to earth and destruction.

The shadow and the great noise in the night sent people tumbling from their beds. They say that the shadow passed so close to the cathedral that it almost grazed its massive facade. They say that the fabric of the machine

PAGE THREE, COL. FIVE.

WHAT WAS THE CAUSE?

AIRSHIP IN PERIL THREE MONTHS AGO.

LEAKING GAS BAGS.

"Daily Express" Air Correspondent.

I am able to reveal the complete details of an alarming experience which overtook R 101 during her flight over Hendon during the Air Pageant last July which now casts a mysterious shadow across the terrible disaster of the airship yesterday morning.

There were elaborate precautions against the publication of the following facts and until now the secret has been kept successfully.

The destruction of the great air liner with its toll of forty-six lives, however, makes it vital that the truth should be laid at once before the public.

PREVIOUS ESCAPE.

Officials have carefully concealed the fact that disaster almost ended the career of R 101 on the day that she made her public appearance at the Royal Air Force display.

The airship began her cruise on that day shortly before noon, and spent a considerable time hovering over the east coast before starting for London, so that she would arrive over the flying field at her appointed time.

Directly on leaving Hendon, after being in the air for more than four hours, the officers in control of the vessel discovered that she continued to lose height.

Members of the crew who examined the gasbags found that the small balloons, which are made of goldbeaters' skin, were chafing against the metal framework. A report was immediately made to the senior officer, who set his course at once for Cardington and ordered all available speed from his engineers.

R 101 returned safely and was moored. The airship, however, was saved solely by the swift action of Flight-Lieutenant Irwin, the captain, who dumped his entire water ballast and two tons of heavy oil fuel, and employed every other possible means of reducing weight.

HOLES WORN.

Shortly afterwards R 101 was taken from the mast and put into her shed for repairs.

It was discovered on examination of the deflated balloons that more than sixty small holes had been worn through the gas containers.

The re-design of the airship's hull, which meant inserting another bay and increasing her length by forty-five feet, was subsequently undertaken, and it was understood in some quarters that a new method had been introduced for the arrangement of the gas balloons.

The theory of yesterday's disaster points, however, to a recurrence of the same trouble which the R 101 encountered on the day of the Air Pageant.

THE PRINCE INFORMED.

Air Ministry officials at Beauvais were in constant touch with the Prince of Wales throughout the day. It is possible that he may fly to Beauvais to-day.

The Last Message: "Going To Bed."

The last of a series of wireless messages received at Le Bourget from the R 101, states Reuter, was timed 1.50 a.m., fifteen minutes before the crash. It read as follows:—

"At the moment, the passengers, after an excellent meal and after enjoying a number of cigars, are getting ready to go to bed."

THE KING'S "HORROR."

The King immediately on hearing of the disaster sent the following telegram to the Prime Minister:—

"I am horrified to hear of this national disaster which has befallen Airship R 101, and the consequent serious loss of life, including that of Lord Thomson, my Air Minister.

"The Queen and I sympathise deeply with the relatives and friends of those who have perished in the service of their country, and also with the injured survivors.

"GEORGE R.I."

LATE NEWS.

REVOLT AGAINST BRAZIL GOVERNMENT.

NEW YORK, Sunday, Oct. 5.

According to despatches received here to-night, Brazil is in the throes of revolution. Practically the entire State of Rio Grande do Sul has revolted against the Federal Government, while outbreaks are also reported in the States of Parana, Minas Geraes and Rio de Janeiro. Owing to the strict censorship the extent of the uprising is not known.—Reuter.

Broadcasting Programmes on Page 13

NUNS' VIGIL OVER BODIES.

BEAUVAIS, Sunday, Oct. 5.

While the rain beat down on the smoke-blackened girders of the wreck of the R 101, the Sisters of Mercy at the Beauvais convent gathered at the little schoolroom where most of the victims' bodies lie, to begin an all-night vigil.

They will spend the night in prayer. The schoolroom is too small to accommodate all the bodies, and eleven are to remain in a field outside, protected by tarpaulins.—B.U.P.

TRAPPED MAN'S ESCAPE.

Mr. A. Disley, wireless operator, told how he escaped by tearing and biting his way through the fabric. "I used both my teeth and nails," he said. He became exhausted; then suddenly fell through "a raging tornado of fire," and found himself safe on wet grass.

Figure 30 Disaster coverage of the day Fleet Street style: the front and back pages of the *Daily Express* for October 6, 1930

Daily Express

MONDAY, OCTOBER 6, 1930.

ADVERTISERS' ANNOUNCEMENTS.

AERIAL VIEW OF GREATEST BRITISH AIR DISASTER.

A "DAILY EXPRESS" PHOTOGRAPH, taken from an airplane, of the wreckage of R 101 at Beauvais, France.

ONE OF THE ENGINE GONDOLAS.

FRENCH OFFICERS EXAMINING A WRECKED ENGINE.

THE AIRSHIP'S FLAG, WHICH BY SOME CHANCE ESCAPED THE FLAMES.

Printed and Published by the London Express Newspaper, Ltd., 8, Shoe-lane, London, E.C.4, Great Ancoats-street, Manchester, and at Albion-street, Glasgow.—Monday, October 6, 1930.

Figure 31 Harry Guy Bartholomew (1885-1962), creator of the modern *Daily Mirror* and architect of the 'tabloid revolution'

none of the existing papers – and certainly not the *Daily Mail* since Rothermere had inherited it from his brother, Lord Northcliffe – truly served the working class market then just emerging from the depression of the 1930s.

Bartholomew was a rough diamond, a cantankerous man, rude and ruthless in his dealings with people, but with a genius for understanding the mind of the ordinary man in the street (he was not heard to mention the ordinary woman in the street) and with the technical skill to devise newspapers to reach him. He could see that a good deal of newspaper content was outside the interests of such readers. A new sort of product was needed. For him this meant a handy tabloid-sized paper, down to earth in its viewpoint, which entertained as well as informed, was easy to read, with competitions

DAILY MIRROR, Monday, September 4, 1939

Daily Mirror

No. 11,152 ONE PENNY
Registered at the G.P.O. as a Newspaper.

BRITAIN'S FIRST DAY OF WAR: CHURCHILL IS NEW NAVY CHIEF

BRITAIN AND GERMANY HAVE BEEN AT WAR SINCE ELEVEN O'CLOCK YESTERDAY MORNING. FRANCE AND GERMANY HAVE BEEN AT WAR SINCE YESTERDAY AT 5 P.M.

A British War Cabinet of nine members was set up last night. Mr. Winston Churchill, who was First Lord of the Admiralty when Britain last went to war, returns to that post.

Full list of the War Cabinet is:—

PRIME MINISTER: Mr. Neville Chamberlain.
CHANCELLOR OF THE EXCHEQUER: Sir John Simon.
FOREIGN SECRETARY: Viscount Halifax.
DEFENCE MINISTER: Lord Chatfield.
FIRST LORD: Mr. Winston Churchill.
SECRETARY FOR WAR: Mr. Leslie Hore-Belisha.
SECRETARY FOR AIR: Sir Kingsley Wood.
LORD PRIVY SEAL: Sir Samuel Hoare.
MINISTER WITHOUT PORTFOLIO: Lord Hankey.

There are other Ministerial changes. Mr. Eden becomes Dominions Secretary, Sir Thomas Inskip goes to the House of Lords as Lord Chancellor, Lord Stanhope, ex-First Lord, becomes Lord President of the Council, Sir John Anderson is the Home Secretary and Minister of Home Security—a new title.

None of these is in the Cabinet, which is restricted to the Big Nine. These are the men who will be responsible for carrying on the war.

But Mr. Eden is to have special access to the Cabinet.

The Liberal Party explained last night that although Sir Archibald Sinclair had been offered a ministerial post, the Party had decided at this moment not to enter the Government.

POLES ATTACK

POLISH troops are fighting on German territory, according to a Warsaw message.

A Polish counter-attack pushed back the Germans and penetrated East Prussia near Deutsch Eylau, it was claimed.

The Polish Embassy in London described a Nazi report that troops had cut the Corridor as "entirely false."

Later (according to the Havas Agency) the Polish Radio announced that Poland had retaken the frontier station of Zbaszyn.

The German News Agency claimed that Nazi troops, operating on the Southern front had taken the town of Radomsko.

Radomsko, north of the industrial region round Katowice, is about forty miles from the Polish frontier.

1,500 Raid Casualties

The Poles' latest estimate of casualties in German air raids was issued last night in Warsaw.

It is alleged that 1,500 people were killed or injured in German air bombardment of open towns and villages during Friday and Saturday. A considerable proportion of the victims were women and children.

(The German Government had secured from

Contd. on Bk. Page, Col. 1

"BREMEN IS CAPTURED"

—French Report

The £4,000,000 German liner Bremen was reported to have been captured yesterday and taken to a British port.

A report from a high French source stated that the Bremen was captured at 4 p.m., but the area in which the liner was captured was not mentioned.

A French Government radio station broadcast the report which was picked up by the Mutual Broadcasting System of America.—Associated Press and British United Press.

Petrol Will Be Rationed

The first meeting of the new war Cabinet took place last night. Mr. Churchill was the first to leave and the crowd broke into a cheer as he walked out. Mr. Hore-Belisha was driven away by a woman chauffeur in uniform.

The Premier went from Downing-street to Buckingham Palace where he stayed with the King for three-quarters of an hour.

It was announced last night that as from September 16 all petrol will be rationed. In the meantime all car owners are asked not to use their cars more than is vitally necessary.

To-day all banks throughout Britain will be closed.

Australia yesterday declared war on Germany. "Where Britain stands, stand the people of the Empire and the British world," said Prime Minister Menzies in a broadcast message last night.

New Zealand has cabled her full support to Britain. There is a rush of recruits in Canada. At Toronto a queue of 2,000 men lined outside the Recruiting Office.

Japan has assured Britain of her neutrality in the present war.

Britain's last two-hour ultimatum to Germany was revealed to the people of Britain in a memorable broadcast from Downing-street by Mr. Chamberlain at 11.15 yesterday morning. By that time

cont'd in Col. 4, Back Page

The King to His People

"The task will be hard. There may be dark days ahead. . . . But we can only do the right as we see the right, and reverently commit our cause to God. If one and all we keep resolutely faithful to it, ready for whatever service or sacrifice it may demand, then, with God's help, we shall prevail."

These words were broadcast by the King last night. And to every household in the country a copy of his message, bearing his own signature facsimile, will be sent as a permanent record. The full speech is on page 3.

Page 2 THE DAILY MIRROR Wednesday, August 15, 1945

DON'T BE SILLY!

IT would be nice to have a lot more clothes; a lot more sheets and blankets; a lot more things of various kinds for the house. But no sensible person, though he may grumble to relieve his feelings, really believes that you can turn over from war to peace in ten minutes. The problem is complicated. It ranges from outworn machines to outworn ideas.

Why is there a shortage of cotton goods? We will not indulge in that old irritating retort and remind disgruntled critics that there has been a war on, but we ask them to try to remember the facts. The cotton industry, like other industries, has its labour problem, and it also has special problems of its own. Machinery is obsolete. Methods are out of date. There is, in short, a vast accumulation of "dead wood" which must be cleared away before any real progress can be made.

The last Government appeared not to be aware that the dead wood even existed! Before this, for a number of years, it was equally indifferent to the failure of this industry to revitalise itself; to keep itself up to date, and to do the things necessary if its place in the world's markets were to be retained.

Home to Roost

All this is having its effect now. The old chickens are coming home to roost. It is not to be expected that the product of years of folly and neglect — of masters, and maybe of men—is going to disappear overnight simply because a Labour Government has been returned to power, and Sir Stafford Cripps has made a train journey to Manchester.

For the time being Sir Stafford has done all that could be expected of him. He has shown that his journey was *really necessary*. He has made the cotton folk aware of what the nation expects from them. He has infused into it a new spirit of enthusiasm, co-operation and confidence. This is a big advance and a necessary preliminary to the modernisation of a declining industry. But this piece of "rehabilitation" will take time, and the country should be satisfied so long as it is clear that the Government is fully conscious of the problem, and is making every effort to solve it. Meanwhile, we have to put up with the old suit and the moth-eaten blanket. It may not, however, be for long.

British industry when it really gets going can do wonders. What was done in war can be done in peace if the right feeling is there and the carpers keep their mouths shut while the workers get on with the job.

B. B. B.

"—And now let's learn to live together!"

The 88,000 waifs of war

THERE are 88,000 orphans in Yugoslavia who cannot remember the names of their parents, or where they lived.

Mr. N. Petrovic, of Yugoslavia, told the council of UNRRA (United Nations Relief and Rehabilitation Administration) about them in London yesterday.

It was announced that the question of broadening the basis of the central council by the inclusion of three more nations would be considered and brought before a further meeting.

Before the resumption of the discussions on UNRRA's programme, the Director-General (Mr Herbert H. Lehman) pointed out that the carrying out of any proposals would depend on getting more money.

Including in a list of Poland's essential need—"rivers of milk and mountains of powdered milk"—Dr. Ludwik Rajchman said: "The time has come when there must be more definite co-ordination. Governments must know what they are going to get from UNRRA.

"They must not be told, 'We cannot give you any clear idea because of the many contingencies.' The time has come for some clear indication of what basic needs UNRRA can supply.

"Let UNRRA become a household word in every family in Europe. Let it be a symbol of practical co-operation.

"We must try to satisfy some of the essential needs in every country."

Dr. Rajchman added: "Let us provide sufficient soap, so that it will no longer be a luxury."

Other basic needs, he said, were transport (including horses), meat and fats

Figure 32 The war in the pages of the *Daily Mirror*: the first day, September 4, 1939, and the day Japan surrendered, August 15, 1945, an edition that carried a famous Zec cartoon

and lots of reader participation. It should get away from wordiness and hit the reader with big bold headlines and bold layout. Above all things it should be about people and the things that affected people – ordinary people.

It was thus that the *Daily Mirror* threw off its staid picture tabloid image and blossomed forth in the mid-1930s in bold new clothes which were to take it to Fleet Street's all-time peak daily circulation of 5,250,000 and make it the most powerful influence of the century on British newspaper design. It appeared on the news-stands distinctively (Figure 32) with thick black type headlines of a size not used before in dailies. The length of the news items shrank, foreign news almost disappeared, pictures became bigger, a page of comic strips appeared, and lots and lots of readers' letters. Human interest stories abounded, and where serious matters concerned the ordinary person they were spelt out in short words and paragraphs and charts with symbols and hurled at him (or her) in a way that could not miss. Into the paper came a vivid political cartoon and hard-hitting by-line columnists who berated the authorities on behalf of the readers and embodied the paper's us-and-them philosophy.

In design terms, the *Mirror*'s style was often referred to as poster journalism. In fact, on one or two celebrated occasions the front and back pages were upended to make one gigantic headline poster to smash a message home to readers on a page carrying the minimum of words.

The new journalism

The 1940s and 1950s were traumatic years for British newspapers. With wartime newsprint rationing reducing them to eight tabloid or four text-size pages the profligate use of space for headlines and text that had characterized the pre-war pages ended, never to return, and papers became leaner and more deliberately planned in order to put space to the best use possible. The *Daily Mirror* news page for May 8, 1945, shows the careful dovetailing and tight subbing by which fifteen items could be contained in a tabloid page (Figure 33).

It was the age of radio, but fears that the spoken word would dampen the appetite for print proved groundless. The immediacy of broadcasting might force newspapers to follow up rather than lead in the more important spot news, but it did not stop people rushing out to buy newspapers to get 'all the facts'. Moreover, there were

Tuesday, May 8, 1945 — THE DAILY MIRROR — Page 3

The King thanks Ike and his men

THE King sent a telegram last night to General Eisenhower, the Supreme Commander, asking him to thank the Allied Forces for a "complete and crushing victory."

"Eleven months ago you led the Allied Expeditionary Force across the English Channel, carrying with you the hopes and prayers of millions of men and women of many nations," said the message.

"To it was entrusted the task of annihilating the German armies in Western Europe, and thus liberating the peoples whom they had enslaved.

"All the world now knows that, after fierce and continuous warfare, this Force has accomplished its mission with a finality achieved by no other such expedition in history.

"On behalf of all my peoples I ask that you, its Supreme Commander, will tell its members how deeply grateful we are to them.

"I would ask you also to convey a special message of congratulation to my own Forces now under your command."

In case you are uncertain, today and tomorrow ARE HOLIDAYS

"TECHNICAL HITCH" CAUSED THE VICTORY DAY HOLD-UP

By BILL GREIG

CAPITULATION of Germany to the Allies was announced by Doenitz yesterday—but a mere technicality in the arrangements made with Russia and America delayed the British people's celebration.

One result of the delay is that Britain's workers get two clear days' holiday—today and tomorrow. This Cabinet decision was disclosed last night. Originally it was intended that the first day's holiday would not begin until the Prime Minister had spoken.

The announcement late last night that today will be VE-Day was the final act in a bewildering day following the German surrender.

Weather—as before

Now that the war in Europe is officially over, here is a weather forecast—for today:

Wind freshening; warm and sunny at first but rain can be expected later. Last night the glass was falling a little—indicating that the sudden warm snap of yesterday would not continue.

2,000,000 MILES

Liverpool minesweepers and trawlers have steamed nearly two million miles during the war.

GOEBBELS, WIFE, AND 6 CHILDREN FOUND POISONED IN BERLIN

DISCOVERY has been made in Berlin of the bodies of Dr. Goebbels, his wife, and their six children.

They were found by the Russians. All had taken poison and this was the cause of death.

Hitler's body has not been discovered and neither has the body of Goering.

There is now some speculation as to whether, after all, the Fuehrer and Goering may not have fled to a place of hiding.

It is pointed out, however, that their bodies may have been burned and the ashes lost in the burning Chancellery.

● *Mussolini's son Vittorio has been captured by Italian patriots near the Swiss frontier, says Paris radio.*

Rangoon curfew: looters to be shot

Martial law is now in force in Rangoon. Looters will be shot at sight. A curfew has been imposed between the hours of 6 p.m. and 6 a.m. Any person breaking the curfew is liable to be shot.

The Japs, before they left, encouraged looters to run riot.

Under Major Tooke, a Londoner, military governor of the city, every effort is being made to restore normal living.

Czech troops fly from Britain to save Prague—"great day"

SOON after a Czech Spitfire squadron left Britain for Czechoslovakia yesterday large formations carrying Czechoslovak ground troops also took off for home—and the battle for the liberation of Prague.

The commander, before leaving, said, "This is the greatest day of my life."

VE TRANSPORT CHANGES ARE ANNOUNCED

Restriction Relaxed

UNION WELCOMES PREMIER'S CHALLENGE

Channel Islanders await call

"Daily Mirror" Reporter

TWO BOMB HITS PUT PAID TO HUN CRUISER

NORWAY AWAITS ALLIED LANDINGS

DUTCH PATRIOTS TO KEEP ARMS

Even Franco kicks 'em now

Spain has severed diplomatic relations with Germany, it is officially announced in Madrid.

Sweden has done the same.

Spain will hand Laval over to the Allies, it was stated last night.

NO COUPONS FOR RED, WHITE, BLUE

Until the end of May you will not have to give up coupons for cotton bunting providing it is in red, white and blue or any of these colours and does not cost more than 1s 3d per square yard.

Figure 33 A *Daily Mirror page* for May 8, 1945, shows the tight dovetailing needed to use every inch of news space in an eight-page wartime tabloid

whole areas of news that had no chance of being among the handful of items which were all that could be included in a news bulletin; and so circulations leapt up in the 1950s as newsprint restrictions were eased. Even television, from the late 1950s onwards, did not have the effect on overall sales that the gloomiest prophets foretold, although it reminded editors of the importance of pictures, and forced them to take background and magazine features more seriously, and to take more care over the visual side of page design now that there was such visual competition. The more perceptive editors could see, even when colour television arrived, that there was a permanence in a good newspaper picture that was denied to something of which you got a quick glimpse 'on the box'.

The wartime period was also the high watermark of hot metal and the craft of the skilled printer, soon to be approaching its swansong. At the height of the London blitz, when Lord Kemsley's Gray's Inn Road plant was hit by a bomb and the underground Fleet River surged into the basement, the composing room continued to set as water reached the lead pots of the Linotype machines and sent clouds of steam swirling. Imperial printer George Eggington shepherded his flock until all was complete and the Kemsley titles came out on time. After the war, on his visits to the vast Kemsley composing room at Withy Grove, Manchester, George would review his men and machines like a general, walking down the long ranks, greeting each operator as he rose from his seat and stood to attention.

The history of newspaper design has a number of watersheds. The launching of the first 'popular' daily by Lord Northcliffe and the ideas that went into it, was one. The birth of press photography in the early 1900s was another; Bartholomew's tabloid revolution was a third, and the traumatic effects of two world wars on attitudes and practices could be said to constitute two more. Another such watershed occurred in the 1960s when, under the new Thomson ownership, the Sunday Times was revamped first by Robert Harling and later by Harold Evans, as a sectional newspaper on the American lines, but with a new type dress employing mainly lowercase light and bold serif faces (Figure 34) in a horizontal format with multi-column single line headlines crossing the pages below the fold, and with the pages tabbed and segmented by horizontal thick and thin rules. The most potent effect of the new approach was that it gave genuine strength below the fold to text-size pages, which was something they had previously lacked.

The style, taken up by *The Observer* and, later, *The Guardian*, was ideal for newspapers who wished to compartment their news and features and to make imaginative use of their text size untrammelled by traditional forms. The pages that resulted were characterized by a more daring and artistic cropping of pictures as design elements, the creative use of white space around headlines and pictures and the liberal placing of logos on regular items. The style also set off a shift

Lord Moran's controversial Churchill diaries

THE SUNDAY TIMES

No. 7457 APRIL 24 1966 PRICE 9d.

Anti-trust threat to Sotheby's US market

By Henry Brandon

WASHINGTON, Saturday

THE DEPARTMENT of Justice is seriously weighing anti-trust proceedings against Sotheby's the auctioneers, who in July, 1964, acquired the leading auction House in New York, Parke-Bernet.

Enough choice

Slipped back

Drift from pits will mean a £200m bill to meet power needs

By Ian Coulter, Industrial Correspondent

BRITAIN faces the prospect of having to pay an extra £200 million a year for increased imports of oil because massive manpower losses in the coalfields are threatening to present the country with an energy gap of at least 13 per cent. by 1970.

The loss of men is running at the rate of 3,190 a month. And a symptom of the decline is the fact that in coalfields as far apart as Llanelli in Wales, Welbeck in Nottinghamshire and Moston, near Manchester some 250 new Coal Board houses are empty beside profitable pits.

Mr George Brown's National Plan should be completed by 1970, but the accelerating run-down of the mining force is bound to jeopardise its chances of success.

No such crisis in the energy field was foreseen when the Plan was published last September. To be sure, mining manpower had been running down for years, and between 1960 and 1965 averaged 27,000 men net each year. But the Plan confidently asserted that in the next few years the fall would slacken off.

Instead, the decline has accelerated. Since the Plan was published the net loss each month has been running at over 28,000 a year, or 11,000 a year more than the planners bargained for.

In terms of coal production this will mean a short fall of 25 per cent. by 1970. In spite of the promised increase in mechanical coal-cutting from 80 per cent to 100 per cent., the industry will be 45 million tons short of the 175 million-ton target set by Mr Brown. Manpower will be about 264,000 instead of the hoped-for 333,000.

This grim prospect will be put before the Government together with a warning that by 1970 the country's balance of payments will have to carry an extra £200 million for oil to bridge the energy gap.

North Sea gas will be on stream in considerable volume by 1970 and many new nuclear electric power stations will be operating: but these are likely to produce only about 18 million tons of coal equivalent energy.

The causes of the dramatic desertions from the mines are many and complex, but the National Coal Board has put its finger on an obvious one: Government-sponsored factories sited near collieries. The most notable examples of this conflict of interests are at Kirkcaldy, Fife, beside the Seafield colliery, and as in the Durham coalfield, where there are six factories.

There are at present 434,200 men in the industry; production is running at around 200 million tons a year. The National Plan's estimate that the country's 1970 requirements would be only 173 million tons undoubtedly affected morale in the pits. Most of the men leaving the industry are in the 25 to 40 age group, and their outlook is bound to be affected by how they see the industry's long-term prospects.

Problems caused by the manpower drain, equal to two men for each mine in the country every week, are heightened by absenteeism, so that frequently men have to be taken off the coal face and put on non-productive jobs, such as maintenance and supply. The consequent loss in coal production is equal to one cwt. for every man on every shift. It is estimated that about 250,000 tons of coal are being lost each week.

The Moran Diaries, from which the second in our series of extracts appears today, are stimulating interest and argument to a degree unmatched by any other book this year.

During the whole of his 25 years as Churchill's personal physician—the years 1940-1965—Lord Moran kept this detailed, perceptive and often deeply moving record.

It is a candid book and a controversial book—one that brings the reader closer to Winston Churchill than ever before.

In the Weekly Review 25-28

Stanley Devon

A Royal titbit for the goose. The Queen and the Duke of Edinburgh toured the grounds of the Wildfowl Trust in Gloucestershire yesterday with the president, the Duke of Norfolk, after opening a research centre

Firm bans LSD 'fantasy' drug

By Dr Alfred Byrne, Medical Correspondent

Government will not allow BEA to buy US jets

By Our Air Correspondent

Continued on page 3

Rhine search for missing Prince

By Antony Terry

Rate relief law is being abused

By Muriel Bowen

Jenkins will speed police mergers

Dearer copper hits cars and homes

Zambia's decision—page 15

3 die, 40 hurt at a cup final

Reports of matches—page 24

The Pope not to visit Britain next year

CID comb yacht for loot clues

The Missing Four—page 2.

Figure 34 Revolution in Gray's Inn Road: the new-style *Sunday Times* of the 1960s as it blossomed under the Thomson ownership

into wholly lowercase type formats across the board in quality and many provincial titles.

In its development, it instituted what might be termed the modern approach to type use and page patterns that was to herald – and to some extent challenge – the next great watershed in newspaper design, the availability of full page composition on the computer screen.

Hot metal technology

Newspapers as they appear today, though every stage in their production has been computerized, are in essence the end product of old printing technology. The concepts of typography, design, page make-up and press work are rooted in methods based upon the use of hot metal at all stages, and the purpose in building computerized systems has not been to evolve a new printed product, or to bring about changes in newspapers, but to reproduce exactly by cheaper and more efficient means the sort of newspapers being produced by the old methods. The facilities built into the new systems are thus orientated to working parameters that already existed.

Pages were made up – and had been from the very earliest newspapers in the seventeenth century – by putting into a frame called a *chase* the lines or slugs of type that had been set by compositors and assembled ready for this purpose in long metal trays called *galleys*. The galleys of type represented the editorial input of news stories and feature articles written by reporters and feature writers for the edition and edited to fit the spaces on the pages. The checking and editing to length of this material was carried out by subeditors who had marked, on the original typed or handwritten text (called copy), the type and setting instructions for the compositors to follow. Once typeset, this material was further checked for setting errors (called *literals*) by proofreaders working from proofs pulled from the galleys of type.

The pages were then put together by a page compositor, or *stonehand* (Figure 35) to an editorial plan, or layout, which indicated the positions of the text, headlines, pictures and adverts. The pictures and adverts, like the type, were of metal, being engraved on to *plates*, or 'blocks', and then mounted on metal mounts called *stereos* to give them the same height in the chase as the lines of type.

In the early days, as can be seen from examples earlier in this chapter, design played little part in this. Often the printer was the editor and, apart from the first items on page one or two and a crude sequence or pattern for the different parts of the paper, the order of items was often that in which they had been received, with the headlines being simple labels scarcely bigger than the type of the reading matter. While design techniques developed rapidly from the late nineteenth century onwards, typesetting, page make-up and printing methods underwent little basic change other than the introduction of keyboard-operated *linecasters* for the setting of body type in the 1890s.

Figure 35 Page make-up hot-metal style

The pages under the hot-metal method just described were made up on a stone-topped bench, later of metal but still referred to as 'the *stone*' (hence stone-hand, stone subeditor, etc.). A completed page was called a *forme*. From the forme, moulds were taken under high pressure using a composition material called *flong*, from which were cast the curved printing metal plates used to print the pages by the letterpress method on the rotary presses (Figure 36). These presses, though greatly refined so that they could comfortably print up to nearly 70,000 copies an hour, remained in essence the sort of press first used by the *Philadelphia Ledger* in 1843, by which curved metal plates attached to a revolving drum picked up ink from a bath and printed the pages by direct impression on to a revolving web of paper as it passed over

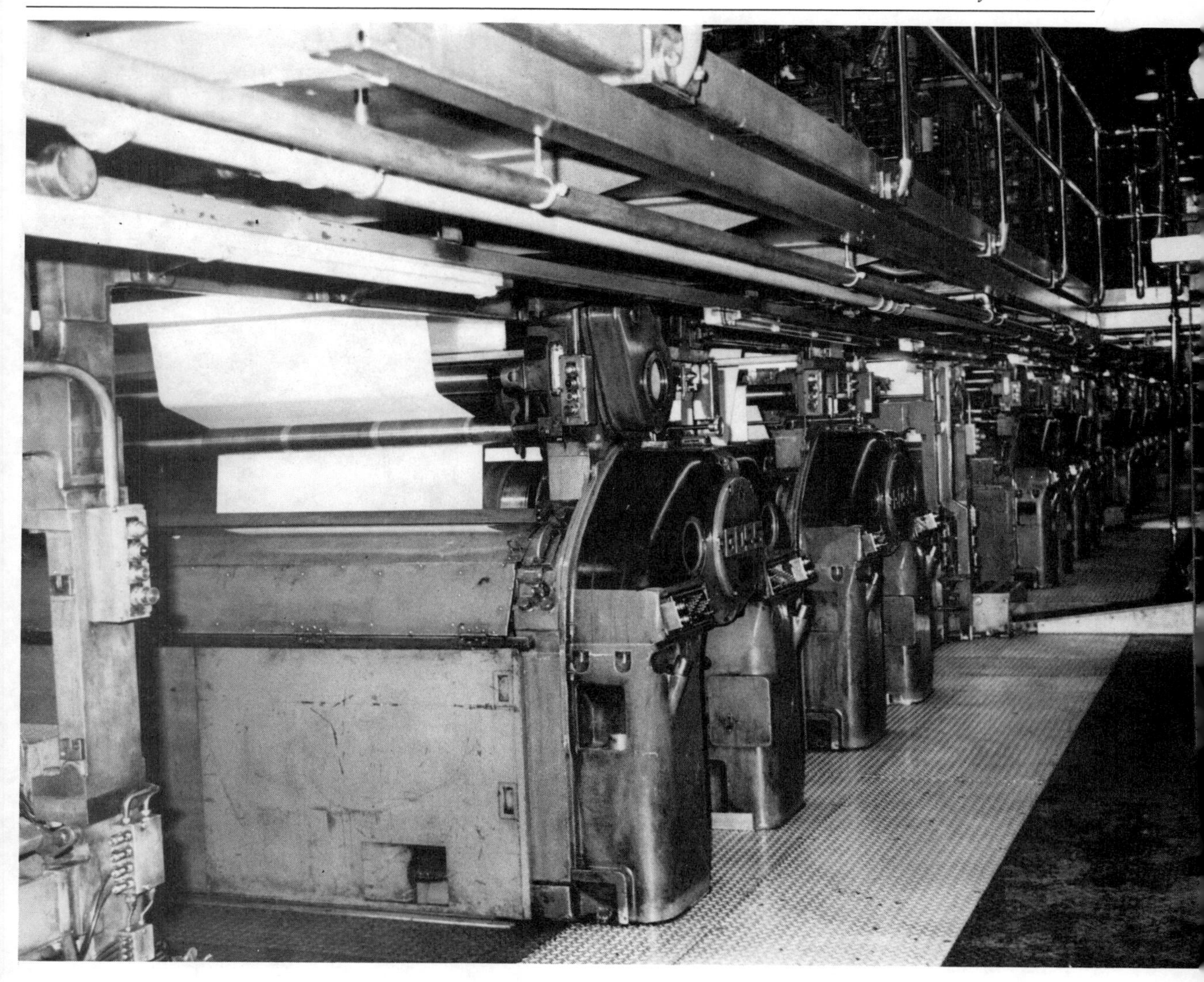

Figure 36 A bank of rotary presses – now mostly replaced by web-offset presses

them, that is, by *letterpress*. The rotary had replaced the earlier flat-bed, or sheet-fed presses.

Modern design techniques grew up in this working environment, as indeed did the whole concept of a newspaper as a product. As markets and revenue expanded so editing and design became more sophisticated and the functions required for them more refined. New types with extended sizes were designed for special roles, in particular some of the versatile modern sans ranges, influenced by designer Eric Gill and, among seriffed types, Stanley Morison's Times New Roman. Type-composing machines had been designed and built since the 1840s, but Ottmar Mergenthaler's Linotype machine, patented in Baltimore in 1884, by which an operator on a keyboard (Figure 37) produced slugs of type from matrices, proved

Figure 37 The workhorse of old technology – Linotype machines

the first practical one. Instead of the laborious hand-selecting of tiny letters it could set lines mechanically at five a minute. Meanwhile, press speeds were being improved and also ink and paper quality, while automatic etching machines replaced the old hand engravers and speeded up the process of blockmaking.

Yet typesetting, picture reproduction, page make-up and printing remained a hot-metal, labour-intensive operation with the basic routines and attitudes little changed, and costing and work practices rooted in earlier days. Typesetting speeds increased hardly at all from the breakthrough with the Linotype machine in the 1890s until the 1950s, when experiments began with computer-fed tape which (against some opposition from print unions who were understandably worried about jobs) increased the setting speed from linecasters to fourteen lines a minute.

Useful results could be achieved by the right cooperation between designer and page compositor. Indeed the cooperation was vital since the tools needed to turn a design into a finished page were in the hands of printers. A disadvantage from the designer's point of view was that any special type effects, or the use of type reversed as white-on-black (WOB) or as black-on-tone (BOT), had to be prepared by photographing the type and creating a metal block or plate. This, together with the growing use of graphics for special display and generally greater demand for halftone and line blocks, including those for adverts, put ever-increasing pressure on the process-engraving department. It was here that the most spectacular bottlenecks occurred in hot-metal production.

The poor quality of labour relations, and the failure of print union agreements to take proper account of the growing sophistication of editorial requirements, plus the general problems of old machinery, outdated methods, demarcation disputes, noise, dirt and slowness inseparable from the hot metal operation, made it harder and harder for editors to get the sort of paper they wanted.

It has to be remembered, nevertheless, that in ideal conditions the hot metal system worked and produced great skills which made possible the advances in design techniques that established the British newspaper in the postwar years as a model to Western countries. When faced with the demand for colour from the mid-1960s onwards, the pre-printing on newsprint of colour advertising gave a high quality result long before run-of-press colour became commonplace with the web-offset presses of the new systems.

The computer

It was not the pursuit of quality that forced the computerization of the printing industry from the 1960s but rather the need to solve the chronic problems of costs, over-manning, poor labour relations and (by contemporary standards) spectacularly old-fashioned technology. Rotary presses had a forty-year life; Linotype machines and page moulding presses had been in use unchanged in some offices for more than seventy years; some Fleet Street composing rooms in the 1970s, with their mallets and space bars and iron trolleys and chain-driven proofing machines, looked like museums of industrial archaeology.

Early problems both with photo typesetting and with printing presses delayed the take-off of 'new technology' but by the mid-1960s most American papers, which were smaller and cost less to re-plant, and a growing number of British provincial papers had changed to the new methods. The development in the 1970s of hard polymer for printing plates and the introduction of faster web-offset presses with better paper control made the new systems suitable at last for the longer print runs of the national titles. By the end of the next decade most British papers had switched from hot metal to 'cold type' to secure advantages in cost-saving and efficiency, once it was clear that product quality could be maintained. The move was helped by the new generation of fast main-frame computers which greatly improved editing and typesetting capability. Through the move into 'new technology' it also became possible to secure improvements in picture reproduction by the use of a finer screen acceptable to the smooth web-offset printing plates.

Early problems of poor match of types on changeover to computer setting have been overcome and now all main manufacturers carry a full range of standard and not-so-standard faces to suit the most fastidious users. There can be, as a result of the American provenance of most of the type masters, slight design variations in popular types such as Century, Ludlow Black and Tempo but this has not prevented papers from maintaining a striking continuity in presentation with even an added clarity of reproduction.

Matching existing body faces has offered more problems, though some papers, addicted to a battered and old-fashioned Ionic or Jubilee long overdue for the bin, have been delighted to find their columns taking on a new sharpness and readability.

Traditional body and headline type sizes, to suit each customer's type format, from 5½ point right through the range to 144 point where applicable, are programmed into the computer for ease of command in editing, and to make type balance more calculable. Yet, in fact, with the computer you can have any size you like down to half a point variation, and the facility to override formatted commands and produce 'bastard sizes' exists in all systems. Used sparingly and sensibly, it is invaluable for getting into a measure a much wanted headline – especially one written by the editor – or avoiding excessive tightness of set or wasted or ugly white.

The command facility to italicize or 'lean' a typeface is also useful, if used sparingly, although not all faces italicize successfully. The best italic types are those that were designed that way, and it is useful to have some stock italic fonts available.

Spacing, both linear and letter, should be programmed in at the outset to maintain style and give visual consistency although, as with type sizes, formula spacing needs to be over-ridden to give special effects or to solve difficult setting problems.

5
Pictures

Newspapers are so dependent upon pictures performing a design function as well as an informative one that it is hard to imagine how editors coped in the days before the camera. In fact, as we have seen, design in early newspapers played such an insignificant part in the marketing of a product that consisted simply of sheets of information for a limited audience that the lack of illustration was not felt. It was when circulations and markets started to expand towards the end of the nineteenth century, accompanied by a broader content and deliberately targeted readership, that presentation began to matter.

Certain news stories, it was found, gained in actuality if the words could be accompanied by a picture. Certain features, even the most mundane how-to-do-it ones, were more intelligible if a point could be made visually. At the same time a picture on a page, it was realized, created a focal point for the eye of the reader and so how the picture was located began to matter.

Role In design

We have seen that whatever the design style or market (the two are connected) of a newspaper, pictures form a fundamental ingredient of the mix along with text, headlines and adverts. In an ideal world every page would have a right size, right subject picture forming a main focal point along with the bigger headlines, around which the text and the other headlines and pictures would be arranged. Such a picture would sit neatly in the editorial space left by the placing of the adverts as if the space had been designed for it. Nothing is so simple. In fact, the choice and placing of pictures is the product of a mixture of planning, good luck, compromise and the sheer inspiration that produces the display idea to unite the ingredients.

Planning a paper's contents, including even some feature pages, is geared to a news coverage dependent on the events of the day and often has to be initiated without some of the material being at hand at the start. While many events (Parliament, courts, meetings, weddings, etc.) can be allowed for in advance, some of the best stories will happen without warning. As a result, the allocation of pictures to some pages might have to be made before they have been received or, alternatively, left to a fairly late stage if the page production slot allows this.

Compromise is of the essence. The most newsworthy text might be the least productive of pictures, and it is unlikely that every page will have a main story supported by a main picture, or that the most wanted picture turns out to be the right shape (that is, horizontal or vertical in composition). A superb picture, or the relative failure of a picture assignment (often a clear case of luck), can cause early plans to change. As a result, editorial material might have to be moved around so that every page gets its share of pictures, and so picture-worthiness as well as news value can become an element in the space and position stories are given. The picture possibility of a feature can likewise govern the space and position it is given.

While change and compromise can play a part in allocating the contents to a page, it does not mean that any old picture – any more than any old text for that matter – will do. The criteria for choosing and using the right picture is as important as the criteria for choosing and using the right story and much care and judgement goes into the process. In addition, therefore, to the basic requirements that a picture should illustrate the text and be an element in the page design, we have to take into account other factors. There are, for instance:

- *Composition*. The grouping and position of the people and main objects in the picture must form a pleasing shape. It has to be the most eye-catching picture of its set or of those available.
- *Balance*. It must balance with the rest of the page and with the picture material of the adverts. It is no use if, for the position it is required to fill, the subject faces out of the page or out of the story it is illustrating.
- *Quality*. There must be sufficient contrast in tones between dark and light for the picture to reproduce properly. A lack of good tonal values can produce a grey effect in mono and woolliness in colour, although a poor picture can be enhanced by the computer.

If the main picture is supporting the page lead or half lead then it must be integrated into the main text and headline area but in such a way that it plays a structural part in the page. If it is with a secondary story then picture, text and headline must be combined to serve a structural function in order to get the best visual effect from it. If the picture is being used for its own news value – say, the first release of a Royal portrait – then with its caption story, rather than just an identifying caption, it must be given its own prominence as a focal point in balance with the main headline area. If there are secondary pictures on the page, even if they are just single-column head shots, they must be balanced against the main picture so as to provide lesser focal points. The examples described below show how these considerations work in practice.

Picture workshop

The news page from the tabloid *Birmingham Post* (Figure 38) shows how to get the best out of a handful of pictures supporting a dramatic page lead. The decision to top the page with the most vivid

THE BIRMINGHAM POST, WEDNESDAY, MARCH 8, 1989 9

One of the carriages derailed in Saturday's crash at Purley, Surrey, is hoisted above the houses before being taken away for examination. Five people were killed in the accident.

Driver 'failed to report Clapham signal change'

A British Rail driver could have prevented the Clapham disaster in which 35 people died by reporting signal irregularities he noticed only minutes before the crash, the inquiry was told yesterday.

But Mr George Christy did nothing, it was claimed.

The reason was that he should not have been driving the train, but had agreed to an "illegal" swop with another driver, said Mr Benjamin Browne, a lawyer.

Mr Christy rejected the allegations from Mr Browne, who was representing Mrs Sue Rolls, widow of the driver killed in the crash, Mr John Rolls.

In an emotional gesture yesterday, she handed back the union badge her husband was wearing when he died.

Mr Tony Staton, district secretary of the Associated Society of Locomotive Engineers and Firemen, told her to keep it, but she said between sobs: "I am sorry, you don't understand. They have let everyone down."

It is understood she was upset at the union decision not to cross-examine Mr Rolls's fellow drivers about the failure to report the signal irregularities.

Mr George Christy: denied 'illegal' swop as driver.

Mrs Sue Rolls: handed back husband's union badge.

The inquiry has heard that a wiring fault during signal modernisation caused the fault, which the driver of a Basingstoke train stopped to report. The signals then allowed a Bournemouth train — driven by Mr Rolls — to run into the back of it.

Mr Christy, aged 60, said he had driven a Bournemouth-Waterloo train past the spot just 10½ minutes before the crash in December.

But he dismissed Mr Browne's claim that the swop he had agreed to — by taking over the train from another driver — was "illegal".

However, he admitted it was unauthorised and a disciplinary offence had the management found out.

He did not report sudden changes in the lights shown by the signals because he thought they were no more than unusual. He thought the reason was that a Clapham signalman had rerouted a train in front.

Mr Browne said: "If you had realised the importance of what you had seen and reported it this tragedy would never have occurred, would it?"

Mr Christy said he did not know how to answer — there was a signal irregularity, but he was not aware of it at the time.

Mr Browne said: "May I suggest that you are so anxious now to say that this wasn't an irregularity and didn't report it at the time because you were well aware that you should not have been driving that train and, had you reported it, you would have got yourself into trouble?"

Mr Christy: "That isn't the case. If there had been an irregularity I would have reported it."

Mr Christy, of Hammond Close, Basingstoke, Hampshire, said he had seen the three signals approaching the one at the centre of the crash, number 138, change to green as he approached the first of them.

Then signal 142, the one before 138, suddenly "flicked" from green to yellow, he said.

A BR official responsible for supervising signals modernisation said he had feared there could be more loose wires like the one that caused the crash.

Mr Alfred Court was being questioned by Mr Michael Spencer, counsel for the bereaved and injured, about why he had agreed to cut off part of the old loose wiring responsible for the disaster after signal engineers found it within hours of the crash.

Mr Court, aged 48, signals supervisor for new works, based at Wimbledon, said he had obeyed orders from senior engineers to make the wire safe by cutting it and had left it in their hands.

Mr Spencer, asking why Mr Court had not left it in place to be photographed and examined later, said: "You appreciated the importance of this evidence, did you not?"

Mr Court said: "Yes, but I had just finished 16 commissions similar to this one and I was a bit worried about wires on my job hanging loose."

Mr Spencer said: "You were worried that some of the wires that you have been involved in could be in the same state?"

Mr Court said: "I was worried. But when we checked there were none."

He said the remaining wire, still connected at the other end to a fuse, was taped up with black tape with a marker to distinguish it.

The next day two senior engineers asked him to remove the wire completely, but he refused. He cut off a short length and gave it to one of the engineers.

The inquiry continues today.

Warning to egg producers

By NIGEL HASTILOW

The Government last night promised tough legal action against food producers who continue to sell infected eggs.

After months of criticism of the Ministry of Agriculture's delays in responding to growing public concern over salmonella in eggs, a clear warning was issued to producers last night.

Mr Richard Ryder, the junior Agriculture Minister, told egg producers: "Clean up your act or face the full brunt of the law."

He was speaking during a Commons debate on the Government's £19 million package of assistance to the industry which came after the slump in sales caused by a remark by the former junior Health Minister, Mrs Edwina Currie, that most egg production was infected with salmonella.

Mr Ryder rejected the claim by the all-party Commons agriculture select committee that his department had been sluggish.

He said the Ministry had put forward 17 new measures to control salmonella in eggs, the most comprehensive package in the world.

Though egg sales were still down by 20 to 25 per cent, the Government's rescue package had restored stability.

He said: "Consumer sovereignty is the key to the market place."

"It can be threatened by monopoly and cartels, trade protectionism masquerading as protection, or neglect of public health standards."

Earlier, Mr John Biffen (Con Shropshire North), the former Cabinet Minister, condemned Mrs Currie (Con Derbyshire South), for her reluctance to give evidence to the select committee and described her as "a political health hazard."

Sir Hal Miller (Con Bromsgrove) said some caterers had still not reacted positively to the Government's measures to improve egg quality. Some small farmers were still finding their supplies to caterers were down by 60 per cent.

Dr David Clark, the Labour agriculture spokesman, accused the Government of mishandling the whole affair by "incompetence, excessive secrecy and delaying tactics.

"By their failure to act speedily, they have allowed the problem to become exacerbated."

Dangerous myths on cancer causes

Myths among children that cancer is caused by eating burnt toast, operating computers or falling off a chair backwards must be destroyed, the Cancer Research Campaign said yesterday.

Survey results published by the campaign showed that children had various misconceptions about the disease and were deeply confused about its causes and curability.

The campaign is worried that children will grow up with false ideas and fail to seek early treatment as adults for possible symptoms.

A hundred children around the country, aged between 13 and 16, were asked their views of the disease.

Dr Anne Charlton, of the campaign's Manchester-based education and child studies research group, found that a third of the children were correctly aware of the dangers of smoking. A fifth were aware that excess sunlight could cause cancer, but many were wildly wrong in their view of the disease.

"Old wives' tales are developing all the time about cancer. Schools can help to project a more accurate message and expose some dangerous misconceptions," said Dr Charlton.

Tesco pays half costs

Tesco is to pay half the costs of the case in which two sisters were awarded £800 damages each last week after suing for unlawful arrest, false imprisonment, libel and slander.

The sisters, Mrs Frances Warby and Mrs Ann Chastell, from Hertfordshire, faced a costs bill which some estimates put at £20,000, because they pursued their court action after refusing a settlement offer of £1,500 each.

Tesco will pay costs up to the time it paid its settlement offer into court. The firm is not paying the sisters' portion of the costs.

Jail for father who caused son's death

A drunken driver was jailed for four months at the Old Bailey yesterday for causing the death of his two-year-old son in a car crash.

Paul Swaby, who was already banned for another drink-drive offence, was two-and-a-half times over the legal limit when he set off home with his son from a party.

Swaby (28) put the boy in the front passenger seat with an adult safety belt over him.

Mr Tobias Davey, prosecuting, said that as Swaby went through the Blackwall Tunnel he overtook on the inside at 50mph.

The car careered from side to side before smashing through a crash barrier on to the opposite side of the road.

The toddler was killed almost instantly.

Swaby had denied causing death by reckless driving.

Mr Andrew Bright, defending, said the boy's death had had an appalling effect on Swaby.

The judge was told that Swaby, unemployed, of Sheridan Road, Manor Park, East London, is already serving 15-months for deception and handling. He ordered yesterday's sentence to be served concurrently.

Figure 38 Handling a news picture: the tabloid-sized *Birmingham Post*

Irish Independent, Wednesday, March 8, 1989 7

Tug-of-love men 'home in 3 months'

By FRANK KHAN

THE Jordanian father of Nadine Nassir — the Dublin-born child at the centre of the tug-of-love drama — said last night the two Irishmen, jailed for attempting to abduct her, should be home "in three months".

And he added: "I've nothing against them."

Dr. Sameer Nassir denied, in an exclusive interview with the *Irish Independent*, that Gerard Flynn (26), of Knocklong, Co. Limerick, and Denis Dennehy (28), of Portlaoise, were being held in primitive prison conditions.

Dr. Nassir, speaking for the first time of his ordeal, revealed how he had visited both men with Nadine's uncle, Seamus McManus, and declared: "They are being treated very well. I've nothing against them."

He said: "I think they will be let out in three months with good behaviour. I'm glad the boys will be getting home soon." He expected both men would be transferred shortly to Sharjah prison where they would be entitled to buy food and have books supplied.

Dr. Nassir said Nadine, who will be six in April, was fine. "She is very well," and had spoken on the 'phone to her mother, Ann McManus, for 20 minutes yesterday.

Nadine's father, who spent 20 years in Ireland, said he would be prepared to let Nadine come back to Ireland to see her mother if sent an Irish court order giving him full custody. Nadine was under her mother's custody when she was taken by her father to Dubai last summer.

Under the "right circumstances", Dr. Nassir said he would return to Dublin with his daughter. Dr. Nassir, who is a surgeon at Khorfakkan Hospital, added: "I never wanted to deny Nadine her mother, or her mother Nadine. Under no circumstances would I allow them be separated forever." He said: "I did not kidnap my daughter — I feel she should be brought up here by me."

Dr. Nassir said he was disappointed by the way the Irish media had treated him, adding: "They tore my private life in pieces." Referring to the two-year sentences handed down by the court, Dr. Nassir said he had not been in favour of stiffer punishment.

US Yuppies to burn up the ould sod

● Exporting a well-packaged product.

By TIM HASTINGS

BORD NA MONA is to make its first major pitch for the American market later this month with the introduction of Irish Peat Bricks as an "upmarket lifestyle product".

The new Yuppie fuel — which is designed to be used for special occasions — will sell at nine dollars and 95 cents a half bale. That's almost 10 times the price of the half bale here — but the difference is largely accounted for by transport costs.

Packaged in an attractive box format, the new "bricks" are being marketed on a pilot basis in the New England area. And, if successful, they will be extended to outlets in Chicago, Boston, New York and Toronto, according to the board's marketing man John Foley.

The bricks, similar to briquettes on sale here, have already been test marketed in selected homes and more than half the households involved said they would be willing to buy.

"We are not selling the peat as a commodity fuel but as something for special use. We view it as a long-term project," Mr. Foley explained. People like the smell of the new bricks and their appeal is not confined to middle-class and upper-middle-class buyers, he added.

Initially, the board is planning to sell 50 tons of the product in the first phase of the US marketing campaign which followed consultations with Coras Trachtala and months of market investigation.

Shares scam firms crackdown

A MAJOR crackdown on "share pushing" firms is being planned by the government as it emerged several such operations have set up here, with one Dublin-based scam now making more than £100,000 a month by conning investors overseas.

Industry and Commerce Minister Ray Burke admitted yesterday the so-called "boiler-room" operations are an international scourge and promised action to stamp them out.

By BRIAN DOWLING

There is mounting fear in legitimate financial circles that these firms operating from Dublin and Cork will damage the country's reputation and may cause major problems for the new Customs House Docks financial centre.

Already the Director of Consumer Affairs has taken steps to warn potential UK investors — one of the main target groups of the share pushers — to be wary about offers of shares or dealings in commodities markets by firms using telephone sales from Dublin.

In a circular to the British media, the Office of Consumer Affairs warned some of them have links to organised crime and international investment fraud. It says: "The office is concerned about the bona fides of these firms and about the dangers posed to inexperienced private investors by their marketing methods

The latest action follows a series of visits in the last week by staff from the Office of Consumer Affairs to a number of firms operating from Dublin and Cork.

One of them, selling shares in a Hong Kong medical supply firm, is running an elaborate scheme from Spain using Ireland as a covering address. Using a city centre office staffed by one person, share sales made on telephone through Spain are transacted via Dublin and investors' money lodged in a Dublin bank.

Once the investor agrees to buy shares, the Dublin office is contacted and sends out the share purchase form and invoices are then transacted here. Initial estimates are that the company is banking in excess of £100,000 a month.

Other firms under investigation are involved in selling shares in an Australian gold mining company and selling commodities — usually metals, which investors are told will rapidly increase their market value.

The Minister said he is examining regulatory structures to ensure such firms can no longer operate here. Asked in the Dail if he would set up a Securities Commission, as in the U.S., to police such operations, he replied this was like trying to "hit mercury with a hammer".

A golden welcome

● Coming home . . . Marcus O'Sullivan waves to the welcoming crowd at Cork Airport. Picture: TED McCARTHY

Homecoming of 'Magic Marcus'

By DICK CROSS

"MARCUS is magic", declared the schoolboy banner as Ireland's fast-track specialist, ex-Cork schoolboy Marcus O'Sullivan was given a triumphant homecoming with his world championship 1500-metre gold in his bag, yesterday.

The crowds thronged Cork Airport, then Carrigaline pipers played him and team mate Frank O'Mara down onto the tarmac and into an airport reception with speeches and presentations.

Then came the spectacular drive downhill through his native Turners Cross and past his old school, Colaiste Criost Ri, as friends and neighbours showed their pride in the local boy who beat the world's best.

Cork City Hall was on late opening as Deputy Lord Mayor, Mr. Corr, rendered honours on behalf of the city. Marcus rang his American-born wife, Mary, in New Jersey to tell about the reception. She stayed at home with two-months-old Laura.

His father, John (57), his mother, Elizabeth, brother Gregory and sisters, Grace and Lucia strove to hold back their feelings at the dimensions of it all. They were, clearly overjoyed.

"It is incredible", said his dad as he and Bishop John Buckley walked out towards the Aer Lingus plane just in from London.

The Taoiseach, Mr. Haughey, sent down a letter of congratulations and said: "It was a tremendous achievement and maintains the magnificent contribution which Irish athletes have made to sport at international level."

BRIEFLY

No bail on gun charges

A Belfast man who allegedly stored a large quantity of guns for the Loyalist Ulster Defence Association was refused bail yesterday by the High Court in Belfast.

The weapons were found during an RUC raid on the Rathcoole home of Leslie Mawhinney's girl friend last week. A lawyer said that after his arrest Mawhinney told police he had agreed to keep the arms because men had threatened to tell his wife about his relationship with the girl.

Not enough preparation

Progressive Democrat education spokesperson Mairin Quill claimed yesterday there was inadequated in-service training for the new Junior Certificate, due to replace the Inter and Group.

At present teachers are undergoing one day's preparation for the new syllabi which are due to be introduced in September but Mrs. Quill said this was not enough.

WP talks 'helpful'

Devolved Govt. for the North, security, education, economic investment and the promotion of a 'peace charter' were among the issues discussed yesterday when a high-level delegation from the Workers' Party met Northern Ireland Minister of State Dr. Brian Mawhinney at Stormont Castle.

A statement later from the Workers' Party described the meeting as being welcomed by the Minister as 'helpful'.

'Trade and Peace' talks

'THE ROLE of Trade in World Peace' will be the theme of the fourth annual international conference of the Shannon-based Centre for International Co-operation.

The event, from May 17-20 at the Limerick Inn Hotel, will be opened by President Hillery. Trade experts from the UN, GATT, the EC, Soviet Union and the US will be among the speakers.

Dail offer to the new radio stars

By JOHN FOLEY

CHANGES are to be sought in Dail rules to enable newly licensed independent radio stations to broadcast proceedings from lively Dail committees.

The Committee chairpeople want invitations sent to Century Communications to cover the four major public committees because of RTE's persistent refusal to broadcast their proceedings. Angry deputies have accused the national broadcasting station of refusing to honour its contractual arrangements to cover the Dail or recordings of its related committees.

And because of RTE's refusal, there are growing demands that the CPP should now change its rules to extend to the new independent network the option of broadcasting committee proceedings or recordings.

Moves to involve the independent network are being spearheaded by Fine Gael's Gay Mitchell, chairman of the all-party Public Accounts Committee.

"Dail committees have continually been refused broadcasting facilities by RTE which maintains that it does not have the technical equipment or resources to extend its radio coverage.

"In the light of RTE's refusal to do what it has agreed, I will be asking the CPP to consider the position and invite Century Communications to take on the task," Mr. Mitchell said.

He added that all members of the various committees had been angered by RTE's refusal to cover their activities, despite several meetings with senior executives from the station.

Student fee row

THE UCD administration block on the Belfield campus was taken over by students yesterday protesting against college plans for a £50 capitation fee.

The college president, Dr. Patrick Masterson, briefly addressed the students explaining the university's financial difficulties — however, the students said they were staying in occupation and the administrative staff were sent home early.

Health hazard row on river water

By TOM SHEIL

AN UNHOLY water row has led to a rift between clergymen and some of their flock in a south Kerry parish.

Members of a newly established development committee claim that water which is being pumped from a nearby river into 123-pupil Killinliath National School, near Waterville, is a serious health hazard. But local priests are adamant that the supply is perfectly safe for human consumption and that frequent testing has proven this.

So strong are local feelings on the issue that a meeting organised by the development committee passed a vote of "no confidence" in the chairman of the Board of Management, Fr. Fergal Ryan C.C.

In a strong sermon on Sunday, parish priest Fr. Joseph Nolan, called for an end to the acrimony stating that he could guarantee that the water was safe.

Development Committee Chairman Donal O'Shea claimed yesterday that the servicing of the system was haphazard, and there was a danger of dead animals — some of them dogs which had been given strychnine — falling into the river. He said that his committee was prepared to raise the money to supply the school with water from a spring nearly two miles away.

School board chairman, Fr. Ryan told the *Irish Independent* last night that "one of the best pump and filter systems in the country" was installed on the school supply six years ago.

"The water was tested on Saturday and found to be perfectly safe for drinking," he added "I can understand the concern of parents but they have absolutely no need to worry."

In his sermon last Sunday, PP Fr. Nolan announced that he was having the system — which includes a chlorine treatment and charcoal unit — serviced tomorrow.

The £8,000 purification system installed in 1983 was "adequate".

Haughey to hire jet 'at full rate'

By DON LAVERY

THE jet which will take the Taoiseach and his entourage to Washington on St. Patrick's Day will be hired at full commercial rates, the Government spokesman said last night.

The use of a private jet should cut five hours each way off the trip, the spokesman added. The plane will probably be provided by the Jefferson Smurfit organisation.

Hiring an executive jet with the range to fly to the U.S. will cost about £10,000, and may even save the Taoiseach's entourage money when compared to the cost of executive class travel. The many landings and take-offs involved would also mean longer flight time for the party.

The use of the Government's own executive jet, the British Aerospace 125-700, has been ruled out. A spokesman at Hatfield, where British Aerospace produce the plane, said the 125 was excellent for trips around Europe by Ministers but it would have to make two refuelling stops at Iceland and Gander if it was used to cross the Atlantic.

GET A LOAN ON YOUR HOME

If you're a homeowner, Bowmaker Bank now lets you unlock some of the equity tied up in your home.

We can lend you £5,000 or more for home improvements, a holiday home or even education fees.

With Homeowner Finance from Bowmaker Bank you can fix the level of repayments for the period of the loan, and choose a repayment plan that has 10, 11 or 12 payments annually releasing extra cash for those special occasions.

Avail of Creditcover and protect you and your family in the event of accident, illness or redundancy.

For more information on Homeowner Finance, simply dial 10. And ask for Freefone Bowmaker, or Tel: 753031.

Typical A.P.R. 14·1%

·BOWMAKER·BANK·

FINANCE ON YOUR TERMS

Figure 39 Pictures on a broadsheet page: the *Irish Independent*

one of a crashed train coach being hoisted past a house establishes an instant eye-catcher, and takes the reader straight down through the headline to the intro. The tonal values are good and it benefits, with its left to right movement, by being placed on the left-hand side of the page. At the same time the two single column pictures of people involved give a human dimension to the text of the story and provide a middle of the page breaker. Panelling in the story with 2-point rules gives unity to the display and makes the page. The cropping in all cases is impeccable, accentuating in particular the composition of the main one. The picture in the advert adds to the visual balance of the page, and enables the chief sub to fill the remaining spaces with text and headline.

The broadsheet *Irish Independent* (Figure 39) has a bigger problem since the page lead has no picture. Here a virtue is made of an attractive picture of the homecoming of a boy athlete which is composed into a picture + headline + text module and used as a centre page fulcrum round which to wrap the lead. The end column of briefs on this busy news page serves as a barrier which prevents the boy from seeming to look out of the page. The remaining picture (top left), gives a necessary focal point to help balance against the headline, the boy picture and the prominent advert down the page.

The location of pictures in a design, it can be seen, draws the eye into and round the page. It is to achieve this effect unimpeded that both in location and subject they must not clash with, or repeat visually, material in the display advertisements on the page, or material on the opposite page. In arriving at the best effect from the materials allocated, taking into account the design style of the paper, the designer needs to ask:

- Is the page being over-pictured?
- Does the picture's quality warrant its size?
- Does the page contain unacceptable grey areas?
- Is the best use being made of the available pictures?

The examples on pages 178–9 show some faults in picture use.

Picture desk

The picture input of a newspaper is channelled through the picture desk. This is controlled by the picture editor who performs a gathering and collating role similar to that of the news and features editors, briefing staff photographers on requirements of news and features pictures to be taken, and tapping other picture sources as necessary. The following (Figure 40) are the main picture sources.

Staff photographers

Newspapers, depending on their size and circulation, carry from three or four to as many as twenty-five staff photographers. Their work is integrated closely into the news and features gathering operation and staff photographers often work on assignments with a

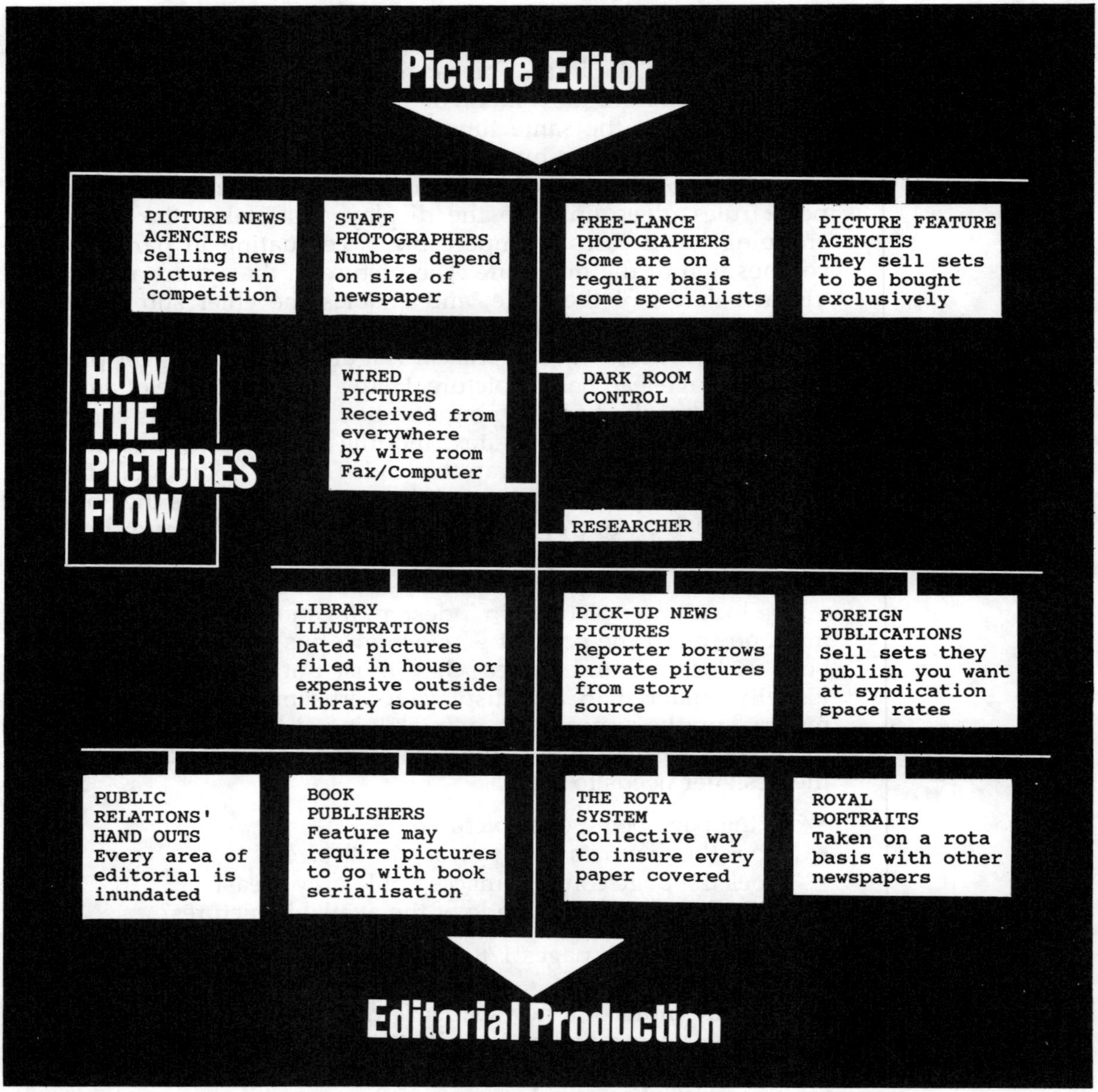

Picture Editor

HOW THE PICTURES FLOW

PICTURE NEWS AGENCIES
Selling news pictures in competition

STAFF PHOTOGRAPHERS
Numbers depend on size of newspaper

FREE-LANCE PHOTOGRAPHERS
Some are on a regular basis some specialists

PICTURE FEATURE AGENCIES
They sell sets to be bought exclusively

WIRED PICTURES
Received from everywhere by wire room Fax/Computer

DARK ROOM CONTROL

RESEARCHER

LIBRARY ILLUSTRATIONS
Dated pictures filed in house or expensive outside library source

PICK-UP NEWS PICTURES
Reporter borrows private pictures from story source

FOREIGN PUBLICATIONS
Sell sets they publish you want at syndication space rates

PUBLIC RELATIONS' HAND OUTS
Every area of editorial is inundated

BOOK PUBLISHERS
Feature may require pictures to go with book serialisation

THE ROTA SYSTEM
Collective way to insure every paper covered

ROYAL PORTRAITS
Taken on a rota basis with other newspapers

Editorial Production

Figure 40 Picture sources: where the daily input comes from

writer. They are aware of the paper's style and requirements and offer the best chance of exclusive pictures.

Freelances

There are many freelance press photographers working in a variety of fields, some heavily specialized into such things as glamour, fashion and industrial photography. They might be employed on particular assignments or on day-to-day or week-to-week arrangements on retainer or fee, or particular pictures might be bought from them. Most freelances are used because of their special experience or reliability in certain types of work, or as holiday relief. Exclusiveness of work depends on the rights bought or the type of contract.

Picture agencies

There are a great many picture agencies serving both special and general fields. Some, such as the Press Association, circulate available pictures to subscribing newspapers to be selected and bought at whatever rights are appropriate. Some such as sports agencies rely on specific contracts for certain types of work, while others are used as required depending on their known specializations. Exclusiveness depends on the rights bought – first or second British, English language, book rights, world rights, etc.

Rota pictures

Major events, such as Royal occasions or celebrity picture calls, sometimes limit press photographers to a set number on the understanding that under the special rota system run by newspaper picture editors and news agencies the pictures taken will be made available to all papers subscribing to the rota. Photographers are usually chosen in turn from the various newspapers and agencies on the rota. The system rules out exclusiveness.

Pick-up pictures

Some news stories and, more especially, some features and series, depend upon pictures supplied by the subjects. They are often bought with the stories and are chiefly pictures of the subject taken on earlier occasions which might not otherwise be available. The question of ownership or rights should be checked before use.

Hand-out pictures

Showbiz pictures, including some glamour, or those from PROs or press officers are often available free because of the publicity value to the owners or subjects.

Picture library

The office picture files, going back in some cases as many as thirty years, are an important source of stock pictures for 'flashback' use or for head or 'mug' shots of politicians, sportsmen and other well-known people.

Briefing

The picture desk organizes the transmission and processing of live pictures and – through the picture library – their storage and retrieval for use. Techniques have greatly advanced in this area and the electronic transmission and storage of pictures has come to be widely used. Where film in out-of-town jobs is developed on the spot, the photographer uses a portable miniaturized mobile transmitter called a 'mobile' (which is connected by modem to a telephone) to turn the film into electronic impulses to send it in negative form down the line to the office. Transmission time takes seven to eight minutes per film of 36 exposures. On remote overseas locations a satellite telephone using a dish aerial might be used.

At the office the images are received directly on an electronic picture desk (EPD), which in fact is a desktop computer, and the contacts are viewed by the picture editor on a television monitor. A choice having been made, the required shots can either be printed out on paper for conventional editing or be passed electronically to the next stage in the production process.

Such things are peripheral to the design function, although they do speed up the availability of needed pictures, and – along with faster page make-up – provide undreamt of facilities for getting the latest news coverage into the paper. Of more immediate importance is the skill of the photographer in getting the right shots and the right 'shape' of pictures, and of the picture editor in locating vital pictures from the various contact sources.

Staff work allows the advantage of a detailed briefing when pictures are being planned as ingredients of certain pages. The photographer might be told to include specific details, or to take particular people together, or to go for close-ups or shots that will lend themselves to horizontal or vertical shapes to fit planned slots on pages. An experienced camera operator, relying these days on 35 mm film, will shoot off a sufficient variety of pictures to allow for such eventualities as a change of mind by the editor or page executive, a change of advertising shape on the page, or the transfer of story and picture for unpredictable reasons to a different page. It is possible that a shot tried on spec by the photographer might yield up a better picture of a different sort than was anticipated to the extent that the projection intended for the page is totally changed to fit it in – such is the balance between planning and pragmatism in picture use.

Picture editing

The task of editing pictures to size and subject falls to the art desk, or to the person drawing the page, and not to the picture editor, who is normally the executive in charge of picture procurement. A necessary preliminary is the choosing of the pictures and this is done usually by the executive responsible for the contents of the page, who might be the night editor, the features editor, the chief subeditor or even the woman's page editor (see Chapter 8). Under the art desk system the art editor would expect to be included at the picture tasting stage to give expert advice on a picture's possibilities and to relate it to the page design.

The electronic way

Few pictures are used exactly as they are taken. There are editorial reasons for including or excluding certain detail and artistic and practical reasons for getting the best possible image out of what is used. To achieve these purposes the picture is:

1. Cropped to mark off the unwanted areas.
2. Scaled or sized to give it the correct size needed in the page.
3. Retouched, if need be, to enhance its quality.

In offices where electronic pagination is in use these processes are done quite simply. The picture, having been chosen for the page, is retrieved from the picture desk computer and is cropped and sized at the same time on screen. The visualizer or page designer enlarges or reduces the required part, excluding any unwanted bits, so that it fits exactly its pre-selected box on the page.

For the purposes of display or for a montage the picture box on the screen, and therefore the picture, can be tilted in any direction. If need be, the entire box can be reshaped to accommodate the selected section of picture, or made larger or smaller should the page design need to be changed for any reason.

This building in of the picture is done after it has passed through the scanning studio where retouching is carried out by adjusting the electronic components, or pixels, of the image under editorial instructions, and any colour correction made.

Adobe Photoshop and similar software programs are achieving remarkable results in electronic retouching and colour correction which go far beyond the facilities available by airbrush.

Cropping: the reasons

In most cases newspaper pictures are used for their functional role in providing a wanted image rather than being displayed simply for artistic merit, and this purpose can often be enhanced by cropping. The purpose might be to blow up part of the picture – perhaps one person in a group or one house in a terrace. Some pictures are cropped to improve the composition of the main image, or to exclude people or things that are not relevant to the story the picture is illustrating. Editorial purpose is the overriding thing. Pictures are not cropped to some abstract aesthetic standards. Some might not even be good pictures in the aesthetic sense; they might simply be the only ones available, with their usefulness depending upon right cropping and the improving of detail by retouching.

What is a good picture from the photographer's point of view is not necessarily so when it comes to its function in the page. It must primarily serve the story it illustrates and the page design, of which it is a part, although it would be a foolish designer who ruins a good picture just to squeeze it into a preconceived slot.

There is usually a sound reason in the visualizer's mind for the way in which a picture is cropped, although this does not stop cropping being a subject of controversy among journalists. There is no doubt that bad cropping can destroy a picture's effectiveness, and the following guidelines should be considered.

A picture should be cropped so that:

1 The relevant parts of the picture's subject fill the main area to be printed.
2 Any distracting background or unwanted detail is excluded or minimized provided it can be done without damage to the picture.
3 People's features and other essential detail are preserved. For instance, in a 'mug' shot giving only a face the crop should leave in the whole of the chin, though it can come down on the hair (unless the hair is a reason for using the picture); ears should be left in. Thus, while achieving the biggest image in the space, the essential features will not be damaged.

(a)

Figure 41 (a) Cropping and scaling a picture. The dotted lines on the cropped image – here shown as on a tracing – indicate how the required width measured across to the diagonal determines the depth

4 Any 'tilting' used in squaring off the cropped area to produce a more balanced picture does not create a nonsense by throwing verticals (lamp-posts, walls, fences, etc.) out of true.
5 Essential features or characteristics of a person or object are not misrepresented by excluding detail.

Manual editing

Where the cut-and-paste system of page make-up and picture preparation is in use or where, for some reason, a picture is required to be cropped and scaled on the print, the following applies:

To crop, mark off the wanted part of the picture on the back of the print in pencil, while exposing it to a light source, so as to exclude unwanted detail and give the shape needed for the page (Figure 41(b)). This part of the picture is then 'shot' to produce the print for

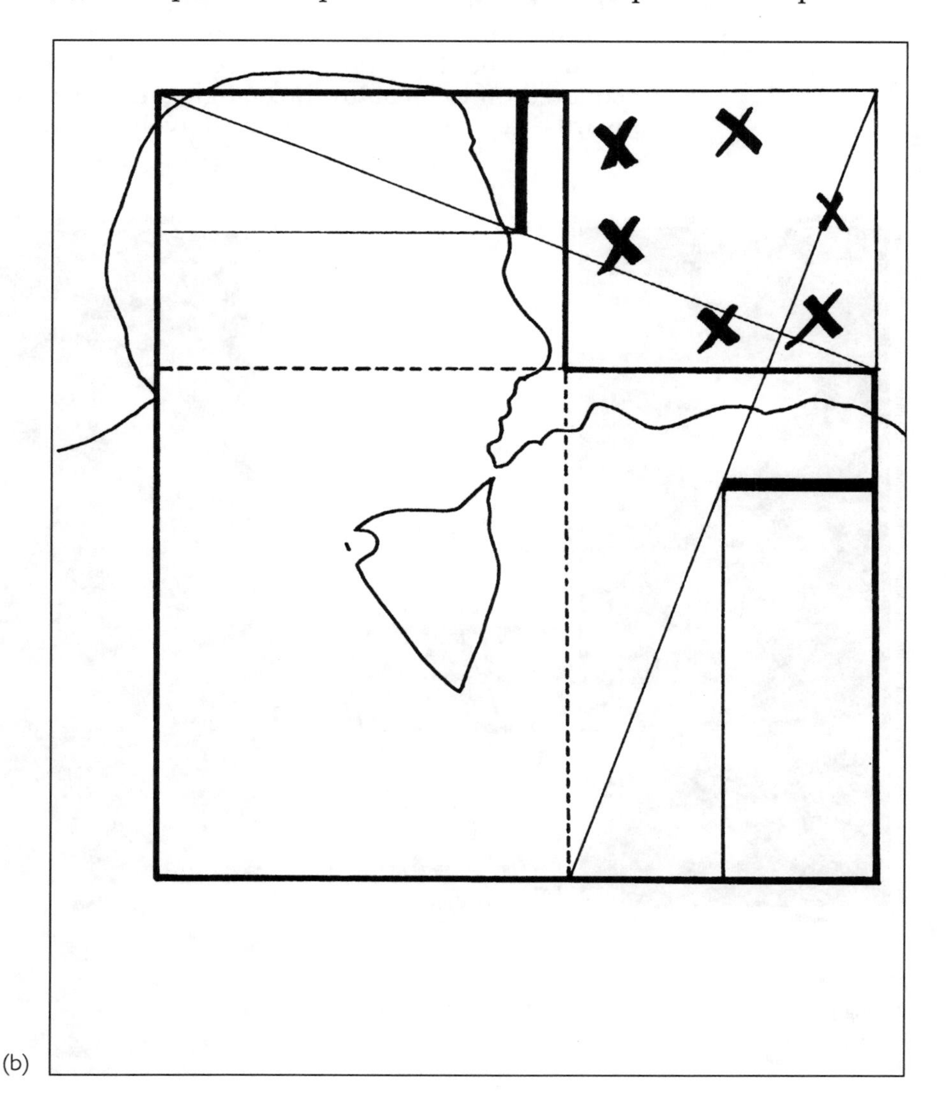

Figure 41 (b) The diagram shows how the traced image is used to draw diagonals to determine the measurements for a step required top right of the picture

(b)

the page. The editing can be done alternatively by pencilling in the cropped area as a tracing on the front of the picture (Figure 41(a)).

To scale, measure the required width – single column, double column, four column or bastard size – across the cropped area on the back of the picture and find the depth by drawing a diagonal line in pencil from corner to corner of the cropped area and measuring down from the point the cross line intersects the diagonal as in Figure 41(b).

If for some reason, such as a proximate advert position, the depth is critical, then the chosen depth should be measured upwards to the diagonal and the picture's width will be the distance from the side of the cropped area to the point of intersection. The method works which ever way the diagonal is drawn, though the usual way is to draw it from bottom left to top right.

If the picture is intended to have a cut out edge or a sloping side to fit a montage or picture composite (see below) it should be squared off to its natural rectangle and the waste area removed afterwards. To scale the picture effectively the working area has to be a

BRITT'S IN FINE FORM

★ BLONDE bombshell Britt Ekland arrives for the premiere of the new Bond film Licence To Kill — and provides her own stunning double feature.

★ Sexy Britt, 46 — who played Mary Goodnight in the 1974 Bond movie Man With The Golden Gun — went on later to a private celebration party. Guests guzzled champagne and downed Mexican food. Hot stuff, Britt.

Picture: NIKOS VINIERATOS

A Britt of all write!

BLONDE bombshell Britt Ekland pops out for the evening — and shows why she's still a big draw for autograph hunters.

The sexy Swede may be 46 but she became a mum again last year and it's obviously keeping her young. Her nifty dance footwork was a real wow in the early hours at a London nightspot. Write on, Britt!

Figure 42 Actress Britt Ekland at a premiere makes a striking picture. The cropping on the left shows her essential sex appeal; the rival paper's version, right, keeps in the pens of the autograph hunters, thus shifting the eye's attention

rectangle. If there is a difficulty scaling it to a size that will fit the space available, then the cropping should be checked to see if the picture area can be widened or deepened without damage to the image. In any case, cropping and scaling are usually done at the same time since requirements of size and space can influence how a picture is cropped.

Calibrated wheels, or even a slide rule, can be used to do the job but the diagonal method is visual, simple, almost as quick, and avoids the danger of a mis-reading.

Retouching: Manual retouching is carried out before the picture is shot for a paste-up page and consists of using an airbrush to sharpen detail or blow back unwanted background. It is done by a specialist artist under editorial control.

Retouching – the dilemma

Whether or not, or by how much, a picture should be improved has always been a source of argument. The electronic means now available, and their utilization for design purposes, have heightened the controversy.

The ability of programs such as Adobe Photoshop to create substantial changes in pixel structure leading to distortions and even inventions of detail has led to tighter editorial control being placed over retouching to avert the lawsuits that would follow such tampering with pictorial fact.

In a technical sense electronic retouching is better than the best possible results achieved by airbrush artists and has put an end to some of the nonsenses and 'works of art' that got into newspapers.

Retouching is invaluable for printing in newspapers pictures that for some reason are not of good reproductive quality. The general rule is that unwanted detail can be removed provided the picture's purpose is not damaged, and poor detail improved by discreet brushing, but that absent detail cannot be created – or should not.

The essence of good retouching is that it is not noticeable on the printed page. Sharp visible lines of the airbrush or the working in of new detail, by whatever means, can make nonsense of a picture and can also land an editor in legal trouble. The discreet removal of background can highlight people and features in the right picture, but retouchers should beware of 'fogging detail' so that the background looks unnatural, or so the subjects appear to be standing in a cloud.

Reversing

Reversing – known in computerized systems as *flopping*, from American usage – means simply using a picture for the page by working from the negative the wrong way round, and is useful to avoid having faces looking out of a page or out of, for instance, a panelled-in features display. The device bristles with dangers such as a well-known person's hair being parted on the wrong side, a wedding ring on the wrong hand, and a jacket being buttoned up the wrong way round. It can be confusing to readers if an identical agency picture has the American President facing in opposite directions in the same pose in two different papers. More extreme dangers, if not spotted in time, are cars turned into left-hand drive and number plates reading back to front. The device should be used only after a careful inspection of the original has ensured that no nonsenses are being perpetrated.

Closing up

Closing up people together by means of a scissors job on the original bromide is sometimes used to dispose of waste in the middle of a picture in cut-and-paste production and can produce a satisfactory page image, but it is fraught with danger if, say, a couple are made to look so ludicrously close as to be out of character. It is a piece of editorial licence to be used with caution when an original is wastefully gappy and cannot be substituted. As with over-zealous retouching it can lead to complaints to the editor, and is ethically dubious since once an original has been cut up its use is open to charges of fabrication. Screen make-up offers a simple solution. The same picture is exposed in two separate boxes each containing one of the two subjects. The two boxes are then brought together with a tiny overlap.

Substituting backgrounds

Outdoor shots and seascapes are sometimes 'improved' by substituting different sky effects where this adds interest and does not clash with the circumstances in which the picture was taken. If this practice is extended to substituting more detailed backgrounds then the danger again is that it can lay the paper open to charges of tampering with a picture. The computer is particularly useful for this purpose.

Montages

An enhanced effect of action or atmosphere can be gained by grouping pictures together in a block or pattern, where there is a linking theme. This is a useful ploy in a features layout where mood or a 'flash-back' element dominates a display. Flashbacks, (that is, happier days or earlier days shots or reproduction of a picture published previously) can look effective as *tear-outs,* or *rag-outs* with a black ragged tear line applied round them. Where, in a blurb or display, pictures are fitted together into a pattern along with type, sometimes in the form of WOBs or BOTs, the device is known as a *compo,* or composite.

Legal aspects

It is possible to libel a subject by the way a picture is cropped or retouched and care should be taken that such work does not misrepresent anyone. And not only humans – a national daily was in trouble in the 1950s when an overzealous retoucher reduced the size of a prize bull's vital member and the owner, who relied on stud fees, sued.

Colour editing

As with a monochrome photograph the tones in a colour picture are reproduced by the halftone process. This means that they have to be broken down into dots by 'screening' the original in order to reproduce the different colour values. The dots of the screened picture, when transferred to the printing plate, accept ink in various combinations of four basic colours: *yellow*, *magenta* (red), *cyan* (blue) and *black*. These are printed in that order from four separate plates (that is, a page containing colour makes four separate passes through the press) which are prepared from colour negatives made by subjecting the picture or transparency to a process called scanning.

Scanning

The scanner (Figure 43) is an electronic device by which an operator, using a set of balances and filters, given total accuracy by the computer, creates the four negatives from the colours in the photograph. The cyan filter creates a yellow negative; a green filter isolates the magenta; a red filter produces cyan, and a combination yellow-orange filter makes black. These negatives, each containing its own colour, are called separations.

Figure 43 Vital machine in colour-picture editing – this Itek 200-S scanner has a microprocessor-controlled keyboard and a colour monitor

To accomplish this the colour print or transparency is attached to a drum in the machine and the operator keys in the percentage of enlargement or reduction that is needed together with the screen (dot size). Laser beams then scan the picture as it revolves, isolating and screening every individual image. Within seconds the computer which activates the scanner converts the information into a screened piece of film for each colour. This information is stored digitally in the scanner for use depending upon whether cut-and-paste or electronic pagination is being used for page make-up.

The versatility of the scanner will allow a complete composed page to be reproduced in the four separations up to its printing size. A single 35 mm transparency can, in fact, be enlarged up to ten times, with the scanner enabling the operator to enhance the sharpness of the image and even the density of individual colours. Modern 'paint box' scanners will allow an art director to change, say, the colour of the sky from night into day, to stretch the image horizontally or vertically, or even take an item out of one picture and transplant it into another. The limiting factor in this sort of operation is the time and expense where page production is on a tight budget.

The operator of the scanner will receive information from time to time from the viewing monitor that the screens are clashing between one colour and another. A crude example of this phenomenon is when an already used picture is used again in a newspaper for a 'flashback' effect. Because the newspaper cutting or old picture has already been screened when originally used, a clash of dots will occur when it is prepared again for the page, giving what is called a *moiré* effect. The same effect can result inside the scanner from the complexity of the colours of a particular colour print or transparency. It can be cured by the operator simply revolving one or more of the screens on the separations, whereupon reproduction will be seen to return to normal.

Picture quality

Picture editing is more critical in newspapers printing in colour. With the advent of the electronic picture desk photographs can be called up in full colour on monitor screens. Sometimes the photographer at the scene of a story will be transmitting, by means of a modem, direct over telephone lines or by earth satellite, into the picture editor's viewer.

With monochrome the dynamics of a picture are traditionally expressed first in its news value and second in its reproductive quality. Faults in focus can, in a curious way, lend a greater sense of urgency to an exclusive hard news picture. More accuracy in focusing is called for with colour since focal 'shake' will confuse the scanner and precious time will be lost manipulating the machine to correct the deficiency.

'Hot' and 'cold' are words that become important in connection with the quality of a picture presented for publication. A hot picture is one that shows too large a proportion of red in its colour range. If red is too strong in a transparency it will 'spill over' into the surrounding colours and the blues will take on a purple warmth inappropriate to the subject. Conversely the cold colour blue, if too strong, will dominate and 'corrupt' the warmer colours. Certain types of colour film are notorious for their predilection for cold or hot colour. Good photographers will balance their cameras to the type of film used to correct this problem.

The subject of the picture will influence editorial choice in colour pictures. A disastrous fire, for example, will benefit in reproduction if hot reds are being produced. Conversely, the effect of strong red would be wrong in a picture of a snow-covered wind-swept mountain. Contrasts, when they occur, can be effective. For instance a picture of a bird like a red cardinal scratching for food in a snow-covered New England garden can deliver a shock to the eye. Achieving colour balance both in the choice of the original and within the scanner is a critical one for the picture editor.

Choosing the picture

The parameters of picture choice, it can be seen, are more complex with colour than with black and white. The editor might want a picture strong on news value and visual impact, whereas production demands are for a range of colour, good tone balance and crisp focus. There are also differences between news and features requirements to be taken into account.

News

News pictures in colour – although this part of the paper remains dominated by black and white – have become commoner since photographers have been encouraged to carry two cameras, one loaded with colour film, the other with monochrome. There are situations where news shots in colour are a standard requirement for a newspaper. In the case, say, of a major disaster story, the picture editor would deploy several photographers, some carrying colour film, the others black and white. This would guarantee pictures for pages that are not programmed to take colour where a story might spill on to several pages.

If the story is at some distance from the office, modern transmission techniques over telephone lines are used to meet tight edition deadlines. Mobile transmitters feeding into computerized systems can now be packed into bags the size of a briefcase and both colour and mono can be sent over the same line at the dialling of a telephone and the touch of a small keyboard.

Computer hardware, as with television, can transmit news pictures from the camera's lens straight to the page make-up screen.

Features

The boom area in the supply and use of colour pictures and transparencies is undoubtedly the features, or magazine, pages. The needs of features colour are best expressed in the perfection achieved by the high quality *fashion* photographer. Every aspect is covered by the person behind the lens. Models are chosen for their ability to 'act' the clothes. Their bodies must be appropriate to the type of garment being photographed, eyes and bone structure being paramount, their movements fluid to produce freedom on film. Successful pictures of fashion in colour should give the reader a feeling of excitement, while subtly conveying the fashion editor's view of the features of the clothes that should be brought out.

The disadvantage from the editorial production point of view is the sheer volume of transparencies resulting from a fashion session. The motorized camera enables the operator to demand continuous changes of stance which, in the end, allows the fashion page designer to choose shapes and poses that are unique. The choice can require much discussion between fashion editor and art editor, and it is a good idea to feed the best of the bunch into a display unit for final choice to get a better impression of size. To produce a short list first it is still common, where transparencies are being produced, to place them side by side on a desktop light box and view them through a magnifying glass. The chosen ones will be those which, while fulfilling their fashion function, allow the marrying of words and pictures to be made on the page in the most visually attractive way.

Some executives handling fashion pages feel the method of choice either with computer screen or with light box and magnifier is simply not good enough, however. Having weeded out the unwanted shots, they examine the short list, many times bigger than required for reproduction, through a projector. This can uncover faults not visible to the eye by normal magnification, and give a better reading of the colour. A knowledge of the dominant colour is vital in preparing a colour fashion page.

Agencies

Features colour will in many cases come from the library stock of agencies. Either these will be supplied on-line or a picture researcher will be sent there to find pictures to fit a given feature. The increased use of colour in newspapers has meant the growth of picture agencies offering colour prints and transparencies, many of them specializing in subjects such as celebrities, gardening, motherhood, country life, money features and so on. A newspaper's researcher will be offered viewing facilities, or the agency will research and send round a selection of transparencies to fit a list of requirements, though for an extra fee.

Computer retrieval of pictorial material on disk has greatly improved the speed at which a wanted subject can be produced from an agency's files compared to the bulky old mono picture libraries. Even transparencies take up less space than prints. A less advantageous development of all this is that prices of picture reproduction are increasing with demand. Whenever a new picture fad occurs it is followed by a price explosion. An example is the appetite for celebrities being photographed at night-spots around the world. Immediately this happens the paparazzi shift into gear and begin producing immense quantities of pictures. The agencies respond by setting up specialist departments to sort the good from the bad. Inevitably this is reflected in the price per square centimetre to newspapers for exclusivity.

Special transmission

Some big organizations can send colour from abroad 'over the wire' as separations which are instantly usable. The quality, unlike the wire pictures of old, is excellent since satellite transmission is used. The high cost of this method is eased if the pictures can be transmitted via a newspaper's own office in the country or origin. A point to remember is that a picture needs four wiring times for the separations.

A system of instant transmitted display, for which the hardware is now available, enables picture editors to simply call an agency which may have a required set of pictures and ask for them to be displayed for viewing on the picture desk VDU. A choice can then be made and a 'receive' command keyed in. An extension of this facility is the colour scanning by agencies of pictures as ready-to-use separations for transmission to subscribers.

A good picture editor balances these costs of picture procurement against the department's budget. Where a staff photographer can be conveniently sent to do a job it will be invariably the cheapest way.

Editing transparencies

Where cut-and-paste make-up is in use and where, for some reason, it is necessary to edit colour pictures from transparencies, the page designer faces a task of particular delicacy: how to establish on the tiny frame of a 35 mm film the specific area required to print.

The tiny film should be placed on the light box that lives below the lens and bellows of the Grant projector (see page 103). Make sure that the correct long lens is in the lens holder, and locate the transparency in the dead centre of the light box. At this point a movement of as little as 10 mm away from the centre will have the designer searching for the image that is to be projected on to a sheet of tracing paper placed on the screen at the top of the machine. Is it to be the whole picture, or is the picture to be cropped and a part of it blown up? Once a decision has been made, the image can be traced off (Figure 44).

A final check can be made on which way round the transparency should be. If it is on the light box upside down then the picture will be traced off in reverse (left to right). A transparency that has been put in its cardboard frame properly will show the maker's name and

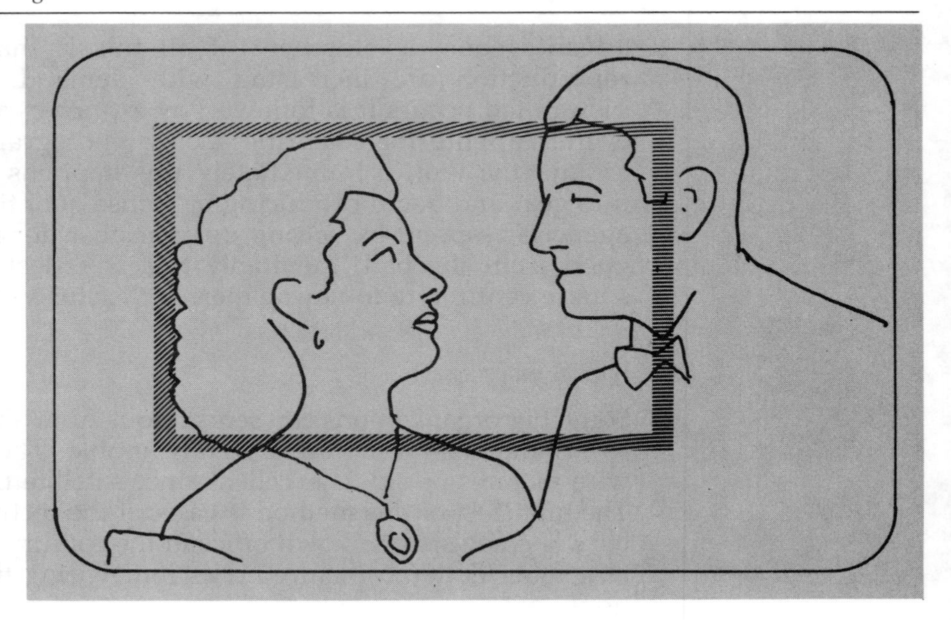

Figure 44 To edit a colour transparency it should be enlarged on a Grant projector or colour enlarger and a tracing made of the required area

other information on the front side and nothing on the reverse. The words on the frame should be facing upwards. If the frame is plain on both sides, the whole thing should be taken between finger and thumb, making sure not to touch the actual film, allowing the light to strike across the horizontal plane of the film. This should show the correct size as shiny and the reverse side as dull and slightly uneven where the chemicals have created the images of the picture.

Sometimes it can be difficult to read the transparency because of the thinness of its emulsion. In this case be ruthless and take the film out of its frame. Peel the cardboard apart and the film will fall out...but not on the floor! Now the sprocket hole area of the film is revealed. Also showing will be the frame number and producer's name between the sprocket holes. You can be certain the film is the right way round since you can read these words. On the reverse side they read back to front. Now put the transparency back in the Grant machine.

The Grant should be focused up to the size needed for the page and the projected image, whether part of the transparency or the whole, traced off and transferred to the layout sheet.

A snag could occur here. The long lens in the Grant might not be long enough to enlarge the 35 mm piece of film to the full size of the page you are drawing – that is, if you are going to make the finished picture that large. The answer in this situation is to trace off the image to the largest size you can, place this tracing below on the light box, and trace up again to the required size. Accuracy in doing this is vital, and so you should be using a very hard pencil or ball-point. Litho houses on publications where page make-up and printing are contracted out will rely on your accuracy since they will be composing colour on screen.

Do not throw away the tracing for your colour picture after it has been transferred to the layout sheet. With a different coloured pen or pencil indicate on it the cropped area of the transparency that you are using. The technique here is the same as when marking up a mono picture. In doing this on no account allow the transparency to be separated from the tracing. Put identifying marks on both to avoid accidents.

A point to note: in cropping a transparency or colour picture be aware of the usefulness of background colour. Not only can the blue of a sky be pleasing when printed – it can also be a useful vehicle for an overprinted headline, especially on a features page. In a special projection a colour transparency could fill an entire page with the essential illustration at the bottom and the rest of it used to take overprinted headline and body text. Remember, the litho house can match trannie colour, so if more sky is needed than is in the film it can be extended – at a price!

6
The modern art desk

Modern newspaper design is centred more and more on the art desk, a production refinement at one time associated only with magazines. Here layout artists, or journalists whose speciality is layout, carry out a range of tasks to do with the presentation and projection of editorial material, ranging from picture editing to the preparation of graphics and the actual drawing of the pages and, nowadays, the utilizing of computer graphics both for paste-up and screen composition.

Many small papers still rely on the chief subeditor, or even the editor, to carry out such design work as becomes necessary, but the increased sophistication of materials and effects, and of computer capability, are resulting in more editors hiving off the design function and establishing an art desk, however minimal, within the editorial area. A common method is to have one art desk serving a number of papers in-house, although this tends to treat design as an assembly line process. Another is to have a suitably skilled senior executive, with the help perhaps of a trained graphic artist, entirely responsible for the design and house style of the paper, commissioning such extra artwork as is needed.

On bigger papers a new breed of journalist/designer, professionally trained in graphic arts, is being recruited to staff art desks under an art editor. This arrangement takes the responsibility for design off the shoulders of hard-pressed chief subeditors who have enough to do with the actual editing. It also ensures that some regard is being paid to evolving a consistent and polished design style that will reflect the newspaper's character and make it attractive to readers. A third advantage is that it saves the newspaper's visual face from being at the mercy of a variety of hands, particularly at times of frequent staff changes.

The art desk system in no way diminishes the control by editors and senior executives over the planning and editing of material.

Rather, if properly organized, it gives more scope and options to editorial decision makers and the means to more imaginative presentation. If, in the chapters that follow, the involvement of an art desk in the design process is assumed by the authors, it is because this is the way editorial practice is moving.

Origins of the system

An awareness by British newspapers of design as a distinct entity, rather than as a loose tradition, owes a good deal to the development of the so-called tabloid layout style by H. G. Bartholomew's *Daily Mirror* in the late 1930s. It was on Mirror newspapers that the first newspaper art desks appeared and functioned as central points for design. In the preparing of pages the duties of picture editor and art editor were initially blurred and newspapers suffered from neither function being performed properly. However, as awareness of the value of centralized design developed, a department gradually evolved that went beyond the simple scaling of pictures demanded by the shouts of a hard-pressed night editor.

The art editor, or with some grander publications the art director, became a powerful influence on the development of the paper's layout style. As the drive for higher circulations and more reader appeal grew in the 1950s and the system, with the easing of newsprint restrictions, began to produce results in improved presentation, the art desk expanded into a fully manned department with a deputy and team of layout artists. The expansion of television at this time and of colour TV in the early 1960s made editors more aware of the need to compete visually with the rival medium.

Trust in the expertise of the art desk took time to develop. Senior editorial executives, dubious of the need for 'visualizers', went on putting their own pages together. Art editors would be thrown 'roughs' of pages by night editors and told to 'put that through the system'.

On the national tabloids the crucial development came in the early 1960s on the *Daily Mirror*, then reaching the peak of its success, and its sister papers. The method demonstrated its value in 1969 when Rupert Murdoch acquired *The Sun* and had it redesigned as a rival to the top-selling *Mirror*. By this time the Thomson titles at London's Gray's Inn Road, principally the *Sunday Times*, had moved into desk-based design, to match the paper's chief rival, *The Observer*. The art desk had arrived.

As the method was seen to work, other newspapers, particularly national ones selling in a competitive market, but also specialist ones and the bigger provincial titles, began to adopt it in the drive to give a professional look to their products. An important effect of the art desk system, with its integrated approach to design, was the teasing out of some of the faults that had bugged the old ad hoc methods in which layouts came with subbed copy from the subs' desk, and picture editors cropped and scaled pictures to their own fancy in between briefing photographers and running the picture library.

A trained visualizer could ensure that headlines did not run into each other from page to page, that editorial material on opposite pages did not fight each other, that pictures balanced properly one page against its partner and stories and pictures did not conflict with the surrounding adverts. Thus the piecemeal approach was replaced by design cohesion, with a working dummy of pasted up layouts or page proofs operating as a master blueprint.

Role of the art desk

Editorial production on a morning or Sunday newspaper is controlled by the editor or the night editor; on evening or weekly papers by the editor or deputy editor, and occasionally by the chief subeditor. For the purposes of this book these executives will be referred to as the back bench. Sitting with the night editor on a national paper (which is the model of this chapter) are the deputy and assistant night editors and sometimes a 'prodnose' whose duty is to read every word in the paper and query any suspect facts or grammar.

From the back bench comes the decision making – the selection of material and pictures that will make up the page contents – as the production cycle unfolds. Feeding into the system are the features from the features department, news stories from the newsroom and agencies, and pictures from the picture desk, with lists of forthcoming material being constantly updated in line with revised timings caused by various obstacles in the way of photographers, writers or machinery.

Production factors on any newspaper dictate that certain pages must be early on the processing list. Stock features pages, including TV listings, at the back of the paper come early, though on a tabloid paper pages at the very back must be kept open since they marry with the front late news pages on the final printing plate.

Now begins the first of the day's debates between back bench and art editor. It might concern a small news page at the back – small because the advertising space on the page happens to be heavy. Therefore the placing of the picture and disposition of stories is critical. The weight of advertising needs to be thought about for the editorial subject matter must be seen by the reader as reading priority. Should the reader's eyes swing to the advertising first, then the creative director of the advertisement is a bigger success than the newspaper's art editor.

As the debates proceed and layout sheets bearing squiggles begin to emanate from the night editor (Figure 45), the art editor is roughly drawing ideas on a layout pad and amending them as back bench thoughts evolve. The art editor might suggest alternative ways of doing a headline to simplify the typographical approach or to carry the length of text better. The wording and general shape are quickly clinched, for time is the master. The chief subeditor is waiting for the page rough to be drawn in detail or to be put up on screen so that instructions can be passed to the subeditors on the treatment and length for stories and the setting and headline sizes.

Figure 45 Creating a page the traditional way: rough, drawn layout, and ready-to-print page one from an edition of *The Sun* using paste-up before the onset of electronic pagination. Reproduced by permission

The development from rough to finished page (see also Plates 2 and 3 of the colour section) will test the accuracy and feasibility of the original concept. It does not follow that because two 'great' minds have produced the first page idea that it will work without some modification. The duty of the layout person is to produce a finished design that will stand several tests. First, the weight and lengths of stories dictated by back bench decision should appear in correct relationship to the page. Second, the type and setting used should come within the house typographical style. Third, the page must be given visual balance as well as a balance of contents.

Even where full electronic pagination is in use (Figure 47) pages, tabloid or broadsheet, usually start life on a layout pad (printed in-house) made up of full-size gridded sheets. The surface should be comfortable and smooth to draw on and paper stout enough to stand rubbing out and redrawing as ideas are tried out and changed.

The sheets should be scaled in centimetres up one side and down the other so that the stories can be measured in length downwards and the adverts more appropriately upwards from the bottom. They should be scaled across in the column widths of the paper with a pica of white space between the columns and showing 18 points of space at the top to take the folio. They should be slightly transparent to allow show-through should a light box or a Grant projector (see page 103) need to be used.

For tabloid papers two sizes are needed – one at single size and the other at double spread size with the grid showing a vertical centre gutter of appropriate width. Broadsheet grids can be printed half-size for ease of use.

Metric measurement is now universal in advert depths (that is, three columns by 20 cm; single column by 14 cm, and so on) while picture sizes, especially those not of standard column width are usually in millimetres (that is, single column by 36 mm; 48 mm by 36 mm, and so on) in place of inches. Page designers, however, still have to cope with traditional print measurements. Type sizes, as we have seen, remain in points. Although variation in size down to half a point is possible in computerized systems, most typesetting is formatted to give the standard series of sizes, which continue to offer the best guarantee of well planned type balance. Setting widths, as a result of American practice, are expressed in the systems in picas and points, and not in ems and ens.

With cut-and-paste the make-up cards look identical to the page layout sheets even to the column widths and centimetre measurements, which are in non-reproducible blue ink. Where picture and setting depart from standard measure, however, a trained eye is of the essence if white space is to be achieved consistently by scalpel in making up the page.

Instructions to subeditors, often with code words for type, which are relayed on screen in the case of electronic pagination, are written on to the layouts for cut-and-paste. By either method, a check with facing pages to make sure there is no duplication of typefaces or pictures should be second nature to an art desk person.

Variations in headline size are not, on the whole, as pronounced in broadsheet styles as with tabloids. While imaginative changes have taken place in the design of many broadsheet papers, as will be seen in later chapters, some base their style on relatively few changes of type size so that the movement of the eye down the page is not so excited by individual treatment of stories. Perhaps a change in body type to bold or italic is permitted, or the panelling in of an item. Yet the technique of creative accuracy in page drawing is even more important on the relatively huge area of a broadsheet page. Picture size and headlines play an enormous part in keeping the interest of readers and helping them to choose what they should read first.

Colour pages

Where a colour element is located on a paste-up page that is otherwise to be printed in mono the mono parts of the page must be accurately made up so that the spaces to accept colour are left precisely located in width and depth for the four passes through the press needed to take ink from the colour plates. A useful practice here in the paste-up room is to apply page-size acetates showing colour location to each page taking colour so that the spaces are seen to be accurate before the page is released to the camera and platemaker. This can avert the serious fault of colour intruding into mono areas of the page through make-up failing to follow the layout precisely.

To this end instructions on page layouts should be generous in what is required to be done in make-up and colour preparation. For instance every piece of colour tint used must bear its Pantone colour number and the percentage of screen needed to effect the colour required.

Type sizes must be marked in detail on the edges of the layout as with mono pages, as also should the thickness and colour of rules. Some contract printing houses prefer type sizes to be in millimetres rather than in the old points system. Differences of this sort should be checked before launching into colour layout and design where make-up and printing are by contract.

With such arrangements it is a good idea, for safety, to mark any original artwork with its instructions as well as putting instructions on the page. If the colour and typesetting are at different locations, make sure that the mono part of the layout with its typesetting instructions is made available separately for this purpose. Leave nothing to chance. Your contract printer might be surrounded by pages from all sorts of magazines and newspapers.

Tools of the trade

The Grant projector

Before the Grant projector's appearance in the late 1950s it was necessary to reduce or enlarge type or pictures for layout purposes by using tracing graph paper. One piece was laid over the picture or type and the other used to enlarge the image by increasing the tiny graph squares by an appropriate number. It was tedious and hard to get an accurate result. The Grant projector (Figure 46) has solved the problem.

Figure 46 Still a useful tool for the creative art desk: the Grant projector

December 31 1995

NEWS
DATA

JOHN MAJOR TELLS OF HIS HOPES FOR IRISH PEACE

CLEESE SAVES WOMAN FROM MUGGER
See Page 5

OUTCRY OVER PARA HELD IN ULSTER JAIL
See Page 5

WATCHDOG ACTS OVER PILL PRICES
See Page 9

TRAVEL BOSS iN RUINED HOLIDAY SCANDAL
See Page 9

WORKING FROM HOME DRIVES YOU MAD!
See Page 11

ARNHEM: THE DAY THEY WENT TO HELL!
See Pages 12 & 13

KENNELS PAID MILLIONS TO GUARD DOGS
See Page 20

SCHOOLS SOLD IN CLASSROOM REVOLUTION
See Page 24

NEWS
THERE'S MORE OF IT!

SUNDAY Success

Price 65p.

The campaigning newspaper - with the accent on NEWS - you can't afford to miss!

Drivers face B-test escape

JUNIE LOOKS A REAL WINNING SUCCESS

Police anger after drink-drive suspect freed by loophole

Si meliora dies, ut vina, poemata reddit, scire velim, chartis pretium quotus arroget annus. scriptor abhinc annos centum qui decidit, inter perfectos veteresque referri debet an inter vilis atque novos? Excludat iurgia finis.

"Est vetus atque probus, centum qui perficit annos." Quid, qui deperiit minor uno mense vel anno, inter quos referendus erit? Veteresne poetas, an quos et praesens et postera respuat aetas?

"Iste quidem veteres inter ponetur honeste, qui vel mense brevi vel toto est iunior anno." Utor permisso, caudaeque pilos ut equinae paulatim vello unum, demo etiam unum, dum cadat elusus ratione ruentis acervi, qui redit in fastos et virtutem aestimat annis miraturque nihil nisi quod Libitina sacravit.

Ennius et sapines et fortis et alter Homerus, ut critici dicunt, leviter curare videtur, quo promissa cadant et somnia Pythagorea. Naevius in manibus non est et mentibus haeret paene recens? Adeo sanctum est vetus omne poema. ambigitur quotiens, uter utro sit prior, aufert Pacuvius docti famam senis Accius alti, dicitur Afrani toga convenisse Menandro, Plautus ad exemplar Siculi properare Epicharmi, vincere Caecilius gravitate, Terentius arte.
Hos ediscit et hos arto stipata theatro spectat Roma potens; habet hos numeratque poetas ad nostrum tempus Livi scriptoris ab aevo.
Interdum volgus rectum videt, est ubi

SAIL TO SWEDEN FOR £5

A COOK FOR EVERY READER - FREE!

Figure 47 The electronic art desk: this mock-up of a Sunday title shows (left) the page one as designed on screen, and (right) the components from which it was made up using Quark XPress tools. Text and pictures are seen in their individual boxes awaiting placement, the picture box being defined by a cross. The eight 'handles' round each box enable you to place it into position – or to change its shape and size if need be – by dragging it with the mouse

SUNDAY NEWS

Success

The campaigning newspaper - with the accent on NEWS - you can't afford to miss!

Drivers face B-test escape

THERE'S MORE OF IT!

JUNIE LOOKS A REAL WINNING SUCCESS

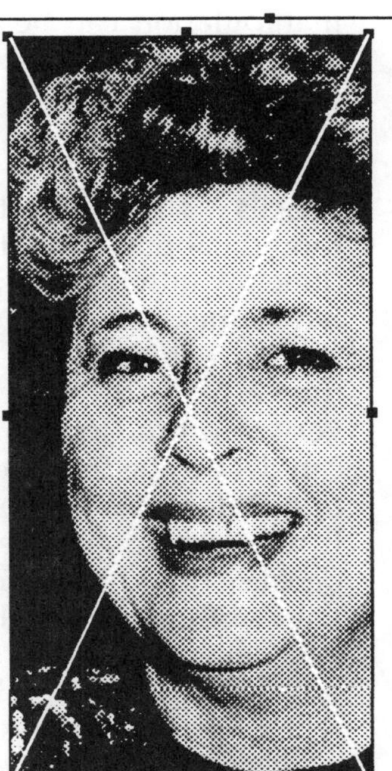

Si meliora dies, ut vina, poemata reddit, scire velim, chartis pretium quotus arroget annus. scriptor abhinc annos centum qui decidit, inter perfectos veteresque referri debet an inter vilis atque novos? Excludat iurgia finis.

"Est vetus atque probus, centum qui perficit annos." Quid, qui deperiit minor uno mense vel anno, inter quos referendus erit? Veteresne poetas, an quos et praesens et postera respuat aetas?

"Iste quidem veteres inter ponetur honeste, qui vel mense brevi vel toto est iunior anno." Utor permisso, caudaeque pilos ut equinae paulatim vello unum, demo etiam unum, dum cadat elusus ratione ruentis acervi, qui redit in fastos et virtutem aestimat annis miraturque nihil nisi quod Libitina sacravit.

Ennius et sapines et fortis et alter Homerus, ut critici dicunt, leviter curare videtur, quo promissa cadant et somnia Pythagorea. Naevius in manibus non est et mentibus haeret paene recens? Adeo sanctum est vetus omne poema. ambigitur quotiens, uter utro sit prior, aufert Pacuvius docti famam senis Accius alti, dicitur Afrani toga convenisse Menandro, Plautus ad exemplar Siculi properare Epicharmi, vincere Caecilius gravitate, Terentius arte.
Hos ediscit et hos arto stipata theatro spectat Roma potens; habet hos numeratque poetas ad nostrum tempus Livi scriptoris ab aevo.

NEWS

SAIL TO SWEDEN FOR £5

A COOK FOR EVERY READER - FREE!

Several versions of projector were marketed for this role but the Grant is undoubtedly the toughest. The principle is simple. Camera bellows and lens are suspended below a flat screen. The bellows contract and expand to give a focal point while a movable platform below the lens supports the artwork, pictures or type specimens. Two spindle handles, one on each side of the machine's front panel, control the two elements. Under a perambulator hood to keep out excess light is a glass screen on which layout paper or tracing paper can be laid. When the spindles have been adjusted together to give a sharp image on the paper, and regulated to give the size required by the layout, the image can be traced off directly on the paper. If needed, images can also be traced from, say, a parcel or a hand.

The original purpose of the Grant projector was for use in advertising studios. While they are still in use in some newspaper offices, newer technology is beginning to phase them out. There are available some efficient replacements such as the high definition colour copying machines made by Konica and Canon. These will reproduce any flat original artwork or transparencies up to the percentage size required in less than the time taken to trace off an image under the hood of a Grant.

Where full electronic pagination is practised using Apple-Mac computers and Quark XPress software (Figure 47) the instant imaging of scanned artwork on to the screen has rendered both the Grant machine and its successors redundant. Picture editing and manipulation can be carried out on the page by the simple finger touch of mouse and keyboard.

Pencils and felt-tip pens

Pencils – still the best tool for drawn layouts – should be carefully thought about. Now that photocopiers and computer print-outs have done away with multiple carbons, lines can be changed and rubbed out as the design comes together, so a soft eraser is a must, too. If all that is required is an accurate mechanical drawing from the artist, then a 2H is the tool. It is precise because the hardness produces a very thin line. Remember, the thickness of a line on a drawn layout can distort the finished total of millimetres even across the full width of a tabloid double spread or the depth of a single broadsheet page. If representation is needed on a layout then a full range of B (soft) pencils should be available. Large typefaces can be fully pencilled in with authenticity and a proper blackness. Where photographs are still being traced off through a Grant machine to give a layout a polished and finished look, a range of pencils of varying softness will allow tone values to be copied at speed.

Felt tips in colour are dramatic to give visual effect to a finished layout but they should be used with care. The colour spreads and if a visual (as when creating a dummy) is used as a mechanical the calculations will not make sense. Some colours will not only spread but will penetrate the layout paper and ruin sheets beneath the one being worked on. A word of warning: this sort of pen can be dangerous when one is creating visuals under hot lights and a whole group of colours are uncorked at the same time. The headache that ensues can be the effect of the chemical which is the base or drying agent of the colours. Treat with care.

Tools of measurement

Every pencil and felt-tip pen feels at home running along the edge of a ruler. The ancient printer's gauge, made of dependable steel, is still the most reliable, usually with picas and 10 points on one side and inches and millimetres on the reverse. However, the shiny surface can reflect light, particularly strip lighting, and many mistakes have occurred through this. Another problem is that some are too comprehensive in print measurements, giving 5, 6 and 8 point as well as 10 points and pica. This can result in the user accidentally reading off the wrong figure. The modern 18 in/45 cm rule is the best for drawing pages.

7
Graphics

Line drawings, scraper-board, charcoal and wash, even woodcuts, were among early forms of newspaper illustration that gave way to the photograph with the invention of the half-tone reproduction process at the end of last century. Since then the camera has dominated page design. Today there is a strong movement back into what has come to be called newspaper graphics. This does not mean that the photograph is in any way an endangered species – just that there is a new awareness of the potency of drawn illustration in conveying certain types of news and information. The programs available in computer software have made the assembly of the various elements of a graphics illustration almost instantaneous.

Graphic artwork has never entirely been absent from newspaper pages. The big political cartoon of the Victorians stayed in fashion as did a variety of thumbnail cartoons by signed artists. Comic strips, both for adults and children, were increasingly used in the popular dailies from the 1930s as circulation pushers. Line drawings appeared with fashion articles on women's pages (Figure 48) and were resorted to when illustrating Sunday series, especially fiction, which was popular with 1930s newspaper readers. Yet, against the camera, artwork was not taken seriously in the main business of informing readers and illustrating the news and features of the day.

The change began with the Second World War in which maps and diagrammatic illustrations were devised to explain the fighting on the various fronts in a way that was beyond the camera. The chronicling of big battles in illustrated maps, with picture break-outs of the guns and tanks used became common in wartime newspapers.

One editor/strategist who was the 'officer in charge' at Kemsley's *Sunday Graphic*, which was noted for its picture coverage, had the entire North African desert war laid out on a sand table containing a whole cubic yard of sand. Models of German Tiger tanks and British

MONDAY, THE DAILY EXPRESS. OCTOBER 6, 1930. 5

PARIS *Peep Shows Of Fashion.*

IN this amusing article BEVIS describes and illustrates her impressions of some varied aspects of life in Paris.

The "Marquise" cap is new. This one is of velvet—with a tiny wisp of very transparent tulle over one eye.

PARIS.

PARIS is a place where one sees many Americans. That is a well-known fact. They sit in fashionable places like the Ritz, Ciro's, Prunier's, and stare at each other; or "artistic" places like the Dôme, and Rotonde, where I saw some American youths, looking very bohemian, selling some awful "works of art" to their compatriots. I have never seen so many Americans in Paris, in spite of Wall-street.

That Water Wave!

RUE CAMBON is very interesting, and if you walked up and down it all day you would see quite a number of celebrities—mostly English and American, it's true.

There are, however, still some French people left in Paris, and they are very quaint—although they seem to find us even quainter. A friend of mine, whose hair is straight, wanted a wave with irons. The coiffeur, an artist in his profession, insisted upon a water-wave.

For hours my friend sat and broiled in an infernally hot encasement with a tight net on her head. The result was not quite successful, and the coiffeur almost wept. So chagrined was he that he refused payment of any kind.

Sleek Mannequins.

I WAS fitting on some new autumn coats; they all suited me, but were far too expensive. (I know they will be cheaper in London.) The assistant kept on bringing more models, and I was becoming embarrassed. I could not honestly say I did not like them, and to mention the real reason was rather delicate. But the girl was beaming: "Really, madam," she said, "it's a joy to put these coats on you. Madam has a real mannequin figure." And when I bravely said I should have to think the matter of buying a coat over, she treated it as something quite negligible, completely overcome by my mannequin figure.

But I think it was a doubtful compliment. At the various shows I found the mannequins much too sleek. The only dresses that really suited them were the full ones. In the high-waisted, tight-about-the-hips evening dresses they looked like shrimps—half-starved and pitiable.

It is amusing to watch French workmen, staid old gentlemen, and policemen stand for minutes on end, gazing, enraptured, at the shop windows of the popular stores. The "Exposition des Modes Nouvelles" seems to be an unending source of interest to them. I have even seen taxicab-drivers leave their cabs to have a peep at the new dresses.

At the theatre, a "répétition générale" of a new play is quite an experience. So much excitement. Such a clatter of conversation. And what a crowd! The entr'actes are almost longer than the play. The atmosphere is torpid with central-heating and many people jostling one another in the tiny foyer.

I think the greatest attraction for the French playgoer is the foyer during the entr'acte. At least this seemed to be the case at the new Dranem show—although it was amusing enough. French actresses grow in popularity and ingénue-ism the older they become.

There are a lot of old, old films here, too. With different titles. No new ones—except those of Chevalier. The French are amused at the "happy end" which English-speaking people seem to insist on. In fact, "appy end" has become a recognised Continental expression in some vague way representing the Anglo-Saxon frame of mind. They find us superficial!

Jane Renouardt in Sip's revuette gives a marvellous imitation of Yvonne Printemps, looking like her twin sister, with the exception of her nose.

Whereas Yvonne's is retroussé, Jane's nose is—well, it's not snub. She wears black satin pyjamas with a heavy fur border from ankles to knees. Also an afternoon dress with the widest of bell-shaped sleeves in green lattice-work effect.

"Yashmak" Veils.

VEILS are to the fore, some even thickly crocheted, and one I noticed looked like a yashmak worn the wrong way. It covered the eyes and had holes at either side to allow freedom for fluffy curls.

Jewellery is getting more ostentatious. One sees enormous cabochons, tremendous lumps of solid stones, gigantic blobs of bracelets and baubles as big as Christmas-tree decorations.

One woman wore as much of one thing and another that she could not walk. She staggered. Yards and yards of heavy parchment satin embroidered with pearls and diamanté, masses of velvet and fur for the cloak, and loads of heavy jewellery. The puzzle was to find the woman, and the effect produced was a draped mass of animated material and fur.

Surely jewellery has reached barbaric limits, or shall we have to wear rings in our noses like African belles?

Pink Tulle Beret.

AS the Parisienne wears gloves from early morning until—early morning, she wears her rings as well as her bracelets on the outside of the glove.

In a shop window I saw a pale pink tulle beret with a light woollen design on it, and a small scarf to match.

Some dresses are extremely haphazard. They look unfinished and awry, buttoned up the wrong way, and with all sorts of odd colour combinations.

One evening wrap has six or seven magnificent fox furs as trimming. Even a Frenchman (who surely should be inured to anything which might happen to "la mode") was heard to exclaim: "Mais c'est fantas—tique, c'est formi—dable!"

"Mais c'est fantas—tique, c'est formi—dable!"

Six or seven magnificent grège coloured fox pelts arranged on a green velvet evening wrap.

Sophie Tucker's "Says I—"

Miss Sophie Tucker — the great comedienne — has a philosophy of life that has won her the affection of the world at large. Its refreshing sincerity and naturalness of expression led the Editor of the "Daily Express" to invite her to become a guest-journalist on his staff. "I'm no writer," was the reply; yet here is the first of her daily PHILOSOPHIES, exactly as she wrote it out in the language she has made peculiarly her own.

WHENEVER I see a woman cry it sets me thinking.

There's something wrong in crying, beyond the cause of your tears. Don't tell me that every woman has something to make her blue. I know that better than most people . . . d'you think because I sing every night and get called a good fellow I never have a heartache?

Heartaches? Maybe I have more than most. But I'm too old and wise to show it. There are some silly folk in this world who hold that tears have a cash value—but don't you believe it.

Tears never got a girl anything but a red nose.

Tell me you can't pay your rent . . . tell me you're fired from your job, tell me your man's left you and you'll never have a new sweetheart. But will Soph prescribe that "good cry and feel better for it"?

Not on your life.

Here's a wisecrack that's part of my philosophy: the tougher the break the broader the grin.

If your sweetheart deserts you it isn't salt tears that are ever going to bring him back. It isn't sticking around and sobbing and letting him see you're crazy for him. Believe me, holding a man is very simple—after you've lost one or two.

So take it all with a smile as part of the training, and put on your smart clothes and your best smile and go out and enjoy yourself.

If your kisses can't hold the man you love, then your tears won't bring him back!

Sophie Tucker

To-morrow: Another piquant "Says I—" talk.

KITTENS and birds and jolly little bears, all in lifelike colours, make toys that the small members of the family can squeeze to their hearts' content without injury, for they are made of rubber sponge. An extra touch of delight is added by the fact that the base on which they stand squeaks upon pressure.

Seen in the Shops :: *Notes by A Housewife.*

HOW often are eggs spoilt by being just over or under done? But now you can get a little egg timer to hang on the wall which will time the eggs quite correctly.

Simply set the controller at the three, four, or five minutes' point, as required for the eggs, and switch up the little sandglass.

When the time limit is up the sandglass falls, striking the alarm bell on the left and calling attention to the fact that the eggs are now done and quite ready to serve up.

Motorists' Reflector.

A MOST useful contrivance has been thought out to enable motorists adequately to attract attention when they require the car behind to "pull up." It is a small silvered mirror reflector with a celluloid backing, which reflects bright scarlet when a light shines on it from the lamp of the vehicle behind.

By means of flat elastic securely attached, and a small buckle, it can be worn upon the hand of the motorist, or used by a cyclist on the hat or arm, or in any position required. A pillion girl wearing one on her ankle would keep traffic from approaching too closely, and thus make for greater safety.

Pedestrians who use the dark country roads where there is no footpath will find them a great safeguard. Being extremely light in weight under one ounce—there is no discomfort in wearing one.

Wire Pronged Brush.

COMBS are often difficult things to wash, especially small shingle combs with fine teeth.

This comb washer will simplify matters, and will quickly and easily cleanse any kind of comb, with either fine or coarse teeth. Quite small in itself, the little wooden handle is fitted with a large number of very fine wire prongs, which do their work well and speedily.

Spick-and-Span Dusters.

PICTORIAL polishing cloths add to the "Brighter Homes" movement. The figure of a Dutch housewife keeping her home spick and span appears on the large side, while smaller ones made specially for shoe polishing are embellished appropriately.

Rupert Gets Stolen. By Mary Tourtel.

No. 35.—Close Quarters.

ALL RIGHTS RESERVED.

Out come the women from her house when two passing soldiers cried: "Have you seen a little bear about? Likely somewhere here he'd hide!" "No," said the woman; "I am sure no little bear's been here, For I've been hanging out my clothes." Little they know he's near And lying, trembling, 'neath the clothes in the basket. "Thanks," said they, As they hurried off, "he can't be far. We know he came this way."

Figure 48 Fashion graphics in the 1930s: the *Daily Express* looks at the Paris scene

Shermans, together with British 25-pounders and German 88-millimetre guns, were laboriously constructed and the battles, from El Alamein right up to the Mareth line, were fought out on the table week by week. Information was brought up to date each day and the contours of the sand adjusted to point up the difficulties under which the troops of Montgomery and Rommel were fighting. Early on Saturday morning the photographers would be sent into the huge 'war' room to photograph simulations of the latest battle scenes 'from the air'. Guns would be seen firing their puffs of cotton wool and bombs exploding. By this means, as they turned to the centre pages, the readers of the *Sunday Graphic* were given an armchair view of what it was like to be in the front line.

The *Daily Mirror* under H. G. Bartholomew, in its tight wartime format, pioneered the use of symbol graphics to explain simply to its readers the effects of government policy, of the latest budget, of the balance of armies in battles and the dietary value of wartime rationing, by presenting charts tricked out with little men and half men, guns, tanks and milk bottles.

Graphics generally means any non-photographic illustration, whether it be maps, diagrams, cartoons, comic strips or drawn symbols submitted or ordered, or provided in-house, used to illustrate the contents of a page. The days when a drawing or cartoon was chosen to lighten a page in the absence of a suitable photograph are a thing of the past. Today, graphic illustration is an art form in its own right that forcefully claims page space because it can do a required job better than any photograph. The rapid growth of the art desk system in newspapers is a recognition that there is more to artwork than drawing pages, scaling pictures and preparing blurbs. Newspaper artists skilled in diagrammatic reconstruction, 'exploding' pictures to simulate an event, caricature work, and the creation of ingenious information charts are in great demand. Nowhere is this more apparent than in the quality Sunday field where the interpretive approach to the week's news can be powerfully enhanced by graphics.

Cartoons and drawings

The use of stock symbols for charts and motifs has been made easier today by the many disks of 'instant' (Figure 49) artwork and stored computer graphics. There is still a wide area covered, however, by traditional line work either submitted by freelance artists or agencies or drawn in house. This includes cartoons, full-size and thumbnail, comic strips and simple diagrams to go mainly with features, and drawn motifs for holiday pages and advertising and other supplements. The important thing in the drawing is that the lines be strong enough to reproduce easily on newsprint, bearing in mind that they are usually drawn to be two or three times reduced on camera. Problems of reproduction are likelier to crop up with work submitted on spec, staff artists and regular freelances being more aware of a paper's requirements. In the case of commissioned illustrative

Figure 49 Instant art: what the stock books offer to enliven sports logos. An example from *Instant Art Book 1*. Graphic Communications Ltd, reproduced by permission

work the drawings must be geared to the mood of the words that are to accompany them and an artist should insist on being shown the text. Contributing artists need to be aware of the need to enhance cartoons and strips with stick-on reproducible tints to give a drawing texture (easily available from art shops in sheets) and for colour washes to suit colour reproduction, now the thing with cartoons and strips in web offset-produced newspapers. A pressing demand exists, and a potentially high income, for comic strips with strong characterization and good gag lines, and an adult appeal, that will stand up to long-term publication in a regular slot.

Instant art

There are now many books and computer programs of instant art available for a variety of editorial and advertising purposes. Simplicity in its use is essential. Symbols and little stylized drawings, catalogued by subject, are often needed for a quick chart or motif. Letraset and other art companies similarly provide stick-on art material (Figure 50).

Art editors, and sometimes picture editors and advertising departments, usually keep a library of such books and materials, which are particularly useful for smaller papers without skilled artists. They

Figure 50 Transfer graphics: some examples from Letraset's catalogue. Copyright Letraset 1989

should be frequently updated since the artwork in them will reflect the trends in design at the time. Unless the journalist-designer is producing a vignette or dated effect, illustrations in a book even five years old can give an unwanted image.

A collection of previously used, or even rejected, illustrations is worth keeping around for copying on one of the sophisticated machines now in use when instant art is called for, although it must be regarded as a supplement to 'live' work produced in-house or commissioned.

Some companies specializing in the production of instant art offer only pieces in mono, though there is some sophisticated work around which is accompanied by colour separations for use in run-of-press colour should they be required.

Type and halftones in graphics

The main input of original artwork comes from a newspaper's own art desk and a good deal of it is concerned with the preparation of blurbs and compos (composites) specified in page designs in which a combination of typography, line work and half-tone might be used.

A *compo* is a useful device to set off a blurb or feature display by combining a type headline with elements of the illustration – a cut-out head or drawn motif, for example, often with the type reversed into a WOB or a BOT. The type can be the main headline or simply a label or logo. This device is used to provide a dominant focal point in all manner of newspapers (Figure 51).

Figure 51 Top left: the two blurbs – serious and popular – the aim is to catch the eye. Below: the use of type, line and halftone is shown to good effect in the selection of news and features page logos

The commonest form of compo is the logo on a regular column or feature, which can consist simply of the writer's name and 'mug' shot in a type and tint panel. Such logos are most effective if the type and graphic devices are made appropriate to the person's style. The Philip Wrack column in the *News of the World* some years ago, which started as a half-tone full face shot with the name in type enclosed in a panel, became at one stage just the name and a stylized version of the writer's characteristic dark-rimmed spectacles. The *Daily Mirror*, too, used Marje Proops's exotic spectacles to good advantage in her long-running 'agony' series.

The quality dailies and Sundays and many of their European counterparts have pioneered the use of graphics in logos. *The Times* identification logos use high quality artwork to give 'colour' to what used to be rather flat features pages, setting the mood and subject of text and writer. Even small support features below the fold have their own effective mini-logos.

Charts and graphs

Charts and graphs are a source of creative work for in-house newspaper artists in which instant art symbols might be combined with stick-on tints or colour tones and line work to produce a main illustration for a feature with a strong statistical content. These can range from milk yields and baby booms to defence spending and soccer hooliganism. Tints from Letraset and other companies are a stand-by in this sort of work and can give freedom and flair to the finished product.

Type

Letraset stick-on (transfer) type is much used for effects calling for type outside the system's normal range of faces. Bizarre examples are available for special display calling, say, for stencil or cursive letters, or the bastardizing of letters to produce logos by ligaturing, or the altering of strokes to transpose type characters into eccentric designs. They give great freedom of thought and movement in one-off pieces of artwork. More inspiring to a designer is using a keyboard and mouse for screen work. For quick results there is plenty of scope in modern computer systems for special typographical effects, such as condensing or expanding letters for a given result, provided that this usage does not damage the page's normal typographical style.

Dr Swee Chai Ang abandoned her NHS career to care for Palestinian refugees injured in Lebanon. More recently she moved to the Ahli Hospital in Israeli-occupied Gaza. She describes the squalor, brutality, terror – and determination – she has encountered.

Pictures

The half-tone process allows pictures to be tampered with for graphic purposes. A dramatic effect can be given by coarsening the screen on a particular picture to emphasize its symbolic relationship to a story. Cut-out pictures will often enhance a feature display or a blurb compo, while a slightly distorted close-focus shot reproduced big will lend a touch of drama.

To give greater authenticity a *bleach-out* picture of a well-known face or landmark, such as a political figure, the clock of Big Ben or a pithead gear, can be used as a motif on a story or long-running series instead of commissioning a drawn one. The motif needs to typify the story and does not need a caption. The effect is achieved by developing the picture on until all intermediate tones have been bleached out leaving essential detail in stark black and white (Figure 52).

Figure 52 A bleached-out photograph can produce a dramatic motif to flag an important story as in these two examples

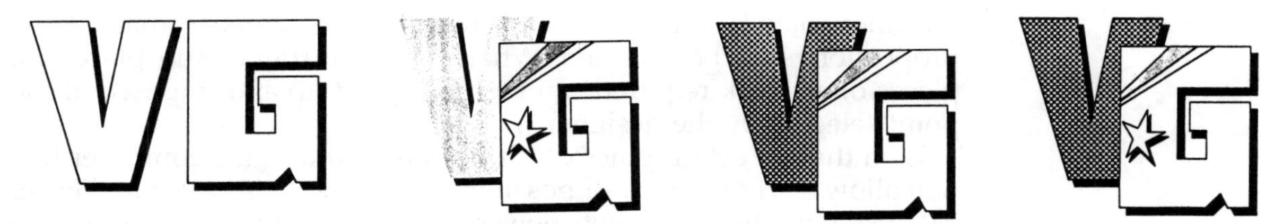

Figure 53 Computer graphics at work: how a page motif is created on screen

Computer graphics

Many pieces of artwork, especially logos, are in regular daily demand in a newspaper office and it is a good idea, where cut-and-paste is still being used, to keep a stock of copy bromides in a box file, and even to recover and keep usable artwork from finished pages. In screen make-up, now more generally in use, original artwork which might be needed regularly can be inputted and stored digitally in the 'pi' fonts of computer setting equipment and retrieved at will. In this way a library of instant art can become even more instant.

The advantages of using computer graphics, either supplied with the system or inputted from originals, are many. To be able to call up at any size a piece of graphics and superimpose it on screen on, say, a financial graph, or any other sort of overprint, has been the dream of designers for decades. The facility is now being widely used.

The more sophisticated word processors have graphics disks of symbols and chart models. If your newspaper cannot afford the more expensive on-screen make-up systems, then examine the possibilities of inputting Bitstream software into word processors that are already in-house. They may be able to supply sufficient graphics for the paper's general requirements either for partial or full on-screen make-up or for printing out for use with paste-up (Figure 53).

Computer graphics can speed the work on a big occasion such as a Budget or a general election. As Budget information is relayed from the Chancellor's lips to the copytaster's screen, graphics staff are searching their computer's memory banks for illustrations for the vital figures. Drink, cigarettes, motor cars and mortgages are high on the list. The prudent old hand will, in any case, have these symbols ready from instant art books, with drawn charts lying ready in permutations of column widths awaiting the figures that will bring them alive. With full screen make-up the corners have already been cut – the artwork is there waiting to respond to the artist's sizing and overprinting.

Information graphics

On-screen compilation of facts in the form of information charts is both fast and accurate and has become one of the great advantages of computer technology. The tedious and time-consuming beginnings of a chart that would require a ruling pen and a steady hand becomes a thing of the past. The keyboard and mouse will call up the lines, panels and boxes on screen with confidence that all the proportions will be accurate. When the basic frame is in place, the pieces of artwork required can then be called up and imposed at the point selected by the designer.

With the correct graphics disk programmed into the computer this will allow a wide variety of possibilities. Rolf Rehe's example (Figure 54), used in the Argentine newspaper *La Nacion*, shows how the country's meat exports are declining. In it he boldly uses the facts reversed on to a large symbol of a bull, a concept that is a classic of its kind.

In most systems, the more advanced and complex charts benefit from hand-drawing being brought in to refine the end product. The computer is used as a tool on the understanding that at any time during the production of the artwork the designer can take a print-out so that embellishments can be applied by hand to the computer's part of the operation. In adopting this method, the chart, when complete on screen, is sized and returned to the page.

Sophistication has grown in information graphics in recent years. The *Daily Star* in 1984, in its hot metal days, was concerned about the traffic of radioactive materials through built-up areas of Britain. It would have been easy to have written a simple news piece about the facts that had been uncovered but the editor knew that a graphic approach would push the story home more effectively. The feature was headlined THE ATOMIC DUSTBIN and the basis of it was that atomic waste from nuclear power stations, foreign as well as British,

Eliminan retenciones a la exportación de carnes

En medios allegados a la industria frigorífica trascendió que el ingeniero Roque Carranza, ministro de Obras y Servicios Públicos a cargo interinamente de la cartera de Economía, firmó ayer una resolución por la que se eliminan a partir del lunes los derechos de exportación que rigen para la carne vacuna refrigerada y que oscilan entre el 15 y el 20 %.

El documento, que sería publicado hoy en el Boletín Oficial, determinaría también algún nivel de reembolso para las carnes con mayor valor agregado incluido (cocidas y enlatadas). Si bien no trascendió este nivel, existía la creencia, con la consiguiente insatisfacción, de que no alcanzaría los porcentajes requeridos por la industria. El reembolso no sería mayor del 5 %.

De todos modos, y pese a la demora en el dictado de la resolución -la eliminación de las retenciones y el otorgamiento de reembolsos habían sido anunciados en Palermo por el doctor Raúl Alfonsín el 12 de agosto último-,la medida lleva cierta tranquilidad al sector industrial, ya que precisamente en estos momentos se encuentra en puerto un buque que viene a cargar aproximadamente 800 toneladas de cortes Hilton para la Comunidad Económica Europea. El incumplimiento de este embarque hubiera puesto en peligro la continuidad de la cuota de 12.000 toneladas de ese tipo de cortes que la Comunidad tiene asignada a la Argentina.

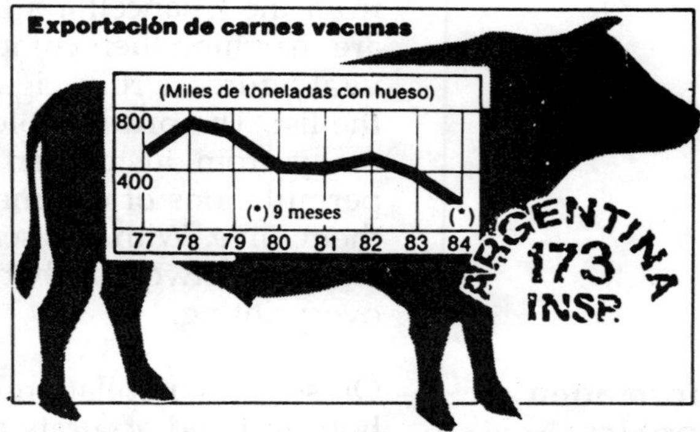

Figure 54 A simple information graphic about beef exports prepared by Rolf F. Rehe for the Argentine paper *La Nacion*

16 DAILY STAR, Thursday, October 25, 1984

A DAILY STAR INVESTIGATION

10 tons of the world's lethal atomic waste travels on British Rail every 24 hours, and all tracks lead to BIG S

● LEFT: Rods of uranium nuclear fuel

By HARRY PUGH and VIC GILES

EVERY working day, three trains haul a deadly cargo through the heart of Britain.

They carry the waste from all our nuclear power stations.

And each load is capable of killing our entire population ... 12 times over.

Meanwhile, off-shore, five ships bring in more atomic waste from power stations all over the globe.

And all 2,000 tons of it are heading for the Big S—Sellafield in Cumbria—which has been dubbed the Atomic Dustbin of the world.

The nuclear waste is sent there for reprocessing—which involves extracting the uranium from spent fuel for re-cycling.

But Sellafield doesn't have the plant to reprocess the more volatile fuel from the latest reactors.

So it has to be put into storage tanks, where it will stay for at least six years until the building of the new plant is completed.

During the summer, the Central Electricity Generating Board staged a spectacular train crash to prove the flasks which carry nuclear waste are safe.

But earlier this week, the environment group Greenpeace claimed the demonstration was "rigged."

Other experts are also very concerned.

Howard Clark, a lecturer at Bradford University, who has made a study of the storage and movement of nuclear waste, points out: "It's a risky business.

Dangerous

"Everybody should realise that these dangerous substances are travelling through their areas."

Jean Emery, head of the Greenpeace monitoring team at Sellafield, adds: "This stuff will still be radioactive in 50 years' time.

"We will be leaving it to future generations."

To avoid the threat of terrorist attack, the actual schedules of these unarmed atom trains is kept secret. But as this exclusive Daily Star dossier reveals, they travel through some of the most heavily populated areas of the country.

MAGNOX
THE earliest nuclear power stations are of the Magnox type, and run on ordinary uranium fuel.

AGR
THE second generation of power stations, called Advanced Gas Reactors, run on enriched uranium.

THE FLASK
SPENT fuel from the power stations — either in the form of uranium rods or pellets — is carried in 50-ton flasks, with steel walls 14½ inches thick.

HPR
BRITAIN'S only Heavy Water Reactor — at Winfrith — runs on enriched uranium.

THE TRAIN
IN a spectacular stunt this summer, the CEGB crashed a 100 mph diesel train into a fuel flask. The flask remained intact, but the train was destroyed in the £1.6m experiment.

CHANNEL DANGER
A FLEET of five ships sail the world to bring in 400 tons of highly volatile nuclear waste every year.

ATOMIC DUSTBIN!

DAILY STAR, Thursday, October 25, 1984 17

IF ...THE NUCLEAR ACCIDENT DID REALLY HAPPEN

THE horrors of what could happen if a nuclear flask ever did leak are not just a science fiction fantasy.

They have been carefully calculated by a team of scientists.

And they predict that if as little as 10 per cent of the contents of a flask escaped, it could kill thousands of people.

Most of them would suffer a lingering death over a number of years from cancer or leukaemia.

But people closer to the leak would die almost immediately.

A cigar-shaped area, two miles long downwind from the fractured flask, and a quarter of a mile wide, would have to be evacuated immediately.

High

And the level of radioactivity would be too high for people to return for at least 10 years.

The survey was carried out by the Oxford-based Political Ecology Research Group, directed by consultant biologist Peter Taylor. He says: "A faulty flask, or perhaps a careless operator who forgets to do something simple like secure a flask lid, could contribute to a major accident.

"But the biggest threat, in my view, is from sabotage or terrorist activity. The right explosive, or an anti-tank missile, could pierce a flask.

"And if any terrorist did this, and used an incendiary device to start a fire, he could create a nuclear cloud with the devastating results we have calculated."

LONDON: THE FLASHPOINT

A QUARTER of Britain's nuclear waste passes through London itself.

And many other major cities are on the routes of the deadly nuclear freight trains.

They move at night when there is less traffic on the rails.

British Rail and CEGB make no effort to route the trains away from dense population areas.

Our map shows how the nuclear trains from East Coast and Kent power stations are routed through the capital.

Often, trains halt in sidings at Willesden Junction to be linked together for the journey north to Big S.

The three-man crews are not armed. But they are instructed in how to alert the emergency services if a flask springs a leak.

Our map shows just what would happen if that leak—or a terrorist attack—happened at Manor Park, with a westerly breeze blowing.

The nuclear cloud would spread for two miles down the line to Ilford, contaminating a corridor a quarter-of-a-mile wide.

More than 10,000 people who live in the Manor Park-Ilford area would have to abandon their homes for at least 10 years. And many of them would die.

LONDON'S NUCLEAR 'RUSH-HOUR'

Figure 55 A tabloid spread projection by Vic Giles on the perils of the 'atomic dustbin' (from the *Daily Star*)

THE STAR EXAMINES BOB GELDOF'S PLAN TO RESCUE THE HUNGRY

We ARE the world!

–AND WE'RE GOING TO MAKE IT A BETTER PLACE FOR PEOPLE TO LIVE IN

BOUNCING POP THROUGH SPACE

THE greatest show on earth may reach outer space. Organisers are trying to link with the Space Shuttle which could then link both stadiums.

KENNEDY STADIUM

● MORE than 90,000 people will pack the John F. Kennedy Stadium in Philadelphia — pictured above during a Rolling Stones concert four years ago. For the Live Aid spectacular, a television crew of nearly 300 will be manning 20 cameras for the global link-up.

● "The Boss" — Bruce Springsteen — and Stevie Wonder will be among the stars whose voices will ring around the world. And as at Wembley 3,000 miles away, the JFK Stadium will explode with noise from the appreciative fans.

BOB GELDOF . . . and some of Ethiopia's starving children

WEMBLEY

BETWEEN each act at Wembley, the 80,000 crowd will see live pictures of the Philadelphia show on giant screens. The link-up with America starts at 5 p.m. our time.

CONCORDE

BOB GELDOF'S "Mr. Fixit" powers persuaded British Airways boss Lord King to lend him Concorde. Now some Live Aid stars will be able to perform at BOTH stadiums.

ITALY

IN Italy, there are plans to have opera superstar Luciano Pavarotti and 500 Italian rock musicians singing together on a Soccer pitch. Special events are also planned in West Germany, Yugoslavia and Austria.

By ROBERT WILSON

WHEN megastar Paul McCartney admits to feeling nervous, you know it's something special.

When a supergroup like The Who decide to get back together for a single show, you begin to realise it's out of this world.

And when you read the rest of the cast list, you're convinced.

For next Saturday's Live Aid concert is the greatest gig in the galaxy.

The sheer scale of the rock spectacular makes it a prime candidate for the Guinness Book Of Records — and a mammoth headache for Mike Appleton, the BBC producer, who is co-ordinating the world-wide broadcast.

"The situation is changing all the time," says Mike, during a brief break from the biggest job he's ever undertaken.

"We are waiting to hear from Russia and Poland and have sent emissaries to India and mainland China.

"We are also hoping that Japan will be making a live contribution."

So it's hardly surprising that even McCartney has had a few second thoughts about the sheer scope of it all.

"Paul wanted to help," explains Live Aid originator Bob Geldof. "But he was a bit nervous because he hasn't played a live gig for ages.

"So I wrote to him, talked to him about the problems of Africa and he said: 'Right—I'm in!'"

But McCartney is just one of dozens of big-name surprises...

Mick Jagger and David Bowie, who have already recorded a single and video together, will be appearing live in the concert—but on different sides of the Atlantic.

For five hours after the sound explosion begins at London's Wembley Stadium, a twin concert will join in from the John F. Kennedy Stadium, in Philadelphia, 3,000 miles away.

Bowie will be in London and Jagger in Philadelphia.

Another duet—this time live—could feature Elton John and Wham! star George Michael.

Just about everybody who is anybody is appearing—or trying to appear.

The Wembley bill also includes Adam Ant, the Boomtown Rats, Phil Collins, Elvis Costello, Dire Straits, Bryan Ferry, Howard Jones, Nik Kershaw, Alison Moyet, Pretenders, Sade, Status Quo, Rod Stewart, Sting, Style Council, U2, Ultravox, Wham! and Paul Young.

In Philadelphia will be Bryan Adams, Cars, Eric Clapton, Duran Duran, Hall And Oates, Temptations, Waylon Jennings, Kris Kristofferson, Huey Lewis And The News, Billy Ocean, Robert Plant, Power Station, Judas Priest, Santana, Paul Simon, Simple Minds, Bruce Springsteen, Tears For Fears, Thompson Twins, Tina Turner, Stevie Wonder and Neil Young.

THE COMMAND POST IS GLC BUS GARAGE

ALTHOUGH the Live Aid concerts involve the most glittering names in entertainment, the headquarters of the mammoth spectacular are in a former bus garage in London.

The old building, donated by the Greater London Council, is the HQ of Band Aid.

There, a team of volunteers handle the huge administrative task of helping the starving continent.

Valerie Blondeau is co-ordinator of the team, led by architect Kevin Jendon.

She says: "Everything is sent out and processed from here. You could call us the pulse of Band Aid.

"For example a month ago, we got a call in from the field telling us they were desperate for plastic sheeting because of rains.

"Within 48 hours we had a plane load [illegible] of it out there."

Valerie has been to the Sudan to see the efforts of Band Aid put to their practical use.

She says: "It's then when you realise that you haven't been wasting your time here.

"I went to a huge refugee camp with 85,000 people in it, and 2,000 coming in daily.

"One little girl was eating Band Aid biscuits, and following us about all day. She smiled all the time.

"That said it all."

AUSTRALIA

TOP Australian band INXS will join the concert live by satellite. There will also be a special "Down Under" day.

BRAZIL

PELE (left), the great soccer legend of Brazil, is expected to appear on screen during the world-wide spectacular. And many of Brazil's neighbours plan to take part in Saturday's show

AFRICA

AFRICA, the continent the extravaganza is all about, is expected to receive the transmission free. BBC chiefs hope to arrange interviews with people in Ethiopia to link up with the concert.

HOW THE MONEY WILL ROLL

GOLDSMITH . . . promoter

"SELLING" the concert is in the hands of some of the world's greatest experts.

Worldwide Sports, an organisation set up for last year's Los Angeles Olympics, was recruited by the promoters.

Sponsors

The money will flow in from the thousands of American TV and radio channels—who have promised half their advertising revenue — and from international sponsors.

At Wembley, the organiser is leading rock promoter Harvey Goldsmith. The concert is already a complete sell-out.

Tickets were £25 a head, but some people have paid £250 for a ticket which includes a four-course meal in a private room.

Similar revenue is expected from the JFK Stadium in Philadelphia.

More money will come in from a Telethon appeal for the charity, plus sales of everything from programmes to special T-shirts.

The BBC is spending £60,000 for 15 continuous hours of TV and radio, and has leased time from satellites.

Total

Radio One will be joining in with a simultaneous broadcast. This means listeners will be able to watch the show on TV and hear it in stereo.

And throughout Europe there are plans to install huge screens to attract mass audiences to help with the cash-raising.

What seems certain is that the total from the original British and American records, plus Saturday's Live Aid concert, will raise more than the £[illegible] million which the British Government donated to Ethiopia last year.

LINING UP FOR THE GREATEST GIG OF ALL TIME

THE THOMPSON TWINS | PAUL McCARTNEY | TINA TURNER | MICK JAGGER and DAVID BOWIE | PHIL COLLINS | WHAM! | THE WHO | STING

TOMORROW: Sting talks to the Star

Figure 56 Graphic treatment for a Live Aid campaign feature from the *Daily Star* – artwork and finished page

was being transported by rail and sea to the nuclear disposal site in Sellafield, Cumbria. The journalist/graphic artist team produced the double-page spread map (Figure 55).

It had to solve two problems: to allow space on the map for text and headlines dealing with the different parts of the story, and also to allow for the presence on the double-page spread of a large advertisement consisting mainly of text. The shape of the available space precluded the first idea of superimposing the map on a black dustbin. Also time was short. The verdict on the spread as it appeared was that it could have been stronger in illustration, and that more could have been made of the shadow on the bottom of the map. Yet, in the short time available, the device achieved the editor's aim of riveting the reader's eye on the subject.

A double-page spread devoted to pop star Bob Geldof's campaign in support of the Ethiopia famine appeal, also in the *Daily Star* (Figure 56) showed an advance in technique. It was based on a huge surrealist guitar as a vehicle for a map of the world showing how countries were responding. At the same time the spread was packed with information for the paper's young readers – an important part of the projection. There can always be criticism: the elements across the top of the page, as can be seen, are fragmented, but the two halves of the spread are held together by the strip of pictures across the bottom.

USA Today, which sells widely in Europe, is a notable exponent of the mass readership graph. In fact, the idea of the paper is to provide the reader with eye-catching pieces of dressed up information. An example is the regular page one colour feature, USA SNAPSHOTS – A LOOK AT STATISTICS THAT SHAPE THE NATION (Plate 4 of the colour plate section) with its dramatic background cartoon work. Figures on pay for nurses in different cities, in this example, reach readers who are drawn to the feature by the humour of the artwork and then find themselves hooked by its geography.

USA Today using keyboard and mouse and light pen, deploys colour cleverly, placing tints to give an impression of a greater range of colour. Yet one of its most successful graphics effects in the edition illustrated, the artwork for the Super Bowl battle between Cincinatti Bengals and the San Francisco 49ers (Figure 57) relies only on mono. The projection, because of the three-quarter overhead view, enables the graphic artist to avoid giving the impression of players disappearing into infinity. The simplicity of the concept is continued into the descriptive captions, each one connected to the appropriate player by a fine pointer line. This is a good example of complicated factual information laid out in such a way that enthusiasts can relate the graphic details with the actuality of the television screen and almost forecast the players' moves.

News graphics

Schooled in the actuality of wartime graphics, postwar editors in Fleet Street began to employ art school graduates to help make news

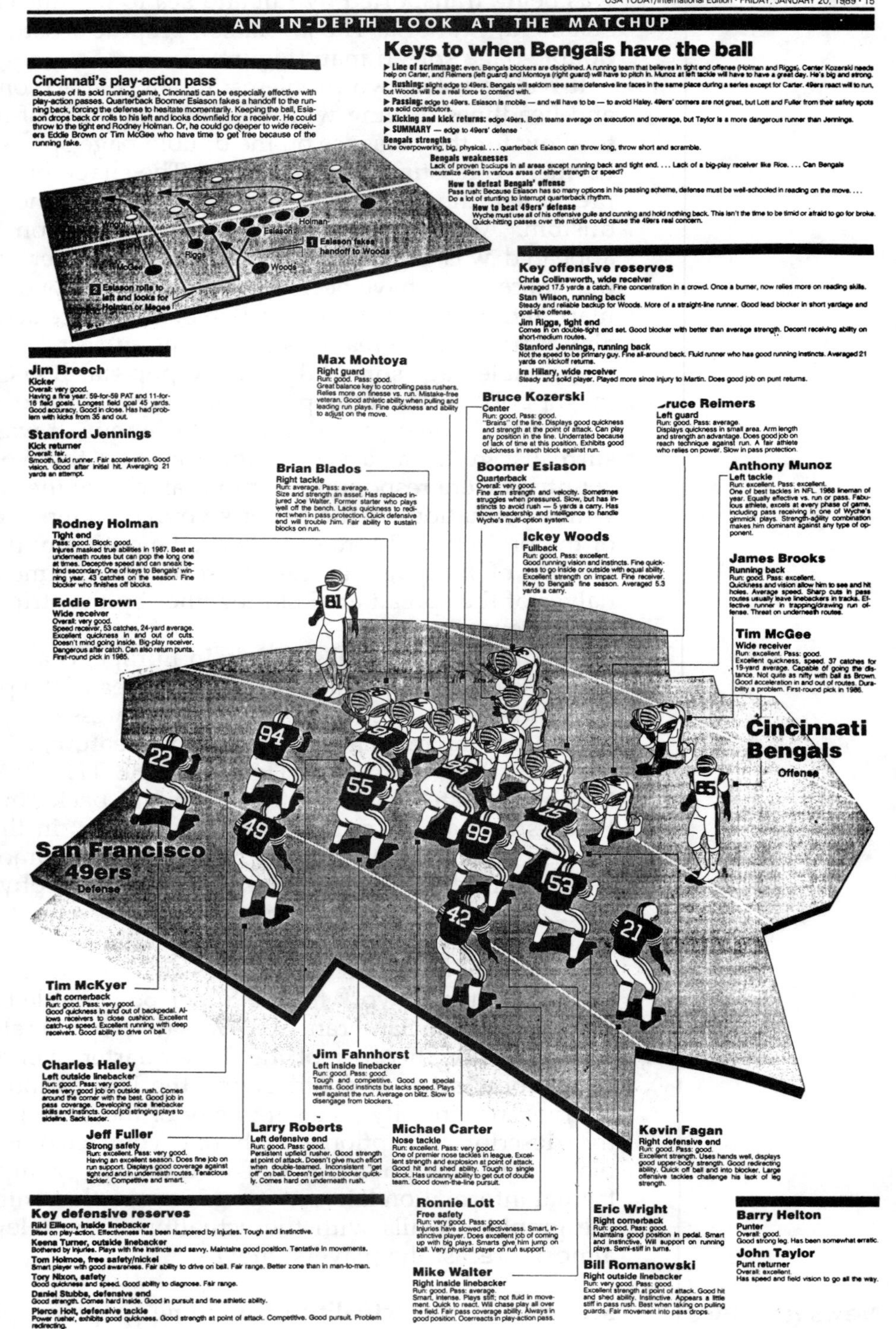

USA TODAY/International Edition · FRIDAY, JANUARY 20, 1989 · 15

AN IN-DEPTH LOOK AT THE MATCHUP

Cincinnati's play-action pass

Because of its solid running game, Cincinnati can be especially effective with play-action passes. Quarterback Boomer Esiason fakes a handoff to the running back, forcing the defense to hesitate momentarily. Keeping the ball, Esiason drops back or rolls to his left and looks downfield for a receiver. He could throw to the tight end Rodney Holman. Or, he could go deeper to wide receivers Eddie Brown or Tim McGee who have time to get free because of the running fake.

Keys to when Bengals have the ball

▶ **Line of scrimmage:** even. Bengals blockers are disciplined. A running team that believes in tight end offense (Holman and Riggs). Center Kozerski needs help on Carter, and Reimers (left guard) and Montoya (right guard) will have to pitch in. Munoz at left tackle will have to have a great day. He's big and strong.

▶ **Rushing:** slight edge to 49ers. Bengals will seldom see same defensive line faces in the same place during a series except for Carter. 49ers react will to run, but Woods will be a real force to contend with.

▶ **Passing:** edge to 49ers. Esiason is mobile — and will have to be — to avoid Haley. 49ers' corners are not great, but Lott and Fuller from their safety spots are solid contributors.

▶ **Kicking and kick returns:** edge 49ers. Both teams average on execution and coverage, but Taylor is a more dangerous runner than Jennings.

▶ **SUMMARY** — edge to 49ers' defense

Bengals strengths
Line overpowering, big, physical. . . . quarterback Esiason can throw long, throw short and scramble.

Bengals weaknesses
Lack of proven backups in all areas except running back and tight end. . . . Lack of a big-play receiver like Rice. . . . Can Bengals neutralize 49ers in various areas of either strength or speed?

How to defeat Bengals' offense
Pass rush: Because Esiason has so many options in his passing scheme, defense must be well-schooled in reading on the move. . . . Do a lot of stunting to interrupt quarterback rhythm.

How to beat 49ers' defense
Wyche must use all of his offensive guile and cunning and hold nothing back. This isn't the time to be timid or afraid to go for broke. Quick-hitting passes over the middle could cause the 49ers real concern.

Key offensive reserves

Chris Collinsworth, wide receiver
Averaged 17.5 yards a catch. Fine concentration in a crowd. Once a burner, now relies more on reading skills.

Stan Wilson, running back
Steady and reliable backup for Woods. More of a straight-line runner. Good lead blocker in short yardage and goal-line offense.

Jim Riggs, tight end
Comes in on double-tight end set. Good blocker with better than average strength. Decent receiving ability on short-medium routes.

Stanford Jennings, running back
Not the speed to be primary guy. Fine all-around back. Fluid runner who has good running instincts. Averaged 21 yards on kickoff returns.

Ira Hillary, wide receiver
Steady and solid player. Played more since injury to Martin. Does good job on punt returns.

Jim Breech
Kicker
Overall: very good.
Having a fine year. 59-for-59 PAT and 11-for-16 field goals. Longest field goal 45 yards. Good accuracy. Good in close. Has had problem with kicks from 36 and out.

Stanford Jennings
Kick returner
Overall: fair.
Smooth, fluid runner. Fair acceleration. Good vision. Good after initial hit. Averaging 21 yards an attempt.

Max Montoya
Right guard
Run: good. Pass: good.
Great balance key to controlling pass rushers. Relies more on finesse vs. run. Mistake-free veteran. Good athletic ability when pulling and leading run plays. Fine quickness and ability to adjust on the move.

Bruce Kozerski
Center
Run: good. Pass: good.
"Brains" of the line. Displays good quickness and strength at the point of attack. Can play any position in the line. Underrated because of lack of time at this position. Exhibits good quickness in reach block against run.

ruce Reimers
Left guard
Run: good. Pass: average.
Displays quickness in small area. Arm length and strength an advantage. Does good job on reach technique against run. A fair athlete who relies on power. Slow in pass protection.

Brian Blados
Right tackle
Run: average. Pass: average.
Size and strength an asset. Has replaced injured Joe Walter. Former starter who plays well off the bench. Lacks quickness to redirect when in pass protection. Quick defensive end will trouble him. Fair ability to sustain blocks on run.

Boomer Esiason
Quarterback
Overall: very good.
Fine arm strength and velocity. Sometimes struggles when pressured. Slow, but has instincts to avoid rush — 5 yards a carry. Has shown leadership and intelligence to handle Wyche's multi-option system.

Anthony Munoz
Left tackle
Run: excellent. Pass: excellent.
One of best tackles in NFL. 1988 lineman of year. Equally effective vs. run or pass. Fabulous athlete, excels at every phase of game, including pass receiving in one of Wyche's gimmick plays. Strength-agility combination makes him dominant against any type of opponent.

Rodney Holman
Tight end
Pass: good. Block: good.
Injures masked true abilities in 1987. Best at underneath routes but can pop the long one at times. Deceptive speed and can sneak behind secondary. One of keys to Bengals' winning year. 43 catches on the season. Fine blocker who finishes off blocks.

Ickey Woods
Fullback
Run: good. Pass: excellent.
Good running vision and instincts. Fine quickness to go inside or outside with equal ability. Excellent strength on impact. Fine receiver. Key to Bengals' fine season. Averaged 5.3 yards a carry.

James Brooks
Running back
Run: good. Pass: excellent.
Quickness and vision allow him to see and hit holes. Average speed. Sharp cuts in pass routes usually leave linebackers in tracks. Effective runner in trapping/drawing run offense. Threat on underneath routes.

Eddie Brown
Wide receiver
Overall: very good.
Speed receiver, 53 catches, 24-yard average. Excellent quickness in and out of cuts. Doesn't mind going inside. Big-play receiver. Dangerous after catch. Can also return punts. First-round pick in 1985.

Tim McGee
Wide receiver
Run: excellent. Pass: good.
Excellent quickness, speed. 37 catches for 19-yard average. Capable of going the distance. Not quite as nifty with ball as Brown. Good acceleration in and out of routes. Durability a problem. First-round pick in 1986.

Tim McKyer
Left cornerback
Run: good. Pass: very good.
Good quickness in and out of backpedal. Allows receivers to close cushion. Excellent catch-up speed. Excellent running with deep receivers. Good ability to drive on ball.

Charles Haley
Left outside linebacker
Run: good. Pass: very good.
Does very good job on outside rush. Comes around the corner with the best. Good job in pass coverage. Developing nice linebacker skills and instincts. Good job stringing plays to sideline. Sack leader.

Jim Fahnhorst
Left inside linebacker
Run: good. Pass: good.
Tough and competitive. Good on special teams. Good instincts but lacks speed. Plays well against the run. Average on blitz. Slow to disengage from blockers.

Jeff Fuller
Strong safety
Run: excellent. Pass: very good.
Having an excellent season. Does fine job on run support. Displays good coverage against tight end and in underneath routes. Tenacious tackler. Competitive and smart.

Larry Roberts
Left defensive end
Run: good. Pass: good.
Persistent upfield rusher. Good strength at point of attack. Doesn't give much effort when double-teamed. Inconsistent "get off" on ball. Doesn't get into blocker quickly. Comes hard on underneath rush.

Michael Carter
Nose tackle
Run: excellent. Pass: very good.
One of premier nose tackles in league. Excellent strength and explosion at point of attack. Good hit and shed ability. Tough to single block. Has uncanny ability to get out of double team. Good down-the-line pursuit.

Kevin Fagan
Right defensive end
Run: good. Pass: good.
Excellent strength. Uses hands well, displays good upper-body strength. Good redirecting ability. Quick off ball and into blocker. Large offensive tackles challenge his lack of leg strength.

Key defensive reserves

Riki Ellison, inside linebacker
Bites on play-action. Effectiveness has been hampered by injuries. Tough and instinctive.

Keena Turner, outside linebacker
Bothered by injuries. Plays with fine instincts and savvy. Maintains good position. Tentative in movements.

Tom Holmoe, free safety/nickel
Smart player with good awareness. Fair ability to drive on ball. Fair range. Better zone than in man-to-man.

Tory Nixon, safety
Good quickness and speed. Good ability to diagnose. Fair range.

Daniel Stubbs, defensive end
Good strength. Comes hard inside. Good in pursuit and fine athletic ability.

Pierce Holt, defensive tackle
Power rusher, exhibits good quickness. Good strength at point of attack. Competitive. Good pursuit. Problem redirecting.

Ronnie Lott
Free safety
Run: very good. Pass: good.
Injuries have slowed effectiveness. Smart, instinctive player. Does excellent job of coming up with big plays. Smarts give him jump on ball. Very physical player on run support.

Eric Wright
Right cornerback
Run: good. Pass: good.
Maintains good position in pedal. Smart and instinctive. Will support on running plays. Semi-stiff in turns.

Barry Helton
Punter
Overall: good.
Good strong leg. Has been somewhat erratic.

John Taylor
Punt returner
Overall: excellent.
Has speed and field vision to go all the way.

Mike Walter
Right inside linebacker
Run: good. Pass: average.
Smart, intense. Plays stiff; not fluid in movement. Quick to react. Will chase play all over the field. Fair pass coverage ability. Always in good position. Ocerreacts in play-action pass.

Bill Romanowski
Right outside linebacker
Run: very good. Pass: good.
Excellent strength at point of attack. Good hit and shed ability. Instinctive. Appears a little stiff in pass rush. Best when taking on pulling guards. Fair movement into pass drops.

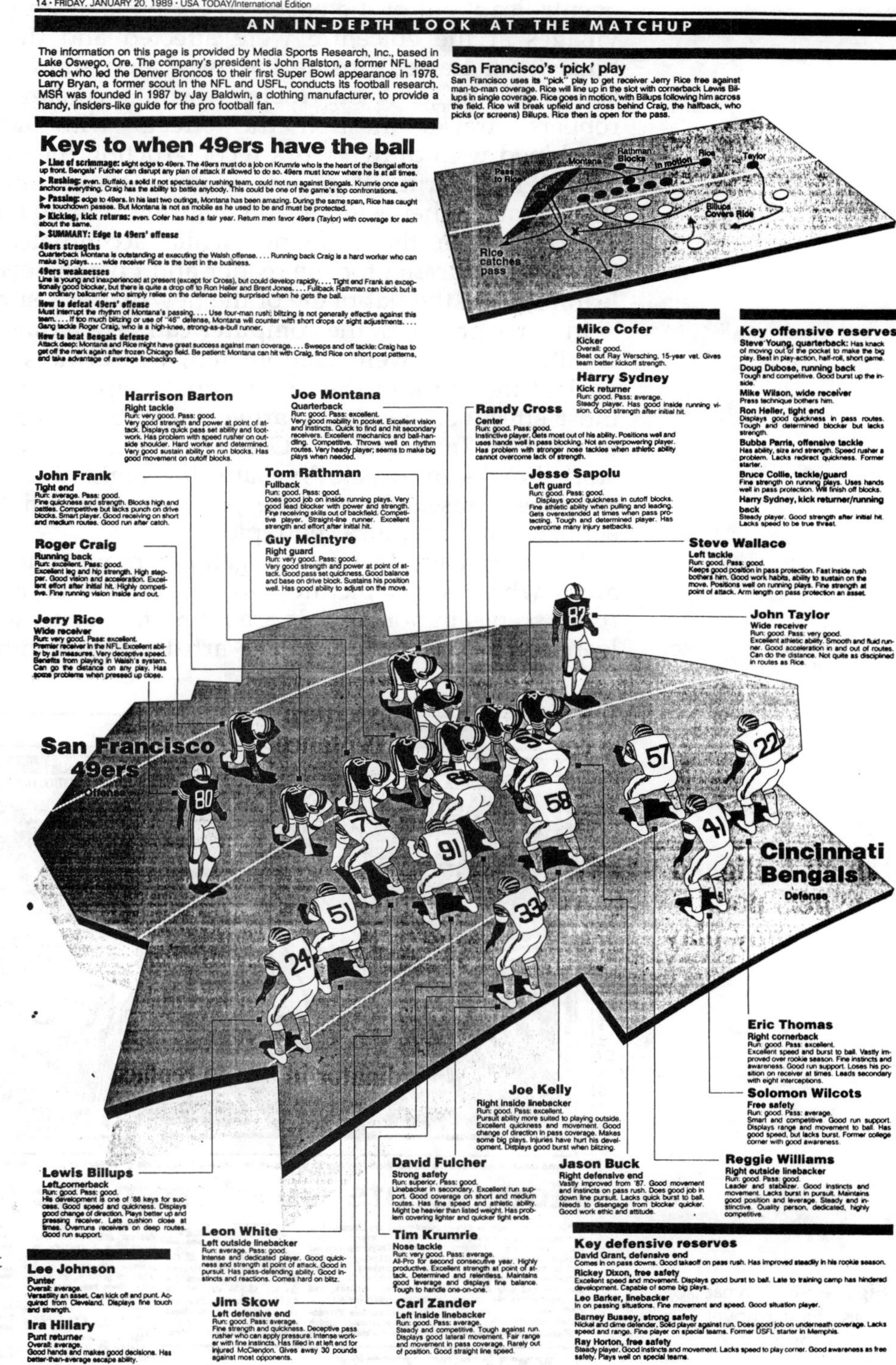

14 • FRIDAY, JANUARY 20, 1989 • USA TODAY/International Edition

AN IN-DEPTH LOOK AT THE MATCHUP

The information on this page is provided by Media Sports Research, Inc., based in Lake Oswego, Ore. The company's president is John Ralston, a former NFL head coach who led the Denver Broncos to their first Super Bowl appearance in 1978. Larry Bryan, a former scout in the NFL and USFL, conducts its football research. MSR was founded in 1987 by Jay Baldwin, a clothing manufacturer, to provide a handy, insiders-like guide for the pro football fan.

San Francisco's 'pick' play

San Francisco uses its "pick" play to get receiver Jerry Rice free against man-to-man coverage. Rice will line up in the slot with cornerback Lewis Billups in single coverage. Rice goes in motion, with Billups following him across the field. Rice will break upfield and cross behind Craig, the halfback, who picks (or screens) Billups. Rice then is open for the pass.

Keys to when 49ers have the ball

▶ **Line of scrimmage:** slight edge to 49ers. The 49ers must do a job on Krumrie who is the heart of the Bengal efforts up front. Bengals' Fulcher can disrupt any plan of attack if allowed to do so. 49ers must know where he is at all times.

▶ **Rushing:** even. Buffalo, a solid if not spectacular rushing team, could not run against Bengals. Krumrie once again anchors everything. Craig has the ability to battle anybody. This could be one of the game's top confrontations.

▶ **Passing:** edge to 49ers. In his last two outings, Montana has been amazing. During the same span, Rice has caught five touchdown passes. But Montana is not as mobile as he used to be and must be protected.

▶ **Kicking, kick returns:** even. Cofer has had a fair year. Return men favor 49ers (Taylor) with coverage for each about the same.

▶ **SUMMARY: Edge to 49ers' offense**

49ers strengths
Quarterback Montana is outstanding at executing the Walsh offense. . . . Running back Craig is a hard worker who can make big plays. . . . wide receiver Rice is the best in the business.

49ers weaknesses
Line is young and inexperienced at present (except for Cross), but could develop rapidly. . . . Tight end Frank an exceptionally good blocker, but there is quite a drop off to Ron Heller and Brent Jones. . . . Fullback Rathman can block but is an ordinary ballcarrier who simply relies on the defense being surprised when he gets the ball.

How to defeat 49ers' offense
Must interrupt the rhythm of Montana's passing. . . . Use four-man rush; blitzing is not generally effective against this team. . . . If too much blitzing or use of "46" defense, Montana will counter with short drops or sight adjustments. . . . Gang tackle Roger Craig, who is a high-knee, strong-as-a-bull runner.

How to beat Bengals defense
Attack deep: Montana and Rice might have great success against man coverage. . . . Sweeps and off tackle: Craig has to get off the mark again after frozen Chicago field. Be patient: Montana can hit with Craig, find Rice on short post patterns, and take advantage of average linebacking.

Mike Cofer
Kicker
Overall: good.
Beat out Ray Wersching, 15-year vet. Gives team better kickoff strength.

Harry Sydney
Kick returner
Run: good. Pass: average.
Steady player. Has good inside running vision. Good strength after initial hit.

Key offensive reserves
Steve Young, quarterback: Has knack of moving out of the pocket to make the big play. Best in play-action, half-roll, short game.
Doug Dubose, running back
Tough and competitive. Good burst up the inside.
Mike Wilson, wide receiver
Press technique bothers him.
Ron Heller, tight end
Displays good quickness in pass routes. Tough and determined blocker but lacks strength.
Bubba Paris, offensive tackle
Has ability, size and strength. Speed rusher a problem. Lacks redirect quickness. Former starter.
Bruce Collie, tackle/guard
Fine strength on running plays. Uses hands well in pass protection. Will finish off blocks.
Harry Sydney, kick returner/running back
Steady player. Good strength after initial hit. Lacks speed to be true threat.

Harrison Barton
Right tackle
Run: very good. Pass: good.
Very good strength and power at point of attack. Displays quick pass set ability and footwork. Has problem with speed rusher on outside shoulder. Hard worker and determined. Very good sustain ability on run blocks. Has good movement on cutoff blocks.

Joe Montana
Quarterback
Run: good. Pass: excellent.
Very good mobility in pocket. Excellent vision and instincts. Quick to find and hit secondary receivers. Excellent mechanics and ball-handling. Competitive. Throws well on rhythm routes. Very heady player; seems to make big plays when needed.

Randy Cross
Center
Run: good. Pass: good.
Instinctive player. Gets most out of his ability. Positions well and uses hands well in pass blocking. Not an overpowering player. Has problem with stronger nose tackles when athletic ability cannot overcome lack of strength.

John Frank
Tight end
Run: average. Pass: good.
Fine quickness and strength. Blocks high and battles. Competitive but lacks punch on drive blocks. Smart player. Good receiving on short and medium routes. Good run after catch.

Tom Rathman
Fullback
Run: good. Pass: good.
Does good job on inside running plays. Very good lead blocker with power and strength. Fine receiving skills out of backfield. Competitive player. Straight-line runner. Excellent strength and effort after initial hit.

Jesse Sapolu
Left guard
Run: good. Pass: good.
Displays good quickness in cutoff blocks. Fine athletic ability when pulling and leading. Gets overextended at times when pass protecting. Tough and determined player. Has overcome many injury setbacks.

Roger Craig
Running back
Run: excellent. Pass: good.
Excellent leg and hip strength. High stepper. Good vision and acceleration. Excellent effort after initial hit. Highly competitive. Fine running vision inside and out.

Guy McIntyre
Right guard
Run: very good. Pass: good.
Very good strength and power at point of attack. Good pass set quickness. Good balance and base on drive block. Sustains his position well. Has good ability to adjust on the move.

Steve Wallace
Left tackle
Run: good. Pass: good.
Keeps good position in pass protection. Fast inside rush bothers him. Good work habits, ability to sustain on the move. Positions well on running plays. Fine strength at point of attack. Arm length on pass protection an asset.

Jerry Rice
Wide receiver
Run: very good. Pass: excellent.
Premier receiver in the NFL. Excellent ability by all measures. Very deceptive speed. Benefits from playing in Walsh's system. Can go the distance on any play. Has some problems when pressed up close.

John Taylor
Wide receiver
Run: good. Pass: very good.
Excellent athletic ability. Smooth and fluid runner. Good acceleration in and out of routes. Can do the distance. Not quite as disciplined in routes as Rice.

Eric Thomas
Right cornerback
Run: good. Pass: excellent.
Excellent speed and burst to ball. Vastly improved over rookie season. Fine instincts and awareness. Good run support. Loses his position on receiver at times. Leads secondary with eight interceptions.

Joe Kelly
Right inside linebacker
Run: good. Pass: excellent.
Pursuit ability more suited to playing outside. Excellent quickness and movement. Good change of direction in pass coverage. Makes some big plays. Injuries have hurt his development. Displays good burst when blitzing.

Solomon Wilcots
Free safety
Run: good. Pass: average.
Smart and competitive. Good run support. Displays range and movement to ball. Has good speed, but lacks burst. Former college corner with good awareness.

Lewis Billups
Left cornerback
Run: good. Pass: good.
His development is one of '88 keys for success. Good speed and quickness. Displays good change of direction. Plays better up and pressing receiver. Lets cushion close at times. Overruns receivers on deep routes. Good run support.

David Fulcher
Strong safety
Run: superior. Pass: good.
Linebacker in secondary. Excellent run support. Good coverage on short and medium routes. Has fine speed and athletic ability. Might be heavier than listed weight. Has problem covering lighter and quicker tight ends.

Jason Buck
Right defensive end
Vastly improved from '87. Good movement and instincts on pass rush. Does good job in down line pursuit. Lacks quick burst to ball. Needs to disengage from blocker quicker. Good work ethic and attitude.

Reggie Williams
Right outside linebacker
Run: good. Pass: good.
Leader and stabilizer. Good instincts and movement. Lacks burst in pursuit. Maintains good position and leverage. Steady and instinctive. Quality person, dedicated, highly competitive.

Leon White
Left outside linebacker
Run: average. Pass: good.
Intense and dedicated player. Good quickness and strength at point of attack. Good in pursuit. Has pass-defending ability. Good instincts and reactions. Comes hard on blitz.

Tim Krumrie
Nose tackle
Run: very good. Pass: average.
All-Pro for second consecutive year. Highly productive. Excellent strength at point of attack. Determined and relentless. Maintains good leverage and displays fine balance. Tough to handle one-on-one.

Lee Johnson
Punter
Overall: average.
Versatility an asset. Can kick off and punt. Acquired from Cleveland. Displays fine touch and strength.

Ira Hillary
Punt returner
Overall: average.
Good hands and makes good decisions. Has better-than-average escape ability.

Jim Skow
Left defensive end
Run: good. Pass: average.
Fine strength and quickness. Deceptive pass rusher who can apply pressure. Intense worker with fine instincts. Has filled in at left end for injured McClendon. Gives away 30 pounds against most opponents.

Carl Zander
Left inside linebacker
Run: good. Pass: average.
Steady and competitive. Tough against run. Displays good lateral movement. Fair range and movement in pass coverage. Rarely out of position. Good straight line speed.

Key defensive reserves
David Grant, defensive end
Comes in on pass downs. Good takeoff on pass rush. Has improved steadily in his rookie season.
Rickey Dixon, free safety
Excellent speed and movement. Displays good burst to ball. Late to training camp has hindered development. Capable of some big plays.
Leo Barker, linebacker
In on passing situations. Fine movement and speed. Good situation player.
Barney Bussey, strong safety
Nickel and dime defender. Solid player against run. Does good job on underneath coverage. Lacks speed and range. Fine player on special teams. Former USFL starter in Memphis.
Ray Horton, free safety
Steady player. Good instincts and movement. Lacks speed to play corner. Good awareness as free safety. Plays well on special teams.

Figure 57 The big show: a double broadsheet spread in *USA Today* given over to the Super Bowl final. Copyright 1989, *USA Today*. Reprinted with permission

events 'leap' out of the page. An earthquake story would show how villages at the epicentre were affected and the artist would put himself (or herself) inside the event and show landslides tracking across the page. A big robbery would conjure up a map of the robbers' tracks through the streets with actual photographs combined with drawn work reconstructing the event. Or a drawing would reconstruct from an eyewitness's account thieves making their snatch, then killing and injuring. Often the headline would integrate with the drawing, while accentuated shadows would increase the drama for the reader and give the eye an impression of leaping from the flat plane of the newspaper's surface.

In news graphics, the computer can give a quick response but the result, however well manipulated, will always appear mechanical against the work of the professional graphic artist. This is why the most effective newspaper artwork can be found in such papers as the *Sunday Times* and *The Observer* (Figures 58 and 59). When a big story breaks midweek, by Saturday night there has been time to produce a researched inside page story based on a detailed and sophisticated graphics centrepiece.

This approach to news is gaining more followers, especially in areas where news pictures are impossible or not allowed. Courtroom scenes in noted trials are a first favourite, with even television leaning to this technique. The artist attends with the reporter and

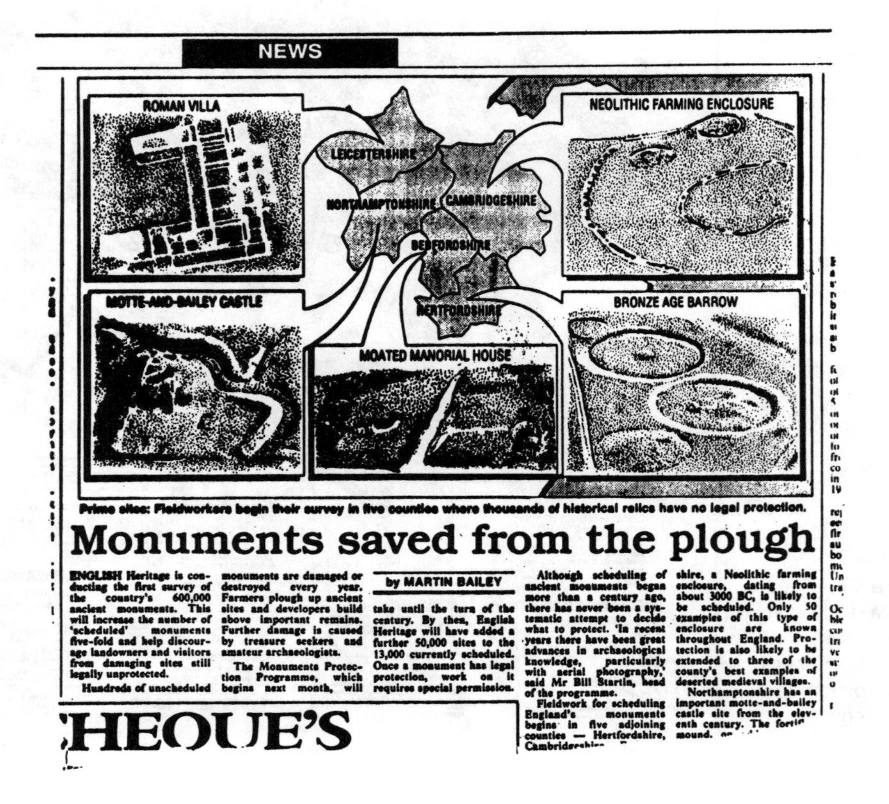

Prime sites: Fieldworkers begin their survey in five counties where thousands of historical relics have no legal protection.

Monuments saved from the plough

by MARTIN BAILEY

ENGLISH Heritage is conducting the first survey of the country's 600,000 ancient monuments. This will increase the number of 'scheduled' monuments five-fold and help discourage landowners and visitors from damaging sites still legally unprotected.

Hundreds of unscheduled monuments are damaged or destroyed every year. Farmers plough up ancient sites and developers build above important remains. Further damage is caused by treasure seekers and amateur archaeologists.

The Monuments Protection Programme, which begins next month, will take until the turn of the century. By then, English Heritage will have added a further 50,000 sites to the 13,000 currently scheduled. Once a monument has legal protection, work on it requires special permission.

Although scheduling of ancient monuments began more than a century ago, there has never been a systematic attempt to decide what to protect. 'In recent years there have been great advances in archaeological knowledge, particularly with aerial photography,' said Mr Bill Startin, head of the programme.

Fieldwork for scheduling England's monuments begins in five adjoining counties — Hertfordshire, Cambridgeshire ...

...shire, a Neolithic farming enclosure, dating from about 3000 BC, is likely to be scheduled. Only 50 examples of this type of enclosure are known throughout England. Protection is also likely to be extended to three of the county's best examples of deserted medieval villages.

Northamptonshire has an important motte-and-bailey castle site from the eleventh century. The forti... mound. ...

Figure 58 Archaeology in graphics: a novel application by *The Observer*. Reprinted with permission

notes the positions and facial characteristics of people and the surroundings of the dock and witness box so as to give authenticity. Care must be taken about the laws of contempt. Juries' features must not be shown. The artist should consult the reporter on legal protocol. The likenesses of well-known people can be researched from stock pictures. Artists who can produce this kind of work at speed can command a high fee and are invariably sought-after freelances.

Political controversy can be pictured in this way, with reconstructions of Cabinet room rows and dramatic debates beyond the presence of the camera. Again, surroundings and well-known people's features must be authentic, with even the kinds of drinking and water glasses on a table liable to be under scrutiny.

An effective example of a news graphic occurred in the *Daily Express* in the edition of July 7, 1978 (Figure 60) when the situation on board a wrecked night sleeper could only be truly rendered by a piece of graphical artwork. The headline INSTANT DEATH BEHIND CLOSED DOORS told the essence of a story which was that people could not break out of their blazing coach. While inside page pictures developed the detailed coverage, the artwork on page one posed the vital question: Were the carriage doors locked? The graphic reconstruction, while something of a comment as well as a question, had hit the point that became most relevant in the subsequent inquiry.

The movement into news graphics to give an extra dimension to illustration has been most pronounced in national newspapers

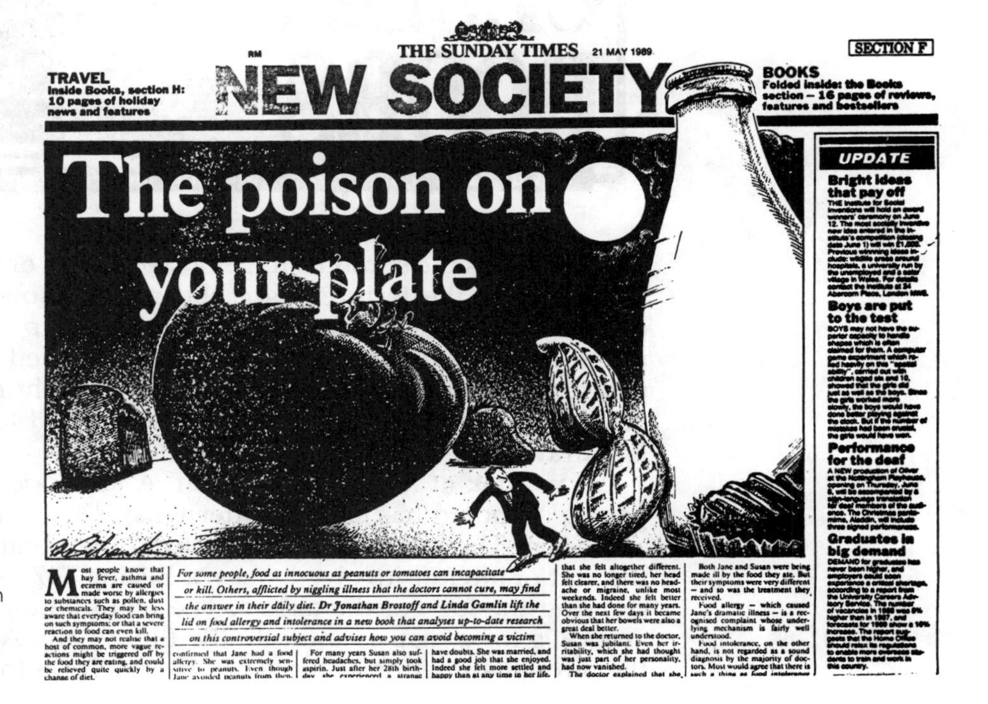

RM THE SUNDAY TIMES 21 MAY 1989 SECTION F

TRAVEL
Inside Books, section H: 10 pages of holiday news and features

NEW SOCIETY

BOOKS
Folded inside: the Books section – 16 pages of reviews, features and bestsellers

The poison on your plate

For some people, food as innocuous as peanuts or tomatoes can incapacitate or kill. Others, afflicted by niggling illness that the doctors cannot cure, may find the answer in their daily diet. ***Dr Jonathan Brostoff and Linda Gamlin*** *lift the lid on food allergy and intolerance in a new book that analyses up-to-date research on this controversial subject and advises how you can avoid becoming a victim*

Most people know that hay fever, asthma and eczema are caused or made worse by allergies to substances such as pollen, dust or chemicals. They may be less aware that everyday food can bring on such symptoms; or that a severe reaction to food can even kill.

And they may not realise that a host of common, more vague reactions might be triggered off by the food they are eating, and could be relieved quite quickly by a change of diet.

confirmed that Jane had a food allergy. She was extremely sensitive to peanuts. Even though Jane avoided peanuts from then,

For many years Susan also suffered headaches, but simply took aspirin. Just after her 28th birthday she experienced a strange

have doubts. She was married, and had a good job that she enjoyed. Indeed she felt more settled and happy than at any time in her life.

that she felt altogether different. She was no longer tired, her head felt clearer, and there was no headache or migraine, unlike most weekends. Indeed she felt better than she had done for many years. Over the next few days it became obvious that her bowels were also a great deal better.

When she returned to the doctor, Susan was jubilant. Even her irritability, which she had thought was just part of her personality, had now vanished.

The doctor explained that she

Both Jane and Susan were being made ill by the food they ate. But their symptoms were very different – and so was the treatment they received.

Food allergy – which caused Jane's dramatic illness – is a recognised complaint whose underlying mechanism is fairly well understood.

Food intolerance, on the other hand, is not regarded as a sound diagnosis by the majority of doctors. Most would agree that there is such a thing as food intolerance

UPDATE

Bright ideas that pay off

THE Institute for Social Inventions will hold an award winners' ceremony on June 12. The most socially inventive new idea entered in the Institute's competition (closing date June 1) will win [illegible]. Previous winning ideas include: wildlife areas around hospitals, a university run by the unemployed and a solar village in Wales. For details contact the Institute at 24 Abercorn Place, London NW8.

Boys are put to the test

BOYS may not have the superior capacity to handle shapes which is often claimed for them. A computer game experiment which relied heavily on this "spatial ability", carried out with children aged six and 10, showed that the girls did just as well as the boys. Since the girls worked more slowly, the boys would have done better playing against the clock. But if the number of mistakes had been crucial, the girls would have won.

Performance for the deaf

A NEW production of Oliver at the Nottingham Playhouse, opening on Thursday, June 8, will be accompanied by a sign-language translation for deaf members of the audience. The Christmas pantomime, Aladdin, will include three signed performances.

Graduates in big demand

DEMAND for graduates has never been higher, and employers could soon experience a critical shortage, according to a report from the University Careers Advisory Service. The number of vacancies in 1988 was [illegible] higher than in 1987, and forecasts for 1989 show a [illegible] increase. The report suggests that the Home Office should relax its regulations to enable more overseas students to train and work in this country.

Figure 59 Highlighting a food controversy: emotive use of graphics in *The Sunday Times* New Society section. Reprinted with permission

Talking back to your TV
SEE PAGE 10

DAILY EXPRESS
THE VOICE OF BRITAIN

No. 24,265 Friday July 7 1978 Weather: Sunny spells, cool 8p ★★★

Where 11 died on the Night Sleeper to London

INSTANT DEATH BEHIND CLOSED DOORS

A SECRET investigation today and a public inquiry later will have to answer the key question in yesterday's rail holocaust: Were the train doors locked?

Eleven people died—most of them asleep in bunks—when the overnight Penzance-Paddington sleeper caught fire near Taunton, Somerset.

They died instantly in a 600degC flash of flame that melted aluminium and glass in their 18-year-old coach.

It seems that the train had been travelling for miles pouring with smoke without the driver knowing until someone pulled the alarm cord.

DELAY

Christopher Helen, attendant in charge of the coaches that burned, told officials: "There was no fire until we stopped. Everywhere was filled with smoke. The fire burst out when we stopped."

By that time passengers were fighting for their lives to get out.

And civil servant Mr Eric Bevan, from Saltash, Cornwall, said: "I was in the second sleeper. There is no doubt the rear two exit doors were unopenable.

"I was awakened by someone shouting. Then another man

Page 3 Column 2

After the inferno: a charred briefcase

List of death: Pages 2 & 3. Dramatic pictures: Centre pages

Figure 60 News in graphics: *Daily Express* handling of a rail crash in a Vic Giles projection

where there is the stimulus to get a step ahead of a rival paper in big stories. How the terrorists got aboard . . . how the ship came to capsize . . . how the bank robbery was set up . . . how the police rescued the hostage. In stories like these a skilled artist can construct and reconstruct so that known facts are brought more vividly to the reader, often through montages in which half-tone combines with line drawing.

Let us look at a typical situation. A jumbo jet with hundreds of people on board has crashed after what is believed to have been an explosion. The picture editor gets from the library stock pictures of the sort of aircraft involved – say, a Boeing 747 in flight – preferably carrying the logo of the airline. The night editor alerts the art desk and calls for a map to indicate the flight path. A designer's plan of the aircraft's interior, it turns out, is available from file. This could help in rendering authentic detail although for the moment the cause of the disaster is conjecture.

Gradually, incoming copy indicates further details – where and how the wreckage fell, the position of the aircraft on the ground. Did it disintegrate in the air, and how?

By this time the artist has been given a new briefing. A decision has been made that the map must include an aerial drawing of the actual explosion, which is now known to have occurred. From crumbs of information in copy of the way the wreckage is lying on the ground, the artist forms a view of the likeliest scene. Should time be short, use will be made of an actual photograph printed to the size needed for the artwork and laser-processed to look like a drawing. The next step will be to literally tear the aircraft artwork apart at the points where it is assumed the explosion happened. The pieces are stuck down on the art board in their assumed order and position. The explosion is then drawn on to the board over the aircraft's flight path on a topographical impression of the town or country beneath. The disintegrated image will show pieces below or well away from the main structure in keeping with the information available.

While the artwork is put together, the subeditor is preparing the story from incoming copy sources, at the same time supplying bits of information to attach to the drawing. Meanwhile the overall dimensions of the job are given to the night editor so that allowance for the shape of the picture, when sized, can be made on the page rough.

Correct sizing and choice of size are the vital keys to the layout. It is better, in a case like this, that the shape of the artwork govern the page design rather than an arbitrary shape cramp the artist's ability to render the scene properly. Each happening demands different considerations of space. In some cases, allowance for sky is important to give correct proportion. Once proportions and effects of distance are lost the reader becomes confused. The design must also make correct allowance for captions and name tabs, which will need to be set separately, perhaps reversed as white on black (WOB) and imposed on the work. Drawn lines and arrows help the reader here. Tabbing and captioning artwork properly is also important to draw the reader's eye to the significance of the drawing. Applied tints can enhance part of it and give it 'lift' from the page through the subtlety of shadow.

Colour applications require even greater care because there will be separations to consider so that the colours can be printed in register. Preparing the separations will need the attention of the colour processing department. An early copy of the artwork should be available for checking by the back bench for accuracy of names and information, and for instructions on possible updating of detail or text from edition to edition, or even for changes of page layout in later editions. If the finished drawing splits easily into two pieces the night editor might consider running it across a spread. In this way

Philip Howard

Times Roman centurion

Today is the centenary of the birth of one of the oddest and most influential of the noble army of *Times* men. For 30 years he was the *éminence grise* of Printing House Square, and keeper of the conscience of *The Times*. A slight figure, in thick, steel-rimmed spectacles, dressed invariably in a black suit with a white shirt and black tie, and outdoors with a silly little black hat, he struck strangers as looking like a Jesuit. He referred to his paper with affectionate mock depreciation as "the sheet", or in the style of an old-fashioned citizen of London, "the House". They say you could hear the capital H.

Stanley Morison was a mess of contradictions: devout Roman Catholic and Marxist and conscientious objector; austere and a keen clubman; shy and a show-off; learned and silly; humble and vain; enjoying pantomime songs and Gregorian plain chant. But the oddest thing about this archetypal *Times* man was that he was not a journalist at all, but a typographer. He introduced the revolutionary notion that *The Times* should be good to look at and a pleasure to read, even if you disagreed with what it said.

The story starts on September 10, 1912, when Morison, a disgruntled young clerk at the London City Mission, read *The Times* printing supplement. This was his vision on the road to Damascus, and concentrated his mind for the rest of his life on the study of letters, written and printed.

He became Britain's greatest authority on letter design, and freelance typographical consultant to several publishers, including the Cambridge University Press. He came to *The Times* as the result of rude remarks he made in 1929 to a *Times* rep about the paper's drab and old-fashioned appearance.

In those days, if you wanted actually to read the paper rather than merely to be seen carrying it under your arm, you needed keen eyes and dogged determination to plough through the unbroken furrows of inspissated thin print, broken rarely by a plonking headline across one column. The rep reported Morison's remarks back to the manager, and the manager appointed the whizz-kid of print as typographical adviser to *The Times*, and asked him to devise a new typeface for the old paper.

In a characteristic memo, Morison wrote (in very small part) that a new typeface had to be "worthy of *The Times* – masculine, English, direct, simple, not more novel than it behoveth it to be novel, and absolutely free from faddishness and frivolity". Typography is a cannibalistic art, feeding off previous types, since only a finite number of faces to represent letters is available to human ingenuity.

Nicolas Barker, Morison's biographer, concluded that the original model for Times New Roman was the "Gros Cicero" type of the French punch-cutter Robert Granjon, dating from about 1568. Whatever its origin, Morison's new typeface became the most widely used typeface of the 20th century, still used *passim* in books and magazines.

The change of type at *The Times*, made over a weekend with no loss of production, involved bringing in 35 tons of new metal and many thousands of new matrices for the machines that set the type. But on Monday morning readers were amazed (and on the whole gratified, except for those who grumble at any change in anything) to see white space around the words, making the page more attractive, and even readable without a magnifying glass.

There were still only single-column headlines, of course; but the new type led the eye easily along the line, yet was strong enough to withstand the pressures of mass production printing. And, good grief, what was this? The Gothic title-piece at the top of the front page had been replaced by Times New Roman. This was the holy of holies, because traditionalists regarded it, together with the Royal Arms (to which, incidentally, the paper is not entitled) as immemorial hallmarks of *The Times*. Morison outflanked them with tradition, by demonstrating that the earliest issues of *The Times* had Roman, not Gothic, title-pieces. He was not allowed to bring the Royal Arms, still quartering the Arms of France and Hanover, up to date.

Morison's position at *The Times* developed into something much more than adviser on printing. He edited and largely wrote the four-volume *History of The Times*, and was for two years editor of *The Times Literary Supplement*. He became the unofficial adviser to two successive editors on organization, appointments of staff, and even for a time about editorial policy. Morison himself spoke (rather too much for his own good) of his "occult influence" on *Times* editorial policy.

After the war, scarred by its support of appeasement, *The Times* had lost its traditional function as noticeboard of the Establishment, without finding a new role. The paper's first (top-secret) management survey by a firm of chartered accountants recommended popularization to attract a larger and broader circulation, including women, the young, and other outsiders. The traditionalist Morison criticism of such a policy was pungently expressed in an outraged anonymous memorandum of 39 pages, labelled "To the Chief Proprietors only" (and written in Morison's fine italic).

It was trumpeting stuff: "Obviously Great Britain cannot function without a strong, educated, efficient, informed governing class. *The Times* is the organ of that class. It remains, and for all we can see to the contrary under a non-capitalist economy, must remain absolutely necessary to that class . . ."

When William Haley became editor in 1952, with instructions to modernize the paper for the new world, he at once removed Morison (without his feet touching the ground) from his position as unofficial guardian of *The Times* tradition. Haley later said: "Sad as it was that Morison so much resented being shut out from anything to do with the editorial side of *The Times* during the last 15 years of his life, the fact is that he should never have been allowed in."

Much turbulent water has flowed under the old bridge since Stanley Morison was keeper of the sacred flame at the House at Printing House Square. But his influence is still potent. You do not have to agree with or even like everything that you read in *The Times*. It would be a very peculiar paper if you did. But the revolutionary idea that it should be a pleasure on the eye and easy to read you owe directly to Morison.

He was the most influential typographer this century, and an endearing eccentric. Lunching a guest at his beloved Garrick Club one day, when the soup (Brown Windsor) arrived, he exclaimed: "To lunch at the Garrick is an act of Christian charity at the best of times, but this is going too far," and swept his guest off to the Savoy Grill.

The great man of print, the true *Times* man, was born on May 6, 1889 – appropriately the feast of St John *ante portam Latinam*, the patron saint of printers.

Commentary • DAVID HART

MAY 6 **ON THIS DAY** 1955

OBSERVER

PROFILE

YITZHAK SHAMIR

Israel's stone-faced stonewaller

LIKE Margaret Thatcher, Yitzhak Shamir, the Israeli Prime Minister who arrives in London tonight on an official visit, is a conviction politician. He is as likely to lecture as be lectured, to hector as be hectored.

He will listen to what his host has to say, but he will insist on delivering a message of his own. Yes, of course, he wants peace with the Palestinians ('the Arabs of the Land of Israel', as he prefers to call them, at least when he's speaking Hebrew), but on Israel's terms, not Yasser Arafat's, not George Bush's, and certainly not Margaret Thatcher's. What he is seeking is diplomatic support for his Government's peace initiative, not gratuitous advice.

If it's up to Shamir, once described in these columns as 'a small, hunched, almost frog-like man who exudes an air of deep suspicion', the [illegible] secret service and politics says: 'He has nerves of ice. He never makes snap decisions. He thinks that time is on his side. His adversaries usually make the first mistake while he waits.'

Two years ago, he killed an opening to Jordan, engineered by his own Labour Foreign Minister, Shimon Peres, by a mixture of attrition and arithmetic. He knew Peres could not command a coalition majority. King Hussein was ready to negotiate. The Americans were pushing. But Shamir sat tight, and prevailed.

The full emotion of his commitment to Eretz Yisrael, the ancestral Land of Israel, emerges most frankly when he is addressing his own, the Likud faithful. 'A foreign State will not arise here,' he promised a party rally in February. 'A Palestinian State will not arise here [illegible]

Behind barbed-wire eyebrows, he calculates, sticks to essentials and bides his time.

[illegible] estine mandate must take precedence, even trying at one stage to make common cause with Germany and the Italian Fascists. After Stern [illegible]

[illegible] known is that their son Yair, an air force colonel, does not vote Likud).

Shamir's last escape was from [illegible]

[illegible] mind that it doesn't pay to undetestimate.'

He is fallible, none the less. For months before the [illegible]

Figure 61 Portraiture in line: Stanley Morison in *The Times*, and Israeli politician Yitzhak Shamir in *The Observer*. Reprinted with permission

the shape and character of the artwork can influence the final page design.

As with chart graphics it is possible to achieve the laborious and less creative part of the work on screen where computer graphics are in use and the outputs transferred to the artist's drawing. Where an advanced graphics facility exists, and where this is the normal entry into the system, it is possible to create the entire job on screen. If the computer's instant art library is comprehensive a reasonable likeness of the aircraft might be featured, explosive symbols utilized, and the topography accurately traced by keyboard or mouse. This could cut down time in the hands of a skilled operator but it is likely that the finished product would appear mechanical compared with the refinement and actuality of a professional graphic artist's work.

Caricature and portraiture

There is a new trend in newspapers towards using graphics for portraiture, some of it stylized into caricature. American papers frequently rely on the artist's pen to devise portraits of their celebrities as fun comments that are also works of art. They are more than cartoons since the work is carefully and artistically executed and generally needs no caption. The distinctive work of Gerald Scarfe is an international example.

In Britain, *The Times*, *The Observer* and the *Financial Times* in recent years have brought cartoon portraiture to an art form in its own right. Peter Brookes's drawing of the great typographer Stanley Morison, commissioned by *The Times* to mark the one hundredth anniversary of his birth (Figure 61), is a notable example of this. Brookes is a superb exponent of the cross-hatch style reminiscent of the copper and steel engraving techniques used in late nineteenth century illustrated newspapers. His method, however, is highly sophisticated and is linked closely to, and extends, the story that goes with the picture, in this case a piece by Philip Howard about Morison's achievements as a designer and typographer. Brookes embellishes the subject by pointing up Morison's splendid Times Roman type and weaving it into the drawing as a pair of spectacles resting on Morison's nose – a lesson to all journalist/designers and cartoonists in the use of caricature and visual graphics.

A similar technique is used by Trog (Wally Fawkes) to give enormous vigour to his drawn portrait of the Israeli politician Yitzhak Shamir in *The Observer* (Figure 61).

8
Essential page planning

Deciding what goes into the paper on a weekly or daily basis is the crux of the editing process and the essential preliminary to designing the pages. The readership market having been defined and accepted, it is thereafter the editor's role to provide the right staff and the content to fill the paper. To this end the news and features coverage are departmentalized (Figure 62) and the work planned so that the input of text and pictures is available in the right quantity and at the right time. Less easy to guarantee is the quality. There are good days and bad days for news, even for specialist papers, and there are great features and not so great features, and features that have to be found space because they fill a guaranteed regular slot. Come what may, out of it all a newspaper has to be produced; it is not the job of design to reflect some absolute standard by which the content is judged but rather to do the best possible job with the material that is available on the day.

The organizing and processing of this material ensures that a flow of copy is constantly entering the production system where the text is being edited and the pages put together. Despite the ongoing nature of all this (some news happenings, for instance, being unpredictable) decisions have to be taken sufficiently early about the overall balance of content and the placing and handling of the main stories and pictures, available or expected, to enable the work of editing to be pegged out to some sort of time scale. In fact, information from the news and features departments and picture desk schedules (though expected stories can fail and better ones crop up) enables an outline plan of the paper to be devised by the late morning conference on a daily paper and by the Thursday conference on a Sunday paper. The actual decision making at this stage is the responsibility of the editor, influenced to a greater or lesser degree by conference discussion. As production unrolls, the changes and modifications to pages necessary in the light of the day's events

BACK Bench
BACK BENCH
A Night Editor who controls the day by day editorial sits with his deputy & assistants on the traditional Back Bench where he is responsible only to The Editor of the newspaper.
ADVERTISING
Every Issue will have a dummy to be started with blank pages in which advertising spaces are drawn. Better if actual proofs are stuck down to enable page comparison.
NEWS EDITOR
He will maintain news contact with the Night Editor at all times.
ART EDITOR
He consults the Night Editor on each page in the paper to provide rough layout ideas for the Night Editor's approval.
ART & DESIGN DESK
The Art Editor runs the group of layout artists and graphics people. He will ask someone to reproduce his rough in a finished form and brief a graphics or illustrative person, should it be necessary.
PICTURE EDITOR
He will keep the Night Editor supplied with the day's picture flow.
CHIEF-SUB EDITOR
Page designs are passed to the Sub-Editorial and the Chief-Sub hands out indivual stories to be fitted into the spaces allotted to them.
SCREEN and/or PASTE-UP ROOM
Layouts or machine commands will be passed from Art Desk on a strict time schedule.
FEATURES EDITOR
The Night Editor must be given constant confidence by the Features Editor that all his pages are in order and up to date. He must also provide blurb material.
PLATE MAKING and PRESS ROOM
These departments depend on the smooth timing of all the other elements, relying, as well, on the marrying of pages at the front end of a tabloid with those beyond the centre pages.

Figure 62 Planning for the paper: the flow of ingredients into the editorial production system and out to printing

are usually left to the executive in charge of them, although a major shift in events can lead to the editor calling a further meeting of executives and reshaping the outline of the paper.

Contents planning

The edition dummy supplied by the advertising department discloses the volume of advertising, the number of pages to be filled and the advertising spaces that have been sold. Newspapers now mostly display all the pages and their advertising on one sheet, called a *flat plan* (Figure 63). While this gives useful at-a-glance information and is easy for ticking off pages completed, it is no substitute for an actual size dummy in which layouts and proofs of advertising are pasted to help establish visual balance as the making up of pages proceeds. Better still, under electronic pagination both flat plan and dummy can be called up on-screen by anyone involved in page production.

The general balance of editorial contents – the amount of space allocated to news, sport and features – is fixed in most newspapers on a broad percentage basis to suit their market irrespective of pagination so that increases or decreases in size day-by-day do not fall unfairly on any one section. A departure from this would be if a heavy sports coverage such as the Cup Final, or a General Election or similar big news story demanded a more than average amount of space.

News pages

Balance of content within the news pages depends on the type of paper. While most newspapers are general newspapers – that is, as opposed to specialist newspapers – they carefully balance out their news space to suit their readership. For instance the national quality dailies such as *The Times, The Guardian, The Independent* and *The Daily Telegraph*, give pages or groups of pages to home news and foreign news and, in some cases, to political and industrial news, giving them a label or tab at the top. Usually the subediting of these sections is carried out quite separately. Financial news has its own pages, as does sport. Town evening newspapers sometimes separate national and local news, while many evenings and local weeklies have special pages for area news.

Not all newspapers have this precise compartmenting. The popular tabloids and more popular based evenings use broad readership appeal as their yardstick for allocating news and pictures to pages, though a balance will be struck on a page between human interest stories and more serious ones so that a page's news 'feel' is not too weighted, and it is not either too serious or too frivolous. Equally, running two stories on a similar topic will be avoided, unless there is merit in tying them in together as a news peg.

In quality dailies and Sundays, where there is less use of what the populars call human interest stories, a balance is struck between 'spot' news and interpretation. Here it is possible for the fine line between news and features to become blurred as byline writers lace their reportage with comment and background.

The planning will have to take account of stories that turn from the page to continue on other pages. Turned stories can be a nuisance to readers and it is best to have a regular system. *The Times*,

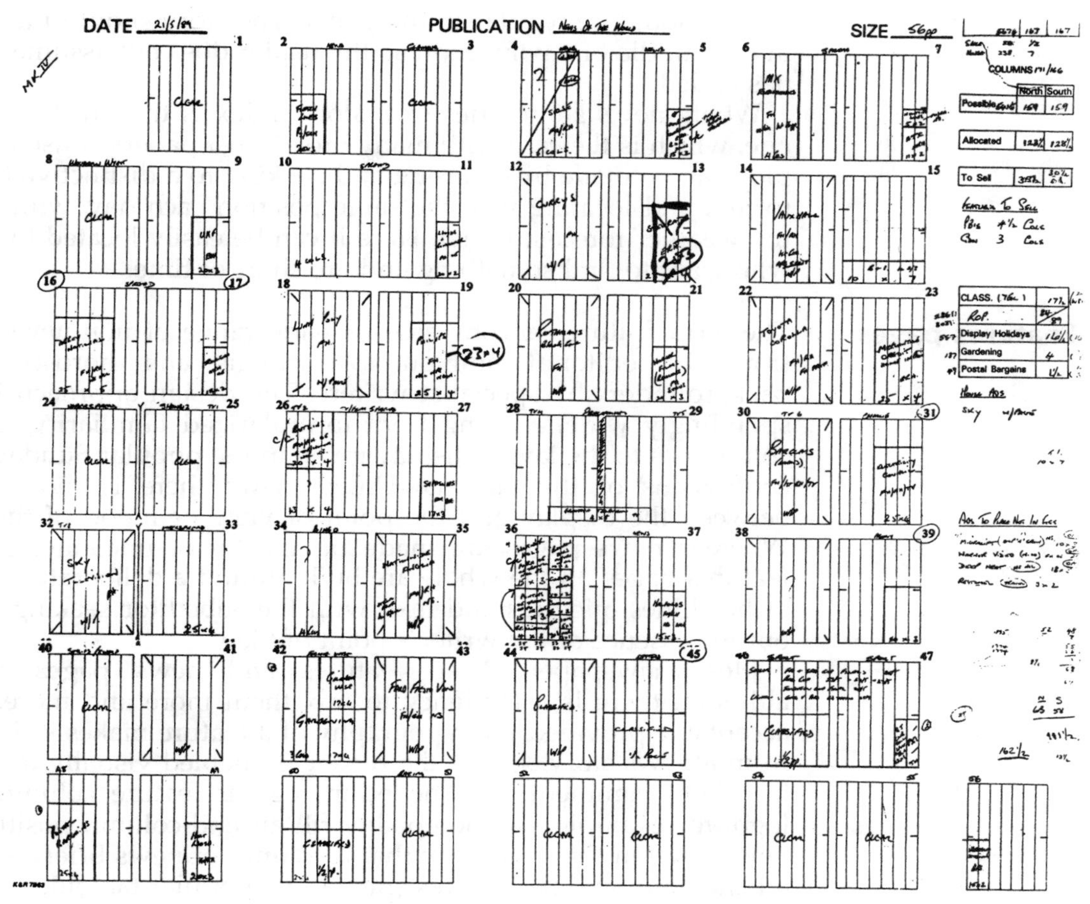

Figure 63 Flat plans, as shown here, are used by advertising departments to indicate pagination and position of adverts. From these, editorial dummies are devised for detailed planning of the edition

for instance, carries all turns of page one stories on page two. Other papers avoid turns as a matter of policy, although it is hard on a tabloid-sized page one not to use them. A way out is to carry a taster of a story on page one with a cross-reference to the full story inside. A design feature can be made of a cross-reference so that, perhaps tied to a small picture, it becomes a blurb of what can be found inside the paper. Such treatment can contribute vividly to the page.

Turn heads should be in a type that connects with the stories' main headline for easy recognition, and should be formatted for instant retrieval on screen to use as needed. It is unwise, for the same reasons, to continue a story on another page in a different body type or setting width. Non-standard setting would make it hard to accommodate the turn, particularly at short notice on screen. The lines TURN TO PAGE . . . or CONTINUED FROM PAGE . . . should be in a distinctive type from the body face for easy recognition. Such usage should be regular in the paper. The story should ideally turn in the middle of a paragraph so the reader does not assume the item has ended.

Where news summaries or a story index is the practice on page one, which is the case with many broadsheet papers, a useful design ploy can be made by choosing and sticking to a distinctive type and format and locating it in the same position each day, with perhaps the weather forecast under. Here it can be easily located by readers who want to find something and are short of time.

Features pages

The term features encompasses a wide range of non-news content from the editorial opinion to cartoons and horoscopes and backgrounders by specialist writers. The design approach is correspondingly wide, ranging from formularized familiarity for name columnists to the lavish display given in the popular Sundays to the confessions of the famous. Unlike news there is more identity between the display and the point of view or mood of the writer. Whereas news pages are made up mostly of unconnected items, features pages on the whole are built around a deliberate projection of headlines, pictures and text, with the subeditor seeking to bring out the essence of the writer's point of view.

Most newspapers keep features and news pages separate, although there is a tendency to mix them more and more. This is acceptable provided the typography of a feature makes it distinctive from a news story so that the reader is enabled visually to separate fact from comment. A panelled-in regular feature column with a logo, for example, can acceptably fill an end-column position on a news page in the same way that a column of news briefs is used in an end column on a features spread in a popular tabloid.

One thing that keeps features separate from news is the longer time span that goes into their planning and the fact that as ordered material, often of a specific length, features are usually ready for page production earlier in the day. Routine ones such as listings and home and beauty columns, often with stock logos, are among the earliest copy into a newspaper office. It is thus that features pages are planned first into the production cycle.

The main difference at the layout stage is that features pages, even in newspapers with a fairly tight design formula, are distinguished by a freer use of pictures and typography, and by more varied layout devices. The broad type format remains, but one or two extra types are introduced for display purposes, and pictures are often chosen and cropped for mood rather than for actuality, and the generally longer texts are lightened perhaps by drop letters, highlight quotes or more elaborate crossheads.

Scheduling the work

Once the editor has decided on the overall plan for the edition and laid down any policy involved, the detailed production work falls to the various executives in charge of pages. On the average town evening or weekly paper this usually means the chief subeditor in the case of the news pages, the features editor in the case of features pages, and the sports editor for the sports pages. With national papers, both morning and Sunday, and the bigger regional ones, there is more likely to be a night editor in charge of overall editorial production, and an involvement in origination of pages by, perhaps, a foreign editor, city editor, women's page editor, or even by assistant editors with overlord responsibility for sections of the paper.

Where there is a variety of hands concerned in originating pages, the art desk system performs a useful function in producing a uniform and acceptable standard of design for the finished product, and the art editor (sometimes the art director) of a big national paper can have a powerful voice in the discussion that goes into the planning of the paper.

The broad blueprint having been set, the drawing, editing and making up of pages is then arranged so that those for which text and pictures are likely to be available first are attended to first and the rest slotted in as the ingredients come to hand. This, as we have seen, usually means that features pages, other than the one containing the editorial opinion, are the earliest, including such areas as the TV and radio programmes, women's pages and serialized material. By the same coin, page one and usually one other news page, plus the main sports page, go to press last so as to include the latest and most up-to-date information.

From this work pattern a production schedule is devised giving nominated press times for each page, ending in the final press time for the edition. This scheduling is not only for editorial purposes but to give an orderly plan to pre-press work such as page pairing (in the case of tabloids) and platemaking. An essential from the production point of view is to stage this work so as to avoid delays to press times than can arise from 'bunching', when too many completed pages are awaiting platemaking or transmission to printing sites.

Drawing the pages

A page begins life as a 'rough' prepared by the executive concerned. It might consist of a 'lead' headline with an approximate note of story length and a main picture which at that moment is still a briefing in the cameraman's head, perhaps one other story that would make a half lead, with a couple of stories still to be found, and perhaps room for another picture. To the outsider it would look like an indecipherable squiggle.

With news pages, particularly papers using electronic page make-up systems, it is possible to alter or switch stories up to just before press time in the light of developments that can outdate decisions taken earlier. A much vaunted exposé story turns out to be worth only half the space; a 'dead cert' vertical picture turns out, on being printed, to be usable only in a horizontal shape; a promised 'pick up' picture, it seems, is not available. There might (although less likely) be a late change of advert shape. There might even be a change of mind by the editor or the executive concerned about the weight to be given to a certain story, or about using it at all. If the page layout has been finished and distributed either on paper or on screen, a revise has to be prepared, or even a substitute layout drawn. If the layout person is still preparing the page, then part of it is erased and started again.

With a running news story of, say, a disaster or an election, a good deal of pressure falls on the art desk to amend the layout right up to the last possible minute to accommodate extra text or late pictures, although a truly flexible news page, by the nature of its design, should allow for these contingencies. For these reasons it is a good idea to build in flexibility as well as balance so that last-minute changes can be made without delaying the edition.

With features pages such dramas are less common, though disagreements and second thoughts on projection can still result in the page being drawn several times. In place of the hassle of speed and revision endemic in news pages the layout artist has to meet the need for sophisticated and detailed presentation of longer texts, and often more complex picture editing.

Making up on screen

Making up pages on screen (Figure 64) has a drama of its own compared to the easier-paced world of cut-and-paste.

The main stories have been chosen, some of the headlines written, the pictures selected from screen-view contacts and given their page code. The page designer sets up the page geometry on the Apple-Mac (Figure 65) from a fairly detailed rough beside the monitor screen. The building work has started....

Each text box accepts a flow of machine-generated dummy words in the correct size for intro and body text. The already written headlines fit into their spaces. Where a headline is still awaited, the page will not be held up; a dummy heading of correct font and size is injected temporarily into the space. The picture box is selected and the coded picture arrives. There is perhaps another text box underneath or alongside to take a caption.

The page appears on the chief subeditor's screen with the dimensions and information necessary for subeditors to be briefed to edit the stories and write the captions and remaining headlines. Every story can be given an accurate fit since the number of words and characters needed appear on the sub's screen at the touch of a key.

The dummy headlines and stories are replaced by the correct wording to fit the character count and typeface displayed. When the chief subeditor is satisfied with the words as well as the fit, a keystroke sends the story flowing into the page to replace the dummy text in perfect fit.

When the page is complete with text, pictures and adverts (which by now have also been called up by the page designer) it can be viewed at 75% up for style and design and up by 200% for words and stories to be checked. After being passed by the editor and legal

Figure 64 Electronic make-up: these pictures taken at the *News of the World* art desk show the pages of an edition taking shape on Apple-Mac screens using Quark XPress to input the ingredients. Used with permission

Palettes

Palettes function like windows in that you can close a palette by clicking on its close box and move it by dragging its title bar. Palettes are always displayed in front of all the document windows. There are seven kinds of QuarkXPress palettes: the Tool palette, the Measurements palette, the Document Layout palette, the Colors palette, the Style Sheets palette, the Trap Information palette, and library palettes.

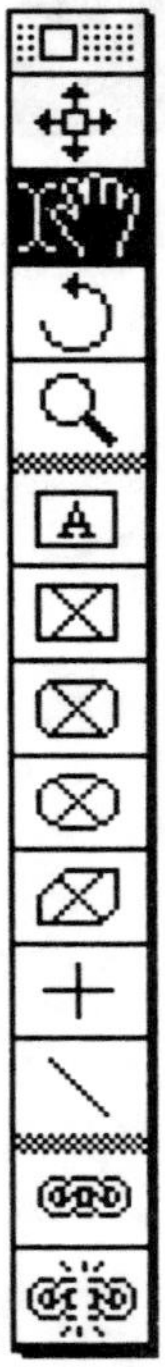

Tool Palette

Select a tool by clicking on it in the Tool palette. The tool you select determines what you can do with the mouse and the keyboard, and which menus and menu entries are available.

Double-clicking on a tool displays the Tool Preferences dialog box, which enables you to specify default settings for the Zoom tool and the item creation tools, +, and ╲.

Keyboard commands:

⌘-Tab . . . . Display Tool palette, or select next tool

⌘-Shift-Tab . . . . Display Tool palette, or select previous tool

Item tool
Enables you to move, group, ungroup, cut, copy, and paste items (text boxes, picture boxes, lines, and groups).

Content tool
Enables you to import, edit, cut, copy, paste, and modify box contents (text and pictures).

Rotation tool
Enables you to rotate items manually.

Zoom tool
Enables you to reduce or enlarge the view in your document window.

Text Box tool
Enables you to create text boxes.

Rectangle Picture Box tool
Enables you to create rectangular picture boxes.

Rounded-corner Rectangle Picture Box tool
Enables you to create rectangular picture boxes with rounded corners.

Oval Picture Box tool
Enables you to create oval and circular picture boxes.

Polygon Picture Box tool
Enables you to create polygon picture boxes. (A polygon is any shape with three or more sides.)

Orthogonal Line tool
Enables you to create horizontal and vertical lines.

Line tool
Enables you to create lines of any angle.

Linking tool
Enables you to create text chains to flow text from text box to text box.

Unlinking tool
Enables you to break links between text boxes.

Figure 65 What does what in electronic page make-up: an explanation of the Quark XPress tool palette. Copyright Quark Express 1996

reader it is then transferred electronically to the production department to be paginated into its place in the paper and paired with its correct partner in the case of tabloid production. A miniature of it will appear along with the other completed pages on the electronic dummy showing how it will look in the paper.

Column formats

The discussion over what is the ideal number of columns from the reader's point of view is now largely a thing of the past. Tabloid papers over the years have been four, five, six and seven columns wide, and broadsheet papers eight, nine and ten columns. Today there are few broadsheet papers that do not print in eight columns, and fewer still tabloid ones that do not print in seven columns, the flexibility required by modern design techniques being best served by these sizes. A factor towards uniformity has been the need for advertisers to have the minimum of variety in advert widths and depths to which to work so that the same, or similar, adverts of the same size can be used in campaigns spread across different newspapers.

Newspapers have always been printed in smaller type than books in order to get the maximum text into the space available, and they adopted a columnar format at an early stage, as we have seen, as an aid to the scanning eye. The 7 point or 6 point across 11½ to 14 picas favoured by the well-filled Victorian papers would have been hard to read across a book's width.

While there is no dispute that narrower columns are easier to read, the limit becomes reached if the column is so narrow – 5½ picas for example – that the words begin to break badly in setting or have to be spread out to justify lines. On a seven-column tabloid paper, panel setting can reduce the lines to 7 picas, but setting of less than this is usually used only in bastard measure caption spaces alongside pictures where problems of bad word breaks can be solved by setting the type ragged right (Figure 78).

With advert widths as a built-in factor on pages there is a strong case in design for sticking to the paper's standard column format, although there is no reason why this should not be varied to permit special effects. If, for example, adverts can be contained in a block across the bottom half of the page, a seven-column tabloid paper could set a special feature across five wider legs in the top half. Where a features series is being carried across a two-page spread it is sometimes the practice to ask for the adverts to be arranged so that the designer has the option of varying the column format, while for design purposes any page setting can be rearranged across legs of bastard measure to fill a given width. Some papers – the *Daily Mail* is an example – vary the column format from page to page to differentiate between news and features pages, or when allowed by advert shapes (Figure 66), but this can involve a complicated juggling of advert positions to avoid setting anomalies.

In the debate that raged some years ago on column formats in design, various aesthetic and practical reasons were adduced as to why a particular number of columns was the right one. On grounds of readability of the text there is no evidence that the eye, scanning and dropping line by line through a story, encounters any greater or less difficulty in an 8½-pica column than it does in an 11-pica column. The only difference is that the narrower measure is read faster because it contains fewer words. Of greater note in the choice of column format,

Page 12 • • Daily Mail, Monday, October 9, 1995

DAILY MAIL / WORLD WIDE

Franc assault could kill single currency

From PETER SHARD in Paris

FRANCE is braced today for a week-long attack on the franc which could savage plans for a single European currency.

Speculators, including international financier George Soros, whose actions helped force the pound out of the European Exchange Rate Mechanism on Black Wednesday three years ago, are said to be preparing to mount the assault.

Beleaguered Premier Alain Juppe faces a huge budget deficit, sleaze allegations and challenges to his authority. But at the weekend he blamed 'the speculators and the gnomes of London' for a run on the franc.

It is seen as highly vulnerable against the deutschmark. And if the Germans do not bail it out and it slips further against the mark, it too could be forced out of the ERM, making the single currency a near-impossibility in the current climate. With France's budget deficit totalling £40billion and unemployment running at 11.4 per cent, conditions for Soros and others to make their move on the franc are ideal, just as they were when sterling sank.

Soros and Co ready for a rerun of Black Wednesday

Tomorrow's one-day pay strike by five million French public sector workers, which could see the country grind to a standstill, will further undermine M Juppe.

There is growing pressure on him to resign as he faces a corruption investigation. He is alleged to have cut his son's rent on a council-owned flat in Paris, as well as offering himself a luxury apartment at a knock-down rate.

Rumours of his impending resignation sent the franc plummeting four centimes on Friday to a new low of 3.51 against the mark.

The strike is seen as a barometer of how his government is faring and, if it is as complete as expected, then the markets will react to what is tantamount to a no-confidence vote.

President Jacques Chirac may have to ditch M Juppe, who shares his federalist monetary approach, possibly in favour of National Assembly president Philippe Seguin, a leading opponent of the Maastricht Treaty.

Bank of France director-general Jean-Claude Trichet knows that if the franc falls further, he will have to raise interest rates, something which M Juppe opposes as it would virtually kill his strong-franc policy.

A Deutsche Bank Research report has concluded that France is crucial to the success or failure of any single currency.

If the franc slips, analysts are predicting that the pound could be promoted — the mark-sterling rate becoming the main European exchange indicator instead of the current franc-mark arrangement. However, British visitors will feel no difference, as the pound is linked with the dollar not the mark and will remain close to its all-time low at 7.86 francs.

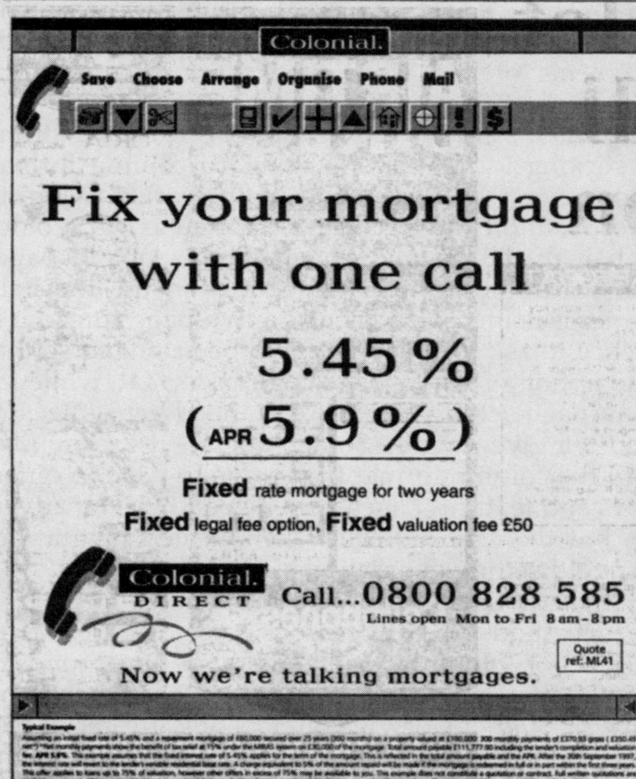

CRUSADER POISED

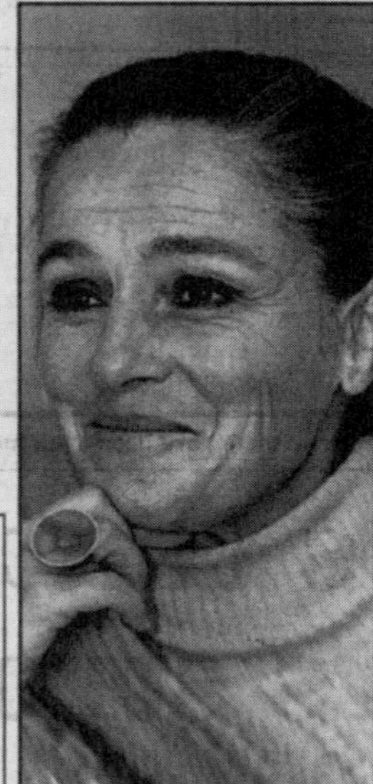

Ritt Bjerregaard: Determined

Ice Maiden may win a freeze on N-tests

A CRUSADER known as the Ice Maiden may force a halt to nuclear tests in the South Pacific.

After a tireless campaign by 54-year-old Ritt Bjerregaard, the European Commission will demand on Wednesday that France proves the explosions have not harmed the environment. The French will be told to hand over all the information on radioactivity levels and geological damage to Mururoa and Fangataufa atolls. A refusal could mean an action at the European Court of Justice, which could forbid further testing.

Miss Bjerregaard, Denmark's European Commissioner and a former teacher and environmental campaigner, was frustrated by the EU's repeated rejection of her calls for action, until she unearthed Article 34 of the Euratom Treaty. This obliges the commission to assess the health and safety implications of nuclear trials before they go ahead. France, as a signatory, had on the face of it acted illegally by not providing relevant information. Miss Bjerregaard sent experts to make an on-the-spot inspection, but the French refused to co-operate.

President Chirac, who insists on seven or eight explosions, is becoming increasingly isolated internationally. In Germany, his closest ally, Foreign Minister Klaus Kinkel, said at the weekend: 'French nuclear tests have made the German people furious.'

Six die as bomb hits refugees

TUZLA: Bosnian Prime Minister Haris Silajdzic threatened to call off peace talks with separatist Serb rebels yesterday after a cluster bomb fired at a Moslem refugee camp killed two women and four children.

At least 30 people, including children, were injured in the outrage at Zivinice, said the UN. It is thought to be the worst incident involving civilians since 37 people died in a mortar bomb attack in Sarajevo in August. That brought Nato air strikes against Serb forces.

Nato said its warplanes were prepared to respond to the latest attack but had not yet been asked.

Clintons toast Ted and Mary

MARTHA'S VINEYARD: Actor Ted Danson, best known as barman Sam Malone in TV's Cheers, married actress Mary Steenburgen at his million-dollar farmhouse at the weekend. The wedding was watched by President Clinton and his wife Hillary — friends of Miss Steenburgen — and actor Tom Hanks.

Tremors shake quake survivors

INDONESIA: Fresh tremors terrified survivors of Saturday's earthquake which killed at least 100 and seriously injured 700 on Sumatra. Five aftershocks were recorded as 800 troops and scores of medics moved into devastated towns.

Cash battle for drug-case driver

MALAGA: Lorry driver Roy Clarke, 47, from Ewell, Surrey, who was freed at the weekend after spending 20 months in a Spanish jail falsely charged with attempting to smuggle £35million of drugs will launch a fight for compensation today. Judges took 30 minutes after a three-day trial to find him not guilty of trying to take 2.5tons of cannabis into Britain.

Hostages on the move

SRINAGAR: Kashmiri guerillas holding four Westerners, including Britons Keith Mangan and Paul Wells, have begun moving camp in the Himalayas. Security forces believe the hostages are safe and ruled out pursuing the Al-Faran group who have held the men for four months.

Jumbo jet's wind of change

JOHANNESBURG: A flight bound for South Africa had to turn back after alarms were set off at 35,000ft. Three hours into the overnight journey, the jumbo jet returned to Heathrow, where the problem was found to be methane gas from 72 flatulent pigs in the hold.

Figure 66 Some papers dispense with a rigid column format when possible. Here, a conveniently-shaped advert allows the *Daily Mail* to set type variously in 11 picas 3 points, 14 picas, and 10 picas 6 points. Reproduced with permission

as with the choice between broadsheet and tabloid size, is its applicability to the paper's readership market and design style. The practice that has grown up in recent decades is that papers that run their stories at greater length and with less display, such as the quality dailies and Sundays and some regional morning papers and county weeklies, use the wider columns in broadsheet size because more words can be got into a given measure. Papers in more popular markets and running shorter items for readers with less reading time, opt for the narrower columns of the seven-column tabloid format. In each case the chosen size and column width suits the display techniques adopted for the length of items and type of readership market – a reminder that content and market shape design techniques.

Column rules

Tabloid papers use fine column rules to separate stories more than do broadsheets. While white space, as style, can look effective and sufficient, fine rules dividing the stories do give the impression on the smaller size that the pages are exceptionally full of things. Strictly, there should be no rule through the centre of a two-leg story. While a cut-off rule marks the end of a story, it is usual where several stories descend a column to separate each with a half centred rule. Where, on features pages, a thicker 1 point or 1½ point separation rule is used as style more white should be left above and below.

The layout

In creating the page pattern, or layout, the designer is imposing a style of presentation on the materials of which the page is to be composed – the text, headlines, pictures and supporting page furniture. Only in the preparing of a dummy newspaper for a new product or a relaunch will the designer have a relatively free hand to try different design approaches and type formats. In the daily task of designing pages for a newspaper running perhaps to scores of pages the parameters are dictated by the paper's accepted style as well as by the number of columns. Thus the designer will know the sorts and sizes of types that can be used and the approximate amount of space on a page that can be given over to display, and even certain things that are permitted or not permitted. Some papers might forbid cut-out pictures; others might allow special types to be introduced for certain feature stories; some styles forbid the use of crossheads or drop letters or keep them only for features pages, while a common practice is to have all headlines set lowercase or to banish the use of italic type. These are style conventions which help give continuity to the paper's appearance.

Yet whether the designer is producing a page within an agreed design format or setting up an entirely new format, and irrespective of any design style or conventions, the page must obey certain guidelines if it is to be successful. It must, for instance, have the following:

Balance

The idea of a page with equal proportions, with left balancing right and top balancing bottom nurtured by some design faddists over the years is a dead letter in modern newspaper design. The arbitrariness of advertising placement always militated against it, but even with a page clear of adverts, or having a block of them across the page, to compel the contents into a symmetrical pattern of 'perfect' balance would lead to distortion of text and headline values and would impose restriction on content selection in addition to those already existing. Instead of the pattern and distribution of highlights being decided by the contents, the contents would be forced into artificial relationship with each other. The result would be a dull page of monotonous appearance.

Thus when we say balance we mean the distribution of headline and picture highlights – of display ingredients – that takes best visual advantage of the page, whatever its shape, and in a way that responds to editorial values. To this should be added that the editorial contents must take account (that is, not clash in shape or mass) with the advertising on the page, and must also create a balanced effect against the opposite page even, and especially, if it is a full-page advertisement (Figure 67). In fact, no page should be consigned to platemaking until it has been checked in good time against its partner. Here a full-size dummy of pasted-up pages helps. Even more useful is the 'electronic dummy' of made-up pages that can be displayed on the computer screen in miniature. It can save a clash of types or half-tone (and even headline words) jumping out at you when you thumb through the first copies off the press.

Variety

Too much variety of type or effect on a page is bad for the eye, which is drawn hither and thither, not knowing where to look first. Yet too little (of type choice and size, for instance) can lead to dullness so that the design fails in its purpose of attracting and holding the eye. Variety is thus a necessary but subtly used weapon – a Century Bold page varied by the odd Century Light headline, or sans kicker; a deep single-column picture livening up a news page that might otherwise be trying to say too much; a long column pierced by a carefully placed cut-off; the bottom of a page saved from fragmenting into bits by a 30 point caps headline across columns that catches the eye just when it thought it had run out of page.

All these are devices that can be used to flesh out the page once the initial balance of materials has been achieved.

Emphasis

The visual highlights of the design should reflect the relative importance of the items. Thus the bigger headlines, pictures and panelled-in

Currys SALE

INDESIT 7.9 cu.ft. Fridge Freezer

SAVE £40

£199.99

FRIDGES FROM UNDER £90

SNOWCAP 5.3 cu.ft. Fridge

AMAZING VALUE £89.99

INTEREST FREE CREDIT ON HOTPOINT

SAVE £30 ON FREEZER

CURRYS – YOUR BEST CHOICE – GUARANTEED

Currys

Theatre

It's sour Grapes from the U.S. stage

Two little faces of courage

Boy who can't eat

Transplant triumph

Lover's tragic stunt

Student falls to death after climbing parapet

As they were: Townshend, Entwhistle, Keith Moon and Daltry

Who in rocking return

As they are now: Daltry and Townshend back in action on stage

CHEAPER MOTOR INSURANCE

DEAL DIRECT FOR A BETTER SERVICE

01-686 2468

041-226 2468

DIRECT LINE INSURANCE

MOTOR INSURANCE MADE EASY

Figure 67 Balancing an editorial page against a full page advert opposite – an example from the *Daily Express*. Note how a column of body type separates the two display areas. Reproduced with permission

stories are not just 'look at me' devices but clear signals as to how the page should be read, how the relative importance of the various items is being conveyed to the reader – in a word, how the page is given emphasis. A page in which the visual highlights that first hit the reader fail to respond to the worth of the items they cover has failed.

Points to watch

To say that good design arises from the nature of the editorial content rather than from the content being submitted to a preconceived pattern is a truism, yet it is not the whole picture. Obstacles, particularly those arising out of advertising, crop up however well-intentioned the designer and can undo all the work:

- *Advert coupons*. Cut-out reply coupons for adverts selling goods and services are sacred. Woebetide the page that has cut-out editorial material such as recipes or competition entries backing on to a similar one, or if one is moved by an editorial hand from an outside page edge where it can no longer be easily cut. It can be a clear excuse for a demand by the advertiser for a refund or a free advert.

- *Clash of interests.* The general charge that advertising influences editorial policy has failed the test of proof in the three post-war Royal Commissions on the Press, yet day-to-day influence of advertising on page planning is inescapable. For example, stories containing criticism of a company or the armed services should not appear on the same page as an advert for the company or for recruitment in case the purpose of the advertisement can be claimed to have been frustrated. Clash of interests detrimental either to the editorial content or an adjoining advert where there is a common subject can be a reason for redrawing a page or moving an advert – for instance to avoid a cosmetic surgery advert appearing next to a story about a bust enlargement that went wrong.
- *Stolen style.* Some adverts, to draw custom, are designed deliberately to look like an editorial item in the style of the paper. These are disliked by editors and sometimes banned as policy or, if used, they must carry a line of type at the top saying 'advertisement'.
- *Edition traps.* Edition area advert changes with entirely different words and display even for the same product, can throw out carefully designed page balance or – if anticipated in time – lead to the same page having to be drawn in two different ways.
- *Spreads.* They should be checked carefully at the make-up stage to see that strap-lines and rules leaping the central gutter are aligned at the exact level to give 'read across'. It will not arise, however, on the centre spread, which allows print-over.

9
Markets and style

Various experts have tried to classify newspaper design into a series of set patterns, or layout formulas, but the dynamic nature of newspaper content and the variability of readership markets (not to say advertising placement) have always proved stumbling blocks.

Layout patterns

Edmund Arnold, a leading American expert, defined six basic layout shapes into which all type styles fitted: symmetrical; informal balance; quadrant; brace; circus and horizontal. Symmetrical – a page of evenly balanced proportions – is, as we have seen, unmanageable and a non-starter in the modern newspaper. Quadrant involves, according to Arnold, treating the page as four quarters and assigning a main display element to each quarter. In the brace, the headlines are arranged, in effect, in steps, with each headline and chunk of text supported by the headline below it, as with a pantry shelf, a feature 'resulting from the very mechanics of newspaper make-up'.

Circus is a layout style which, according to Harold Evans's description, appeals 'to the visual senses, emphasizing appearance over content. Circus . . . aims for layouts with variety, contrast and movement, and are happily prepared to let order and a scale of values go to the lions. The drama or comedy of the layout is as much part of the message as the news'.

Horizontal layout has the text presented in a series of multi-column units descending the page as in many present day equality tabloids (see examples in Chapter 10), and giving – in reaction to the old vertical approach – strength below the fold. This also offers the facility of hiding longish texts in a series of easily readable legs which can be a useful ploy in certain situations. This method is now regularly used in American papers, although being horizontal merely for the sake of it can lead to pages as dull as the uncompromisingly vertical ones.

Of the six layout formulas, informal balance, in which the items are placed in relation to each by means of a series of strategic focal points arising out of the content and shape of the page, has the merit of general application, even if it departs from the notion of precise shapes. In fact, it is not surprising that Arnold, in a later work on newspaper design, went off the idea of six basic formulas. The American Bruce Westley's view that layout is based on the four key elements of balance, contrast, focal points and motion is a more valid summing-up and is closer to the views of the authors of this book which are that design, provided that certain general principles are understood and applied, and the skills properly learned, requires a pragmatic approach depending on content and readership market.

The yearning for pure form which some of the American designers, such as Peter Palazzo with his revamp of the *Sunday Herald Tribune* in the 1960s, have shown in their dummies is perhaps an unconscious wish to elevate newspaper design to an art form of balanced shapes which it is not. However sophisticated the system and imaginative the designer, projecting the material has to do with selling the paper and attracting readers within its market. A page pattern must relate to editorial purpose, which is geared to this function, and the ideas that go into layout must start with this in mind. Text length should relate to story value, headline size to importance on the page; a picture should have editorial usefulness as well as being an eye trap; a piece of rule or type decoration must have some demand to make; the size, or variation in size, of column setting a purpose. Examples in the next three chapters demonstrate this.

In all these tasks – which are carried out within the newspaper's type format – the reader must be drawn in and guided round the pages without being aware of the psychology at work. To be successful a page pattern cannot be a case of anything goes. Nor can it be a rigid geometrical shape. It requires the skilful use of visual devices to achieve the designer's end.

Editorial targeting

We have emphasized the importance of market and content in newspaper design. The precise 'feel' of a paper, although its design is a contributory factor, has to do with what it contains, what it has to say, what package the editor has been able to put together. Most of all it has to do with whether these things are in keeping with what, from your experience of it, you as a reader expect to find in the paper. The way in which a newspaper's content and opinions are put in tune with its readers is the principal aim of its editorial policy; and in its turn it is editorial policy that governs the targeting of the paper on to its readership market.

Few papers share an identical market; some give greater weight to news of the day, some keep the needs of women readers in mind, some go for the younger reader (a diminishing percentage of the population statistically but one still much sought after by editors); some stay up-market with news coverage, giving a wide segment of political, cultural and world news and detailed background features; others go for quick easy-to-read titbits on familiar subjects for readers with little interest in the big world outside. Town evenings go for local news and features primarily with perhaps a bit of national news at the front of the paper and some pages of sport at the back. Free newspapers have a special task of persuading readers that the paper is not just being run entirely on behalf of the advertisers. Quality Sundays, tailored for a day of leisure, go for the long read, plenty on the arts and lengthy pieces of investigative journalism, all of which can find good space on a Saturday production day when there is little hard news about; the popular Sundays solve the problem with double-page exposés or revelations about celebrities.

The proportion of its space that a paper gives to news and features, and the way these are presented, depend on these variable factors of readership targeting, and they are an explanation of the variety of layout devices that are used in newspapers as designers try to develop styles of presentation that correspond to a paper's editorial aims.

Typographical style

The principles of design (which we examined in Chapter 2) might be universal but, for the reasons we have just discussed, typographical style can vary a good deal, as can be seen by looking at any newsstand. It would be convenient for textbook writers if it divided into two main sorts – into popular and quality, for example; or tabloid and broadsheet. Unfortunately there will always be titles somewhere that cut across such neat theories: a broadsheet that has adopted tabloid ideas; a tabloid that has gone upmarket in presentation; a quality paper that has thrown away all the rules and struck out into new territory. So in examining typographical style, we are looking at *trends* to be found in contemporary British papers rather than seeking to put newspapers into stylistic pigeon holes. It is possible, with this caution in mind, to isolate a number of current trends in type style and relate them broadly to markets. For example:

Traditional

The nineteenth century journal of record – *The Times* and *The Daily Telegraph* are surviving examples – bequeathed to the twentieth century an austere, orderly style of presentation based on seriffed headlines of a modest 14-point to 36-point size, mostly in Bodoni or Century; sparing use of pictures, with the headline weight mainly at the top of the pages in the old vertical fashion. The concentration on text was ideal for readers seeking a wide and detailed account of the day's news. Today, with discreet innovations, this type style enjoys a continued use in *The Times* and *The Daily Telegraph*, the *Financial Times* and a number of regional morning papers such as the *Northern Echo* and *Eastern Daily Press* and also London's *Evening Standard* (Figure 68).

Town Final
Express & Star
Tomorrow Full TV listings and much, much more
Low Level station may become leisure centre
MULTI-SCREE
CINEMA PLAN
FOR RAIL SIT
A HUGE leisure development including a multi-screen cinema is being planned for part of Wolverhampton's redundant Low Level railway station site.
Stubbly staff breach dust safety rules
Job hope for 60 sacked workers
Race ace badly injured in smash
Gunman mugs a woman in street
Pava – grotty
Paramedic inj racing to hoax

FINANCIAL TIMES
Boeing wins $12bn Singapore order
Deal is blow to Airbus but boost for R-R engine sales
Electricity charges may increase by 2% in south
Euro-bank urges slower change to single currency
Major hails 'commonsense' agenda
ICI chairman '99% certain' to head Cadbury successor
WHEN WE
A CO
WE'RE NOT
IN ITS WORTH
ONLY YOURS.

Evening Standard
CITY PRICES
AND ANOTHER £50,000 WINNER
Our latest lucky winner: Page 2
LONDON ARTS WIN £42m ON LOTTERY
Sadler's Wells and Globe Theatre share bonanza
Kate's hustle and bustle

Figure 68 Traditional type style at work: the more imaginative use of pictures has given it a new dimension

The principal innovations are that bigger pictures now compete for space and focal points, intros are wider than in the old days, though still of standard measure; the odd wide headline is used down page to give 'strength below the fold', and body setting is varied by the occasional use of a panelled or bold item.

The *traditional* type style does not have to be static. With careful use of a good stock type face such as Times New Roman or Bodoni, the designer can combine authority with elegance, preserving the best of the traditional form on the news pages while introducing discreet typographical novelty on features pages. Eye boredom on the news pages, always the danger with this low key approach, can be averted, as in the examples above, by the careful balancing of main headlines and pictures as asymmetrical focal points, the use of artistic white space and of graphics as breakers.

Though mainly a broadsheet preserve, the traditional type style has remained popular with many city tabloid evenings, of which the *Evening Standard* is an example.

Tabloid

The most commonly used alternative to traditional type style is the *tabloid*, so-called not because it is used by all tabloid papers but because it originated in the typographical and display ideas first

Daily Record
FREE
FEST
20 PAGE
TV GUI
1995 SCOTLAND'S CHAMPION 1995
DEADLY: The face of a monster
Rose looks just like an ordinary housewife, but she is the most evil woman in Britain...
SPECIAL REPORT: Pages 2, 3, 4, 5, 6, 7 plus 12-page pull-out

DAILY Mirror
PLAY TODAY
GOLDEN SCRATCH
SHOWBIZ EXCLUSIVE
Liz Hurley's family secret revealed
Fergie was in it with Diana
IT'S ALL IN YOUR MIND
Mirror EXCLUSIVE
83% You say Di's right
Tory trade chief's amazing insult as jobless figure rises

LIVERPOOL ECHO
The best-selling newspaper on Merseyside
THURSDAY, OCTOBER 19, 1995
Keith's a prize
JOBS and HOMES
12 pages of vacancies • Top property guide
WIN a holiday to Jamaica — Page 24 WIN £10,000 tonic
Parents' agony as girls lose their batt
SIAMES TWINS
Farewell old friend
COLORVISION
INSIDE Weather 2 Crossword 14 Business News 33 TV Guide 35 Letters 44

THE Sun 25p
SAM'S NEW PAGE 3 POSE
You the jury special: Should Diana appear on Panorama?
DI-NAMITE
PALACE RAGE AS SHE KEEPS IT SECRET
I'd have taken back cheat Charles
He and Camilla deserve each other
I'm strong and I live for my boys
FURY OF THE QUEEN
I'VE BEEN KNIFED BY DI

Figure 69 The tabloid type style with its heavy sans headlines has reached far out from Fleet Street

introduced by H. G. Bartholomew in the tabloid *Daily Mirror* in the late 1930s and 1940s. Its characteristics are the use of big sans caps headlines for the main stories, bold picture display, liberal use of print rules, panels and reverse WOB and BOT type headlines to emphasize selected stories, and a general shortness of items and strong magazine content to suit a popular readership not looking for detailed news coverage.

Leading examples are the popular national dailies *The Sun*, *Daily Mirror* and *Daily Star*, together with the popular Sundays, *News of the World*, *Sunday People* and *Sunday Mirror* (Figure 69). In a modified form the tabloid style has been adopted by a wide range of regional evenings selling to a broadly down-market readership in industrial conurbations, of which the *Liverpool Echo*, and the *Irish Press* are examples. As can be seen from the pages shown in these chapters, the tabloid type style does not necessarily mean less news but rather shorter items selected more specifically for the paper's readership market, whether national or regional, and more boldly displayed with bigger headlines than in papers of traditional format.

Quality broadsheet

From the mid-1960s the *Sunday Times*, under the Thomson ownership, and also *The Observe*r began a series of changes in typography and layout which took them away from their modified traditional and mainly vertical type format into new page shapes dominated by well-whited square and expanded types, of both serif and sans families, horizontal single-deck headlines and multi-leg setting separated by heavy solid horizontal rules. These changes were accompanied by a

Scotland's Independent Newspaper
THE HERALD
HEALTH EXTRA Facing the bald facts of life
CINEMA Girls who shop till they drop
OPINION A case of recognising the enemy
SPORT Rangers' Latin lesson
The last post for a racing hero
Forsyth dragged into row
The screw is turned on Howard
Clarke back in Shadow Cabinet
High levels of toxic waste found in Loch Lomond
SUNBELIEVABLE
THIS OFFER EXTENDED...
£350
NO DEPOSIT ON ALL SUITES
NO PAYMENTS UNTIL EASTER 1996
NO INTEREST UP TO 2000 AD
reid
Gays consider vigilante force

THE SUNDAY TIMES
AMERICA ON TRIAL
MARTIN AMIS
Four Classic CDs FREE
WEEK 1: BEETHOVEN & SCHUBERT
Children 'drugged' by computer game's hidden messages
Former Tory minister joins Labour party
Wandering poet returns to claim Nobel prize
Mad police gunman shoots dead two Britons
Taxman's blunders leave millions overcharged
Council sells off spy camera film for sex and crime video
37.5 extra cm.™
virgin atlantic

Figure 70 The quality broadsheet style as seen in pages from *The Sunday Times* and *The Herald*, of Glasgow

bold close-cropped 'artistic' treatment of pictures, chosen more deliberately for design purpose, with fine black border lines, and an increased use of symbol and working graphics slotted into design shapes. By the 1970s both papers, followed in style by *The Guardian*, had moved into a wholly lowercase headline format, eschewing italic type and the use of cross-heads, and relying for eye comfort on shallower horizontal shapes, and on a mix of bold and light versions of their stock types of up to 72 point, with built-in strategic white space.

The pages, and sections into which quality broadsheet papers had now divided, were flagged with matching WOB motifs, often sandwiched between 2-point rules at the tops of pages, while bleach-outs or drawn motifs were used as labels for special presentations.

The components of this type style, which might usefully be called the *quality broadsheet* style, and their uses are demonstrated in the titles shown above (Figure 70).

Transitional

Typography, as can be seen, knows no boundaries. Serif or sans serif type families can be utilized for either tabloid or broadsheet design

Daily Express
THE GREATEST GUIDE TO MOTORSHOW '95
Saving o
village l
Rate cuts and jobs boost will secure future
Manchester Evening News
96 PAGES TONIGHT
TOO GORGEOUS TO DIE
THE SHAPE OF SHOPPING TO COME
MANCHESTER'S BEST ENTERTAINMENT GUIDE
SHUTDOWN SHOCK FOR A SCHOOL
EXCLUSIVE
Home swe
WEATHER 2 DIARY 50 TV 51-54 BMDs 64 POSTBAG 66 BUSINESS: 4 PAGES
Daily Mail
LORD BLAKE: Why I now believe they should divorce
TV interview 'was no mistake'
DIANA: WHY I HAD TO SPEAK
Mawhinney walks protesters' paint
By RICHARD KAY
Evening Mail
YOUR LOCAL EXTRA
£38,000 CAN BE WON ON PLACE THE BALL TONIGHT
150+ PRIZES GUARANTEED EVERY WEEK
Two terror men held over 300lb explosives in van
IRISH SWOOP ON MASSIVE BOMB
Man is knifed seven times in pub
Murder Briton to hang
1500 XMAS PUDS TO BE WON

Figure 71 The tabloid shape but with bold display clothed in traditional dress – what might be called the transitional style

styles, with both families offering a wide range of condensed and expanded variants, and bold and light versions. With some regional titles going for a sans format in broadsheet or a serif in tabloid and some, in recent years, switching from one to the other and even back, it is not difficult to find examples that fit into none of the above categories. Uncertainty of, or changes in, readership market, or worry over falling sales, can be reasons for type format changes, although it is wise to precede a drastic change of type style with some detailed research into market, since this should determine content which, in turn, should determine style.

Nevertheless there are papers undergoing no such trauma which draw successfully from several typographical trends to create something distinctively their own, and which present to the eye a style which, for want of a better word, one might call *transitional*. Such a paper is the *Daily Mail*. Declared a 'compact' paper by Vere Harmsworth in 1971 when it became the tabloid-sized successor to the old *Daily Sketch* and the broadsheet *Daily Mail*, it opted for an elegant mix of Century Schoolbook type, the legacy of the old broadsheet dailies, and of Gothic and Rockwell slab-seriffed type. With items not quite as short as the popular tabloids and a strong features content it was able to combine elegance and busyness in a handy page size which suited its middle-of-the-road readership – middle of the road socially, that is. Following closely alongside it in content, its market rival, the *Daily Express*, a reluctant convert to tabloid size after its successful years as a Century Bold broadsheet, adopted a similar approach with a mixed Century and sans type format which avoided what some considered to be the 'poster page' excesses of the tabloid style (Figure 71).

This middle-of-the-road approach has also suited the content and market of a number of regional evenings who favoured the handy tabloid shape, though not the tabloid style of journalism and presentation.

Design – the creative basis

A discussion on typographical style is to some extent a statement of the obvious since the differences in type use between papers can be quickly demonstrated by a few illustrations. More fundamental to an understanding of modern newspaper design is an awareness of the creative parameters underlying it. We now come to this important aspect.

Whatever the type style in use, design can be reduced in its essentials, to two sorts:

1 The static approach.
2 The dynamic approach.

The term *static* broadly covers the patterns established by many older newspapers in which an 'informal balance' on a page is achieved by locating headline mass and pictures as asymmetrical focal points within a routinized design formula. It is an approach dominated by the use of accepted headline and picture shapes. This does not mean that it is confined to a series of fixed page patterns. The patterns can be varied to suit story length, picture size and advertisement shape and position, but only to the extent that they comprise the same basic headline and picture shapes and the same standardized body setting. Nor is the static approach confined to any particular type format or page size.

Stories, under this method, are placed in the page position and given the headline size that suit their importance or length, these matters being determined by the editor or chief subeditor. If a story grows in importance from one edition or another, then it is moved to a bigger or more important spot, thus taking the place of another story. Yet the concept in this sort of page display remains static in that the headline is written to fit the type and measure, and the story (unless other items have been cleared out to make way) is cut to fit the space on the layout, the page pattern being more or less preserved. Obviously an unexpected picture would require some reorganizing but the simplicity of static layout usually allows even this intrusion to take place without too traumatic an effect.

Static layout need not be dull. There is always room for a bigger than usual picture, and writing headlines to fit the type and measure need not rule out good headlines – the discipline can be even a spur to some subeditors. The routinized shapes are also handy for quick edition changes, when there are minutes to get a story in and out and away. And if the static approach has come under threat on newspapers faced with the tempting freedom of electronic editing and cut-and-paste, it has also had a new lease of life on those taking the move into full page composition and looking for speed and simplicity.

The *dynamic*, or what is sometimes called the *freestyle* approach, is a relative newcomer in newspaper design in which headlines and picture shapes – and therefore page patterns – are created in response to the words and content of the material. In other words, the headlines are written first and the cropping and placing of the pictures decided upon before the page is drawn.

It is still done within an accepted type format, and the freestyle element tends to apply mostly to main stories and 'projections', but it does result in more varied and 'visual' pages than the static approach, and in a more deliberate targeting of material at a paper's readership. It is an ideal way to get the best out of the bigger story and the out-of-the-ordinary picture. It makes greater demands on editorial skills and judgement, but it can bring a story – and a page – alive. It also does not preclude keeping slots and positions for easy edition changes. In fact a freestyle projection works better in the context of typographical familiarity, and is as dependent on typographical and visual balance as is any other sort of layout.

In the next two chapters we consider the creative effects of these two approaches to newspaper design and look at some examples.

10
The dynamic or freestyle approach

The spread of dynamic or freestyle page design can be traced back to two quite different sources: the new tabloid journalism of the *Daily Mirror* of the late 1930s and 1940s under H. G. Bartholomew, and the reborn *Sunday Times* in the 1960s under the Thomson ownership.

Two more different papers could not be imagined. The *Mirror* was slugging out war news and virtual propaganda material for the troops in great poster headlines, with readership stunts and hectoring features by flamboyant columnists; the *Sunday Times* was creating news in depth with its Insight column, and presenting significant 'long reads' in two and three-page spreads, while moving over to an elegant version of the great American sectional newspaper. Yet both, in their use of headline type and page shapes, were breaking out of the old up-and-down 'mortice and tenon' mould and letting the material call the typographical tune. Both were discovering how to project stories using the full resources of typography, pictures and graphics, instead of just getting the material into the paper.

On the *Mirror* in the brilliant 1950s, the marriage of words and typography, and the combination of easy reading and earth-shattering events, was creating a new world for subeditors who were being hired not only for their ability to write headlines that fitted, but ones which shouted 'Read me!' The *Sunday Times* began using type to project and extend the text and not just to provide a comfortable format in which to drop the stories. Each paper in its way was turning upside down stock notions of what a newspaper should look like. And each, in its field, was to become a top-seller.

The original concept behind freestyle thinking in newspaper design was to stretch the visual parameters of the page, to break out of shapes made by horizontal and vertical lines in which every story began with a short double-column intro turning into single column, with the adjoining story fitting neatly into the shoulder, as if put together by a carpenter. Down-page multiple stories were introduced to break up the verticals. The hatchet-shaped headline appeared with one long top line and a second deck of three or four short lines underneath. Pictures were used to give a feeling of display down page. Variable and bastard setting was tried to break the tyranny of standard measures. Headlines reversed as white-on-black or as black-on-tint were introduced. The full area of the page began to be used.

The freestyle method is far from a licence for anarchy, and bears no relation to the contemptuous term 'circus' layout given to it by some American newspaper designers. It is not an 'anything goes' ragbag of type and pictures but a considered technique of presentation working within a typographical framework of chosen faces and sizes and an accepted columnar format. Its practitioners are aware that a newspaper needs to have a continuity of style and appearance that makes it instantly recognizable to its readers. Yet within these parameters it uses the deliberate projection of selected stories as a pivot round which a twin assault of surprise and familiarity is launched upon the reader.

Building the page

Each page is built round one or two main ingredients, whether it be a features or news page. In the case of a news page the ingredients might be the lead and the half lead, or the lead and the main picture (which may or may not be connected to it) which are the natural focal points of the design. But instead of the designer/journalist roughing out a banner head, an intro running round the main picture and a half lead with three lines across two columns, or some other such basic layout shape, the lead story is weighed carefully by the page executive in terms of angle and projection, and might be discussed with the art editor. The headline is then written first. The words of this are the main selling lines for the page (if the story is properly chosen). The main picture, whether connected with the lead story or not, is likewise considered as a piece of projection in its own right, and its position based upon its composition, required size and best shape. The news value of the half lead might also be considered similarly and the headline written first.

The projection might call for an explanatory strap line for the lead story, or for a deliberately wordy headline that takes several lines of a smaller type size than usual. The headline might need to closely integrate with the picture, or the picture might have its own headline and caption. Perhaps an exceptionally wordy second deck is needed to exploit an important quotation that is the fulcrum of the story.

The lead story, alternatively, might consist of three distinct elements with a main statement backed up by separate headlines on the other aspects which need to be knitted together by panel rules – or the story might have good pictures that demand to be run across a spread of two pages. A quick chart or map might have to be worked in. The required length of the story and how it can best be run in relation to the pictures and the space available on the page has to be considered. It might demand a more than average length, while at the same time be lacking in suitable pictures. A separate picture with self-contained material might therefore be allocated to the page to give the layout visual balance.

Whether there is an art bench involved or the page is being drawn by a designer/journalist, the essence of the freestyle approach is that the shape the page takes on the design pad is dictated by the material. This means that time must be allowed for discussion and evaluation. The best projections are often the product of a committee of minds, but essentially imagination and a knowledge of design techniques must play some part in the solution.

Out of this discussion a rough, or several roughs, is drawn to test the projection against the advert shapes and content. On an important page a change in advert shape and position might be negotiated to help the display, although there is usually time for this only in the earlier part of the production cycle. Once this initial idea juggling has taken place and the rough pattern approved, the page is then drawn in detail with appropriate type and lengths marked in. Any single column tops, lesser important doubles or fillers needed for the rest of the page are tasted and slotted in by the chief subeditor in the normal way. Some papers, intent on story balance as well as visual balance, scheme down to the smallest one-paragraph filler.

In the case of features pages, freestyle projection has more obvious advantages since it enables the page executive and subeditor to base a whole page display on an amalgam of text, headlines, pictures and quotations. On a page or spread consisting of one or two stories this makes for very unified treatment with type and picture effects not to be found on news pages. Dramatic or very personal material gives the designer a unique opportunity to 'sell' a page visually to the reader. Even here, however, an element of continuity is helpful to the eye, and it is customary to place stock ingredients such as regular columns and service features in their fixed format alongside a freestyle display, where they help to provide context for the rest of the page.

The virtue of pages conceived in this way is that, far from showing untrammelled freedom, they display deliberate purpose instead of looking as if type has simply been dropped in to fill the spaces left by the adverts.

Disadvantages?

The freestyle method might appear time-consuming, especially for evening papers, or those with limited staff and facilities, or on any paper where edition deadlines are near. Can an editor afford to carefully project particular material as an integrated design when stories are waiting to be placed and subedited?

The answer to this is that it can take more time to think up a headline of perhaps two decks to fit an arbitrary type size and measure and to edit a text to a fixed slot than it does to properly assess a story and 'see' it as a projection comprising a given headline/quote/picture/length. If the back bench is geared to the freestyle approach rather than to filling 48 pages of set patterns and news slots, the projection of main page elements will come naturally and effortlessly as part of the cycle from copytaster to make-up. Each page will take shape round a given projection that stems from the headline, text and picture shape of the main ingredients. If time and page material are properly allocated to take account of edition deadlines there is no reason why a newspaper should not use a dynamic approach to news and to features. The bonus will be pages that tease and excite the reader's eye. The computer has made the mechanics of this approach a lot easier.

The modern tabloid

The Sun has been the most successful example in Britain of the popular tabloid approach to freestyle and the lineal heir, in this sense, of the *Daily Mirror*, which started it (Figure 32). Its readership market is suited to the boldness and inventiveness of this method, and its vigorous page layouts, even if the contents are criticized by some purists, are backed by tight subbing based on a terse distillation of the written word, and a willingness to descend to the colloquial and sloganizing when necessary to communicate. The aim is not only to display big and hit the reader hard but to fill up every last corner of the page with material in which every word counts. Story counts per page are examined at 'morning after' conferences, and the bold simplicity of the concept belies the careful planning of content and projection that goes into each edition.

The purpose at its relaunch under the Murdoch ownership in 1969 was to go for the readership being vacated by the *Daily Mirror*, which was then going through an up-market phase. The typographical approach was derived from the ideas of H. G. Bartholomew to which was welded a flair for self publicity that would have delighted Lord Northcliffe, whose life-long aim was to get his papers talked about.

Its type style and layout ideas have been refined since then rather than changed. The page one shown (Figure 72) is typical of the electronically made-up, well-filled designs of the mid-1990s with a cluster of blurbs (not, alas, seen here in their full colour) round the title-piece, a boldly presented exclusive splash, and a strong supporting column one story keyed in at the bottom with a crisp cross-reference.

The elements of projection *Sun*-style are all there – the pretty blonde, the 144 point sans caps main headline (reversed on to navy-blue!), the eye-catching emphasis on 'free', 'today', 'tonight', 'exclusive'.

The blurbs are typical *Sun* art desk compilations of games to be played, money to be won, free beer to be quaffed 'for every *Sun* reader'.

The Sun, despite its boldness, is closely conscious of the need for continuity and manages to combine free-style thinking with formula.

THE Sun 25p

Monday, October 9, 1995 25p Thought: Just wood friends

'Traitor' forces an early election

By TREVOR KAVANAGH, Political Editor

THE Tory high command was last night braced for an early election after "vindictive" MP Alan Howarth's sensational desertion to Labour.

Premier John Major is furious over the way his options have been cut by what he sees as treachery by an ex-minister.

Mr Howarth, 51, whose true-blue seat is one of the safest in the country, switched to Labour in protest over "harsh" Government policies.

Bombshell

And in a TV interview he claimed that between "30 or 40" backbench Tories shared his disillusionment.

Now party chiefs fear the PM's five-vote majority is too slim to risk waiting for a full economic recovery in 1997.

A senior minister said: "We are on full election alert from now on."

Mr Howarth, who holds a 22,892 majority in Stratford-upon-Avon, Warwicks, resigned in a bombshell letter to Mr

Continued on Page Two

JOHN MAJOR WRITES FOR Sun – SEE PAGE 6

FALDO'S BIRDIE PAL, 20

EXCLUSIVE

From NICK PARKER in Tucson, Arizona

A BLONDE student talked for the first time last night about her friendship with golf ace Nick Faldo.

Brenna Cepelak, 20, has enjoyed candlelit dinners with the superstar.

And since they met at a tournament in Tucson, Arizona, nine months ago, she has flown to join married Nick, 38, at other top American golf events.

Busty Brenna, a rising star of the ladies' game, said: "Nick and I are very dear friends and we share a passion for golf."

Asked if her friendship with the Ryder Cup ace was intimate, she replied: "That's a very personal question. I'm not going to comment." But

Continued on Page Four

Brenna . . . 'a passion for golf' Picture: HARRY PAGE

Ace Faldo . . . candlelit dinners

Figure 72 *The Sun* page one: an example of their electronically made-up designs of the mid-1990s

An ideal arrangement with a spread (Figure 73) is to have a good photographically supported feature flanked by a column of news on the right with a secondary four-column news or features piece on top of the advert on the left-hand page, thus getting the advantages of display width on the main subject, while keeping up the story count and trapping the extra reader looking for snippets.

A particular creation that has lasted has been the Sunspots, tiny tightly subbed pieces of spot news contained in milled rule boxes a paragraph long, which can be used potently as cut-offs between stories. Its pages are also helped by the sharp headline writing in which story points are well taken in the paper's own style, and the lines filled out to the allotted space so that it all looks easy.

The paper was particularly careful not to change its typographical style following the trauma of the switch to computerized setting at its new plant at Wapping, in London's East End.

The *Daily Mirror*, which invented tabloid journalism, moved confidently into web-offset colour on news and features pages under the editorship of Richard Stott in 1988. In its search for new layout shapes embodying both mono and colour elements the Mirror displays the excitement of the freestyle approach when the editor's aim to grab the reader's eye is freed from the constraints of an accepted type formula. The integration of the subjects into the page illustrated in Plate 2 of the colour plate section is cleverly executed.

Tesco arson storm

DISASTER COPS' HQ WAS CHAOS SAYS KENNY

TONIGHT ON SKY MOVIES

Tom Conti in SAVING GRACE

TONIGHT 6pm

Buy or rent your satellite dish now

Name ban by Colonel Crackpot

MINISTER'S GRIEF

BATMAN BATTY!

The Caped crusader is massive hit

BAT PRICES

HERE is a typical Bat gear shopping list:

Bat and Joker T-shirts . £9
Bat children's pyjamas £9.50
Bat sneakers and Joker sneakers £28
Bat nightshirt and Joker nightshirt £14
Batmobile model £8
Bat hat £5
Bat watch £14
Bat halter-top £21
Bat mini-skirt £24
Bat stretch pants £25
Bat lunch box £20
Bat book £3
Bat baby T-shirt £5
Bat movie soundtrack .. £8
Bat enamel badges £3

The Joker T-shirt

Boom

Complaints

Bulging Batsuit

Millions

Del's TV barman in holiday sea terror

EXCLUSIVE

Raced

SEX BEAST GETS LIFE 5 TIMES

PIRATE RAID

Figure 73 Combining boldness with busyness: a typical *Sun* centre spread

PAGE 6 The Post Friday, December 9, 1988

World's biggest gossip column

Far out and fab, Cynthia

CYNTHIA Lennon is opening what promises to be London's grooviest nightclub — for ageing hippies.

The ex-Mrs John Lennon realised there was a need to provide a nostalgic hideaway for the growing army of '60's ravers, who have had more than enough of Acid House and Disco.

And this latest hangout, which is described as a "fun restaurant and rock-and-roll bar", promises to be a breath of fresh air, tucked snugly between top nightclubs Stringfellows and The Limelight in London's clubland.

While you sip your Strawberry Field cocktail and tuck away Sergeant Pepper steaks, Cynthia plans to dash between tables to see that everyone gets some of her magical attention.

Cynthia, the saleswoman,

CYNTHIA: Cook

says: "We are steering clear of the burger style joint — all our food will be fresh every day, imaginative and cooked to perfection."

And she's keeping it all in the family.

"I phoned Julian today to invite him to the launch.

"Unfortunately, he can't come as he's moving home at the moment.

"But he promises to drop in every time he's in London because he misses his mum's cooking."

Holloway hits the road

BUBBLY TV presenter Alison Holloway seems to have got over her six month marriage to Cockney comic Jim Davidson.

While Nick-Nick and his latest bride-to-be, Page Three girl Tracie Hilton, were discussing wedding plans, Alison was out for a romantic night on the town.

She sat laughing, joking and gazing into the eyes of a rather good-looking young man in one of London's trendy wine bars — Joe's in Clapham.

But her smile soon vanished when she spotted yours truly. Before you could say Up the Elephant and Round the Castle she was off into the night with her mystery hunk.

KEN SCORES OVER CABBIE

A COCKY Manchester taxi driver met his match the other day with Chelsea chairman, Ken Bates.

"You're...er, whatshisname, aren't you?" said the chatty cabbie. "Why haven't you been controversial lately?"

"Because," said bully boy Bates, "everyone's been doing what I say."

"Turn right — now."

Edited by LUCY ROBINSON

Busy days ahead for Mrs Planer

MR and Mrs Nigel Planer left their young one, three-month-old Stanley, at home and nipped along to the premiere of High Spirits this week.

But why was Anna Lee looking so doubtful? — perhaps she'd just discovered her husband's views on having babies.

"I think women should have all their babies before they're 30," he told me, "and I'd like them to be very close together — that way Anna Lee will still be young enough to go back to work."

BREEDING: The Planers

POST CLIPS

Bob tells hack to hop it!

BOB Hoskins is clearly not eager to talk in detail about his financial affairs. He received one million dollars for his Roger Rabbit movie, plus an unspecified percentage of the profits.

But, when pressed for additional information by a cheeky journalist, he let fly with TWO four-letter words, including one which indicated to the reporter that he should go forth and multiply.

CROSS: Bob

John's Dynasty

NOW what was Dynasty hero John Forsyth doing at a Beverly Hills bash with TWO wives?

Could it be that Mr Nice Guy had suddenly done something interesting in his old age? 'Fraid not — it appears that the blue-haired actor was with his wife, Julie, and screen wife Linda Evans, who had tagged along for the evening. He's been married to Julie for 45 years.

Bab news

BARBRA Streisand says she earns £600 every hour of every day, of every week, of every year. But she still feels insecure.

Apparently her £50 million fortune doesn't make up for the fact that she can't sustain a relationship.

Must be a moral there somewhere.

Dirty linen

CLINT EASTWOOD lived up to his Dirty Harry name while staying in a very superior New York hotel.

The actor washed his dirty linen in the sink and hung it out on the balcony to dry. For some reason this didn't go down very well with the hotel manager, who asked Mr Eastwood to take in his washing immediately — and check out.

A cool Clint put in a phone call to a pal of his, who happened to own the place, and completed his stay.

The manager is now working elsewhere.

King of the butty

KEN DODD lent his tickling stick and toothy grin to the Greeting Card and Calendar Association's promotion aimed at getting the public to write to each other — as if we needed any encouragement at this time of year.

Anyway, while Ken was having his picture taken in London's St James's Park, a tramp interrupted proceedings to wish him a happy Christmas.

"Merry Christmas, Sunshine," replied Doddy. "God bless you," said the man — and ploughed straight into Ken's prop, a plate of real jam butties from the mines of Knotty Ash.

DJ Tony Blackburn revealed on his show this week that he'd been to see the film Who framed Roger Rabbit — and fallen asleep.

"I thought I'd repeat this fascinating titbit, in case Tony's listener had nodded off and missed it the first time round.

YOU wouldn't normally associate Sir John Mills and lager. But the veteran actor is chuffed that extracts from his 1958 film Ice Cold in Alex are being used in certain TV commercials for the stuff.

Mainly he's pleased with the royalties.

Says the 80-year-old actor: "We don't get any money when the old movies are shown. But these clips are a nice little perk."

HAS Robbie Coltrane been auditioning for a part as Doc Holliday? Or do they just have a rather bizarre dress code at his local Indian takeaway?

Who's a lucky girl?

PLAYMATE Lysette Anthony, the actress and pin-up who told the Post she was "terribly pretty", gets re-acquainted with Dutch boyfriend Luc Leestemaker after being parted for a while by the cold North Sea. Some girls get all the Luc.

Denver booted into orbit

A NEBRASKA pillow salesman is trying to raise funds to send country singer John Denver into space — and keep him there.

Apparently, the money is pouring in.

Brush off

COUNCIL boss Wilf Dumble is sending squads out to sweep the streets of Leeds on Christmas Day...with no bonus and just 30 minutes off for lunch.

Blind drunk

GERRY GAFFNEY put his contact lenses in a glass of water after a Cardiff stag night...but room mate Robert Huckle was so thirsty

Freezer killer jailed for life

A HUSBAND who butchered his wife and stored parts of her naked body in their freezer was jailed for life yesterday.

Colin Buckoke, 30, had battered his wife, Samantha, 24, with a

And when he was stopped by police on the way they found the dismembered remains in the back of his transit van.

Although Buckoke cleaned the house up after the attack, a forensic scientist found a bloodstained

THE PAGE THE STARS CAN'T PUT DOWN

Figure 74 A misreading of the freestyle method: convoluted page design in Eddie Shah's ill-fated daily, *The Post*

Malling KM Kent Messenger

January 22, 1988

Hospital crisis is easing

116 PAGES

Can house hunters afford to buy? PAGE 18

BIG ROW OVER PLANS TO MERGE MAIDSTONE SCHOOLS — PAGE 13

The gift of si… for baby Kar…

by TRACY SCHA…

BABY Karl Smit… his eyes to a w… world this week… operation allowed… see for the first time.

Picture: Roger Vaughan

KING OF THE JU…

DRAKE & FLETCHER London Road, Maidstone Tel: 0622 59161

PROPERTY: Kent's best guide in 28-page pull-out supplement

Brad's the boy

CHAMP'S SPARKLING DISPLAY

Pictures by BARRY HOLLIS

Leanne's best

Green rally too late

CUP HOPES DESTROYED

Valence stalemate

Thieves dump pick-up in the Medway

Helping to stub it out

Cot death

No more washday blues for Freda

by PHIL SCRIVENER

Picture by JOHN WARDLEY

Conmen prey on elderly widow

by BOB BOUNDS

Dithering over merger vote

by LIZ OLIVER

Royal visit

Armed hunt for man fails

Millwood Kitchens

Figure 75 Three pages from an award-winning edition of the *Kent Messenger* show freestyle methods used effectively in a provincial weekly paper

The separations of items are clear in this example, defined by liberal use of solid and tinted rules of varying thickness, yet the eye is well guided through competing claims for attention.

Mr Eddie Shah's national tabloid *The Post*, produced in Warrington, in North-West England, which had a short life in 1988, shows a misreading of the freestyle approach. It is rendered almost unreadable by its convoluted page design (Figure 74). Its failure to attract readers is also a reminder that the athletic power of the design-and-production machine is not enough if a paper has not got something to say and compelling words in which to say it.

A number of provincial papers have successfully adapted the freestyle tabloid approach. The *Kent Messenger*, which circulates in the conurbation of the Medway towns, won the *UK Press Gazette* British Regional Weekly Newspaper design award in 1989 with a 116-page edition containing the pages illustrated (Figure 75). Page one is built around a brilliantly cropped and displayed picture of a baby boy who has had his sight restored. The bold masthead, which prints in yellow and black with the county heraldic symbol, enables an important local story to be given top of the page treatment

Liverpool Echo, Wednesday March 8, 1989

A grand life for Stan and his piano

Graeme Kay discovers that Emmerdale's old Seth is not the country clod he appears

Golden rule

Stan Richards — a cultured man

Showbiz in the blood

Tony Austin meets a rising star from Liverpool hoping for his biggest break so far . . .

Experience

Danger from the deep

Wargames

Gerard Henderson reports on the Irish Sea controversy

Janet's diary day in the life of TV

By Judith Moss

Janet Edge

Figure 76 Letting contents shape design: a fine example of the freestyle approach in this double-page spread from the *Liverpool Echo*

THE SUNDAY TIMES 1 JANUARY 1989 B3

LOOK

SALE AWAY!

Although sales started earlier this year, the best are yet to come, says MARION HUME

A DAY out at the January sales used to be one of the traditions of the British winter. After frantic festive shopping and a few days' respite, we prepared for the battle of the bargains.

This year, it has all been rather confusing. Red and white sales signs merged with pre-Christmas Santas as shops panicked in anticipation of lean times, and shoppers who bought early to avoid the rush were appalled to see discounts appearing before the 25th.

But the best bargains are still to come. Most designer shops have held out against early sales and will not start cutting their prices until this week. Discounting before the year turns is uncivilised, according to Bernadette Rendell, Chanel's elegant ambassadress in England. "An early sale is a sign of desperation," says the designer and retailer Edina Ronay, who will hold out until January 7.

Reductions are dramatic at top designer level – some lines will be sold at half-price – so this is the time to buy that Yves Saint Laurent jacket, body-hugging Donna Karan dress or signature Chanel suit. The quality will be superb, and, if you choose a classic, you will be wearing it for years to come.

Along with the specialist designer shops, Harrods is keeping to the January-sale tradition. "It is physically impossible to prepare a sale so vast any earlier, and anyway people like to pop in before it starts to plot their routes to bargains," explains Andrew Wiles, the press and publicity director. The Harrods sale starts on January 4, so you have precisely three days to work out whether to invest in a Lacroix flame-red riding coat, reduced from £740 to £370, or a Thierry Mugler fitted suit in vibrant lilac, down from £598 to £299. Some of the best bargains are in vibrant colours, such as Isaac Mizrahi's nursery-bright jumpsuit and Arabella Pollen's tweed suit in bon-bon shades.

Out of London, there are also excellent buys, and Alistair Cameron of Ambers in Amersham is even expecting Londoners to travel out in search of his "little treasures", which include Jasper Conran's lace jacket, organza skirt and blouse reduced from £1,300 to £500.

And in Manchester, Richard Creme of L'Homme will be offering bargains in everything from Yamamoto to Romeo Gigli. "The sale is a chance for everyone to be educated into designer clothes, but I don't offer silly prices," he says. "You don't get designer clothes for nothing, even in a sale. And I don't want to be responsible for anyone camping out for a Gaultier jacket and getting pneumonia instead."

Sleeping out in New Bond Street may not be the done thing either, but those in search of the snip of the season should make sure they get to the Giorgio Armani shop early. The price of being elegant and understated will be cut by half in an "everything must go" clearance sale to make way for the two new Armani shops opening next month. It promises to be this year's most stylish rummage sale.

● Above: Multi-coloured wool suit with satin trim, £600 (reduced from £1,200); gilt asymmetric earrings, £40 (reduced from £75); leather gloves (not in sale), £75; all by Yves Saint Laurent from Saint Laurent Rive Gauche, 113 New Bond Street, London W1, and 33 Sloane Street, London SW1 (sale starts January 7)

● Right: Black wool crepe dress by Donna Karan, £195 (reduced from £445), from Browns, 23-27 South Molton Street, London W1, and 6C Sloane Street, London SW1 (sale starts January 14)

Photographs by John Bishop; fashion by Caroline Baker

Edited by KATE CARR

● Above: Beige wool double breasted suit, £591.50 (reduced from £845), navy silk shirt, £238 (reduced from £340), gilt and pearl earrings, [illegible], gilt medallion bracelet, £110, all by Chanel, 26 Old Bond Street, London W1, and 31 Sloane Street, London SW1 (sale starts January 9). Gilt chain belt, £112, gilt and pearl bracelet, £88, both by Butler & Wilson, 20 South Molton Street, London W1; 189 Fulham Road, London SW3. Princes Square, Buchanan Street, Glasgow. Glasses by Ray Ban, £50, all good opticians and leading department stores

BEST OF THE REST

LONDON

Ally Capellino 95 Wardour Street, W1. Starts January 4

Anoushka Hempel 18 Pond Place, SW3. Starts January 10

Antony Price 34 Brook Street, W1 Starts January 14.

Bazaar 1 and 4 South Molton St, W1. Jean Paul Gaultier Jnr, Giuliano Fujiwara, Dirk Bikkenbergs. Starts January 4

The Beauchamp Place Shop 55 Beauchamp Place, SW3. Edina Ronay, Dries van Noten, Workers for Freedom, Ally Capellino. Starts January 7

Boyd & Storey 12 Newburgh St, W1. Helen Storey and Karen Boyd. Starts January 3.

Caroline Charles 11 Beauchamp Place, SW3. Starts January 3.

Chelsea Design Co 65 Sydney St, SW3. Starts January 11.

Comme des Garçons 59 Brook St, W1. Starts Jan 4.

Edina Ronay 141 King's Rd, SW3. Starts January 7

Giorgio Armani 123 New Bond St, W1. Starts Jan 7

Harrods Knightsbridge, SW1. Thierry Mugler, [illegible], Betty Jackson and [illegible]. Starts January 4

Hermès 155 New Bond St, W1 and [illegible] (no sale at Sloane St store) Starts January 21

Issey Miyake 21 Sloane St, SW1, 31 Brompton Rd, SW3 and **Plantation** 270 Brompton Rd, SW3. Starts January 3

Janet Fitch 2 Percy St, W1. Dirk van Saene, Georgina Godley, John Flett. Starts January 5.

Jigsaw 65 Kensington High St, W8 and branches. Starts January 5.

Joseph 77 Fulham Rd, SW3 and branches. Joseph Pour la Ville, Joseph Tricot, Azzedine Alaïa. Starts January 7.

Karl Lagerfeld 173 New Bond St, W1. Starts January 7

Katharine Hamnett 20 Sloane St, SW1 and 38 Princes Square, Glasgow. Starts January 7.

Kenzo 17 Sloane St, SW1 and branches. Starts January 6.

Loewe 25 Old Bond St, W1. Starts January 6.

MaxMara 32 Sloane St, SW1. Starts January 3.

Pallant at the Pantechnicon 19a Motcombe Street, SW1. Starts January 4.

Paul Smith 41-44 Floral St and branches. Starts January 14.

Workers for Freedom 4 Lower John St, W1. Starts January 7.

Xavier Danaud 30-31 St Christopher's Place, W1. Starts January 9.

Yohji Yamamoto 165 Sloane St, SW1. Starts January 4.

OUT OF LONDON

Ambers The Mill Stream, Old Amersham, Bucks. Frank Usher, Caroline Charles, Nicole Farhi, Jasper Conran, Roland Klein. Starts January 6.

Ichi Ni San 123 [illegible], Glasgow. Joe Casely Hayford, Betty Jackson, Creative Stable, John Richmond. Starts January 6.

Jane Young Chain Lane, Newark Market Place, Nottingham. Caroline Charles, Roland Klein, Terence Nolter, Louis Feraud. Starts January 4.

L'Homme 15 St Anne's St, Manchester. Yohji Yamamoto, Comme des Garçons, Romeo Gigli, Paciotti. Starts January 14.

Pollyanna 12 Market Hill, Barnsley. Jean Muir, Romeo Gigli, Callaghan. Starts January 9.

Stone 7a Market Place, Nottingham. Emilio Cavallini, Roser Marce, Wendy Dagworthy. Starts January 3

The Warehouse 61 Glassford St, Glasgow. Jean Paul Gaultier, Sybilla, Sara Sturgeon. Starts January 14.

Yvonne Damant 2 The Square, Richmond, Surrey. Nick Coleman, John Galliano, Ann Demeulemeester. Starts January 12.

LOOK AGAIN ➧

Figure 77 Giving pictures their head: a striking fashion page from the *Sunday Times*

without detracting from the eye-catching splash. The contents index-cum-blurb in column seven, also printing in yellow and black, links the two focal points together. The inside news page is an example of how a panelled-in picture story, in this case in double fine rules, can help a page in which the splash has no illustration. The down page single column picture is correctly placed visually to break up the legs of reading matter across the page, the only discordant note being the unnecessarily heavy three-column rule under the splash headline.

The sports page is an object lesson on how to let the contents dictate the design when the picture desk has turned in a brilliant set of action shots of a local boxing hero. This visually effective spread still leaves room for five sports stories down page. The pages show effectively how a Century bold and light lowercase type style can be made to work in the tabloid format.

The double page spread from the *Liverpool Echo* (Figure 76) shows how a successful town evening paper has harnessed the freestyle approach on its features pages. An all-lowercase sans dress in Futura bold and Helvetica light is used cleverly in a well-balanced display of four items, of which the main one, an interview story, SHOWBIZ IN THE BLOOD, is keyed together across the two pages by an effective 18-point tone rule. The eleven-line set-left standfirst in Helvetica, bastardized to about 33 point to fill out, acts both as an effective break between the two main pictures and as a lead-in to the main headline which is used to tie in the spread across mid-page.

The lack of adverts and the need to give length to the feature on the left has resulted in a daring decision to turn it to run under the first three legs of the main story. The text pivots on another Helvetica lowercase headline which provides correct eye contrast with the main heading, but the master stroke in this risky typographical ploy is to run the text against a beautifully cropped picture of the subject of the piece, Stan Richards, standing up from the bottom of the page in columns four and five. This totally convinces the eye of the cleverness and logic of a design idea that could have failed.

Both wings of the spread in this wholly freestyle concept are well supported without the display detracting from the impact of the main story. The map on the right gives ideal contrast against the five halftones, the only visual blemish on the page being the shy second line of the headline DANGER FROM THE DEEP. As is generally the case with the *Echo*, the pictures are particularly well cropped to achieve their maximum purpose as focal points.

Quality freestyle

What could be called the quality version of freestyle is rooted predominantly in the use of seriffed types and artistically cropped pictures and it tends to rate elegance above boldness, while being otherwise just as free-ranging. The *Sunday Times* fashion page (Figure 77) with its well-balanced interlocking pictures dominating the page, is a vintage specimen. The headline consisting of the

12 BIOGRAPHY The Herald, Wednesday, October 18, 1995

TOM JOHNSTON: A SCOTTISH SECRETARY WITHOUT QUARTER

The man Churchill called The King of Scotland

Tom Johnston, journalist, socialist and statesman, was the man who without question did more for Scotland than any other politician this century. He was described by Churchill as 'the King of Scotland'. The apogee of his remarkable career came during the Second World War, when he was a visionary and highly effective Secretary of State for Scotland. In this, the first of our exclusive prepublication extracts from Russell Galbraith's important new biography of Johnston, we look at the man's outstanding achievements during these dark years.

DAY ONE

Safety first in the city

As Secretary of State for Scotland, Tom Johnston examines a model of the Union Street and Argyle Street crossing in Glasgow at the wartime road safety exhibition he opened in Kelvingrove Art Galleries

TOMORROW

Whisky galore, and how Johnston invented Scottish tourism

SUNDAY Sun
The North's Own Sunday Newspaper
BRUNO LOSES!
Ref. stops fight in 5th round
City Hall fans demand their money back

DRINK-DRIVE CID MAN TOLD TO 'GO, NOW'
Move prompts anger...
North-EastEnder!
Mr Cycle dies, 75
Steam loco takes to the road
The Territorial Army in the North East is ready and waiting to use your skills.
First-class whistler
Contrast on show
Stricken crewmen rescued
New time-share warning

The man who fired a war's first shot...
PARIS SUPER SAVER
SIX DAYS
£99.00

STYLISH QUALITY FIREPLACES
JOHN WILLETTS
FINEST QUALITY PINE FOR YOUR HOME
NEWFERN FURNISHING
Scotland's No. 1 fitted Bedroom company now in England
INTEREST FREE CREDIT ON ALL FITTED BEDROOM AND KITCHEN FURNITURE UNTIL THE END OF MARCH
EASTER OFFER
THE METRO CENTRE GATESHEAD
TEL: (091) 493 2099
The Sunday Sun investigates...
MISSING CHILDREN
A Sunday Sun challenge: FIND THEM!
A tragedy of North East England...
MADDISON JOINERY PRODUCTS
WAKE UP TO SPRING AND SUMMER WITH A BULLDOG CONSERVATORY OR SUN LOUNGE
RING BULLDOG WINDOW CO TYNESIDE (091) 495 0055
Bulldog Window Co.
PHOTO Sales
WE'VE GOT YOU IN THE PICTURE!
Would you like a copy?

A SUNDAY SUN TRIBUTE
Tears for the loss of a Geordie legend
Woman, 86, dies in house blaze
Pullman hostess service by coach!
JOE
New bingo game starts today
Castle Conservatories
PRE-SUMMER SALE
£3,475.00

Figure 79 Six specimen pages from Vic Giles's 48-page revamp of the *Sunday Sun* carried out for the Thomson Organisation at Newcastle upon Tyne in 1989. The designs are a rework, with new headlines and pictures, of material in an existing edition with body setting 'junked' in the columns to fill. They are intended to demonstrate how the freestyle approach to headline and picture use can be used to give vigour to both news and features pages of a provincially-based Sunday tabloid competing with national tabloids in an industrial conurbation

Figure 78 How dynamic design can give power to a long read: a features page from Scotland's *The Herald*, published in Glasgow

simple legend SALE AWAY! is given interest by being mounted neatly on a drawn carrier bag slung through the first 'O' of the word LOOK, which establishes the page, and is an example of the paper's technique of integrating graphics into its display. The pictures help by being superb photographs in their own right. The fine keyline round them protects the low-key backgrounds from fade, while the typey adverts down page act as a foil to the design, at the same time containing useful information for the readers. Like all good freestyle design the page scores with the bold simplicity of its concept.

The *Glasgow Herald*'s serialization of an important biography – *The man Churchill called The King of Scotland* – shows how a long read and quite wide setting of 16 picas can be made eye-catching (in Figure 78) by giving the designer a free hand and a page almost without advertisements. It also shows the strength monochrome can have on a features page when used properly.

The massively enlarged cut-out of Tom Johnston bulges into the folio area as it leaps from the page to engage the reader eyeball-to-eyeball. The headline is a label that for once demands to be read. The eye is tipped into the stand-first and thence into the text. Attention is secured; the designer has won hands down. The bastard setting round the shoulder, the down page picture of cut-off and tinted NEXT WEEK line are all that are needed to break up the legs of type and help the reader comfortably across the page and home.

11
Static and modular design

The shift towards freer design influenced by the tabloid poster styles and the revamping of the quality Sundays in the 1960s still left a large section of the press, particularly in the provinces, committed to traditional presentation. Worries over circulation in the 1960s, however, and concern about increased television viewing, spurred some papers to try to make themselves more attractive and eye-catching. In the main the movement was towards modular patterns, experiments with lowercase type formats, a greater magazine content and the more imaginative use of pictures.

It is tempting to attribute these changes to the expected availability of electronic make-up on screen as a result of the computerized printing systems which began to be introduced in regional production centres from the early 1960s. Yet, apart from the union problems of job allocation, it aroused little enthusiasm at first. The problems of graphics generation looked to be insoluble except possibly in the long term, and even in America the method was not making a great deal of progress. In Britain in the late 1980s, after two decades of 'new' technology, only a handful of papers, national and provincial, had moved into screen make-up and even they were printing out on to bromides to take pictures and adverts by paste-up into the 'windows' that had been left.

Two things influenced the move towards modular design in the more traditional papers. First, once pictures began to be a more highly rated part of the mix, the old established styles of papers like *The Daily Telegraph, Yorkshire Post, Birmingham Mail* and *The Scotsman* lent themselves easily to modular pattern-making, with headlines and pictures sharing the role of focal points formerly occupied almost exclusively by the headlines. Pictures, in fact, rather than headline shapes have become the key in modular design.

Second, the fashion for lowercase type formats, fanned by the successful re-styling of *The Times* and *The Guardian*, required a more segmented page to gain the best eye appeal from headline location (see examples in this chapter). It also required some care to be given in the pages to the juxtaposing of different weights and sizes of the chosen headline type.

A third factor probably weighed with some provincial editors. Where a paper's traditional appeal was valued, the move towards an imposed modular or semi-modular type style with defined parameters gave protection against the dangers of excessive innovation unleashed by the new freedom of cut-and-paste make-up. Some early examples of web-offset-produced local papers were unbelievably garish and messy as papers cut loose from the restraints of hot metal, and young journalists with little training and a determination to experiment were given their head on paste-up pages.

What occurred was that papers with the more modular formats, led nationally by *The Times* and the revamped *Guardian*, were the first to gear themselves up for electronic pagination since, whatever the outcome of the problems with graphics, they felt the change was unlikely to damage their design format.

Influential in their plans was the fact that two national titles that came out in 1986, *Today* and *The Independent*, both settled for screen make-up and modular design from the start, though initially they had to laser-print their pictures and adverts for pasting on to bromides of the pages.

As it happened, the Apple-Mac/Quark XPress solution, when it arrived in the early 1990s, ensured that all styles of page design were now possible on screen and that fears of imposed modularity were groundless.

Modular patterns

Modular, an old word given new currency by computer-speak, simply means in segments – in this case geometric shapes devised from rectangles. In screen page make-up such shapes can be edited and set as complete text + headline segments (within panel rules if need be), and moved into position on screen by mouse or key controls. The stories are given to a subeditor to edit to the required length and shape to fit the layout – say, a headline across five columns with five legs of text running under it to fit a panel 14 cm deep – the same way as with any other layout. The editing screen shows when a precise text length and fit has been achieved and the story then passes through the revise subeditor and is sent to the page. The method imposes no greater burden on the subeditor, once keyboard commands have been mastered, than in hard-copy subbing on screen to fit a space on a paste-up page.

An advantage of modular page design over free style, whether paste-up or on-screen, is that it means a good deal less difficulty for editors with small staffs. It makes page changes simpler since it allows module-for-module replacement. Stories of appropriate shape as well as length can be prepared and held for fitting at edition or slip changes. Where used in conjunction with stored computer graphics it is user-friendly to a newspaper with minimal art desk facilities.

The disadvantages are that unless a determined effort is made and creative use of type and pictures brought to bear it can cause a newspaper to degenerate into formularized pages in which familiarity triumphs over any attempt at surprise. Dull, boring, repetitious design can be the result.

Yet there is no reason why modular design, given the use of a reasonable type range and well-cropped pictures of good size, cannot be the basis of imaginative page design. The need for effective focal points to grab the reader's eye remains and it is up to the visualizer, within the parameters of a paper's type style, to make the modules work for the page. Pictures are the vital ingredient, even though they may be unconnected with the main stories. They give essential contrast to the vertical and horizontal elements in the design, and against the size of the stories. Columns of briefs flanking or centring on the page can give the eye some relief from setting that might otherwise look grey. It is thus the way the modules are put together that really matters. There are plenty of good examples of how this can be done.

Updating the traditional

There have been three revolutions in the twentieth century at *The Times*. The first was the commissioning in 1931 of a new typeface for the paper from Stanley Morison which presented to the world the now familiar and much used Times New Roman. The second was the removal in 1966 by Sir William Haley, the then editor, of the classified adverts that had filled page one for more than a hundred years. The third was the updating of the paper's type format in the 1980s in a bid to widen its readership under Rupert Murdoch's ownership and the editorship of Charles Wilson.

A comparison of the two page ones in Figure 81 with the edition of August 5, 1976 (Figure 80) shows the visual effect of the changes that took place by slow stages through the early 1980s, not the least of which was the restoring of the Royal coat of arms to its position in the title piece. More dramatic, as can be seen, has been the increase in size of the splash headline type from 36 point to 60 point and sometimes to 72 point, and setting headlines centred instead of being nearly all set left.

The sectioning of the paper on American lines and the use of preprinted gravure as part of the modernization were successfully used to draw in both new readers and new advertisers. Most impressive of all, however, has been the way the paper retained its authority while undergoing progressive changes in type and picture use by which it has merged its traditional approach to page patterns with the best of modern modular techniques.

An effective part of the modernization was the reduction of the type area width from 92 picas to 81 to align production with the company's other titles when the paper was transferred from Gray's

Thursday August 5 1976
No 59,774
Price twelve pence

THE TIMES

Time to dump the import dumpers, page 21

Police open fire in new Soweto rioting

Police opened fire yesterday on a crowd of 20,000 blacks, mostly students, in the South African township of Soweto, scene of one of the riots in June in which 176 people were killed. Eyewitnesses reported three deaths yesterday, but this was denied by the police. The students were marching on Johannesburg to protest against the detention of colleagues.

20,000 blacks march on Johannesburg

From Nicholas Ashford
Johannesburg, Aug 4

Lord Thomson of Fleet dies in London hospital after chest illness

By Roger Berthoud

Chemical company takes poison precautions

By Pearce Wright
Science Editor

Government to stock swine flu vaccine

By a Staff Reporter

Employers and TUC settle differences in talks on economy

By Malcolm Brown

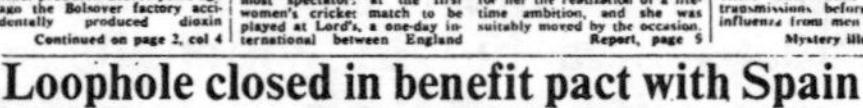

Loophole closed in benefit pact with Spain

By Pat Healy
Social Services Correspondent

Five students reported shot in Uganda

From Our Correspondent
Nairobi, Aug 4

81 executed in Sudan over abortive coup

Teenagers blamed for resort bombs

Dr Kissinger's visit

Wounded evacuated

'Torture' in Ghana

Councillors criticized

Drink-driving law

'Arms minister' is suggested

Figure 80 *The Times*, August 5, 1976, before the changes of the 1980s

Inn Road to the new plant at Wapping in 1986. The number of columns remained at eight but came down to 9½ picas wide, with a subtle gain in readability. The changes entailed a bolder use of pictures and the clever placing of colour and graphics both as page decoration and as breakers. Generally there is an attempt to connect the main page one picture with the splash or half lead. While lengths in relation to the pattern, and the placing of items, have remained broadly traditional there is a willingness to break out of this in type and picture display on the day a big story breaks.

The broadsheet format has been well-developed for busy users for ease of reading and finding. The three-column splash headline on page one is now usually in 60-point Times Bold lowercase, the setting almost always single-column. Despite the changes and the bolder type and picture use on all pages an essential discipline of placement remains with maximum effect being derived from varying sizes and weights of headline type.

All page one stories needing turns go to page two so that readers do not have to dive all over the paper to find story continuations – or, worse still, give up and not bother. The back page forms an essential easy-to-find news and features index with cross-references, built around a bold picture and weather and travel information.

Yet there is a continuing revolution. While remaining a newspaper of record, the old *Thunderer* has broadened its coverage to give its pages an almost populist appeal that would have been inconceivable under the editorship of Geoffrey Dawson or Haley. Human interest has crept in on a modest scale; left-wing writers are given space; easy to follow graphics embellish stores. *The Times* regularly runs full colour, either editorial or advertising or both, on eight pages of a 22-page main section, and sports a strip of two-colour blurbs over the front title-piece.

The editions illustrated (Figure 81) show how *The Times*, despite its broadly modular approach, varies its page one by letting content – and especially pictures – dictate design. Length as much as a headline signal the importance of an item. Shorter legs below the fold show how crossheads can be dispensed with. A basic colour-tinted subject index ties in the bottom of the page.

The revamped *Observer* of the 1990s shows the extent to which a modular features page can look strikingly bold. The example in Figure 82 shows how a picture can be made to dominate. Standfirsts, 72-point drop letters and a by-line compo are among the many devices that are used to remind us that electronic pagination can make all things possible.

The *Northern Echo* page one produced by the team commended in the 1989 UK Press Gazette awards (Figure 83) is an example of how a leading provincial morning paper utilized graphics in adapting its traditional format to modular design. Here, an information chart is used as a centrepiece in the paper's coverage of the 1989 US Presidential election enclosed, with text, in a huge 8 point black

ALL IN THE MIND
Gods that sip milk satisfy mankind's religious thirst
WILLIAM REES-MOGG, P 18

SPORT ON 13 PAGES
Europe regain Ryder Cup in thrilling style
PLUS RESULTS: PP 23-35

ALLA CARTA
Why we all eat like Italian peasants
RACHEL KELLY, P 17

25P

THE TIMES

MONDAY SEPTEMBER 25 1995

No. 65,381

'We are not slaves' storm before agreement

Arafat gains self-rule for West Bank

Europe

Tory associations rebuff Major to select more women for saf

Scottish victory

Clinton's us

TV & RADIO 42, 43
WEATHER 22
CROSSWORDS 22, 44

UNDER THE WEATHER
How to make light of the seasonal gloom of autumn, page 7

SITTING PRETTY
The perils facing artists who paint the famous, page 16

FREUD ON FRIDAY
Why the catwalk that matters is at Newcastle United's ground, page 39

25P

THE TIMES

No. 65,385

FRIDAY SEPTEMBER 29 1995

Fresh views of royalty: exhibition provides a first public showing for photographs from the 'family album' at Windsor

Windfall row mars promise of £50 rebate

Amis fails to make Booker list yet again

'School reports' may replace degree grades

SATURDAY IN THE TIMES

Cook shifts Labour ground on EMU

Education

Carlings announce a temporary separation

Raised in the Highlands

THE FAMOUS GROUSE
FINEST SCOTCH WHISKY
Quality in an age of change

TV & RADIO 42, 43
WEATHER 22
CROSSWORDS 22, 44

LETTERS
OBITUARIES 19, 27
WILLIAM REES-MOGG 18

ARTS 32-34
CHESS & BRIDGE 39
COURT & SOCIAL 20

SPORT 38-42, 44
EDUCATION 35
LAW REPORT 38

Figure 81 *The Times* of the 1990s: two examples show new approaches to page one. Headlines have doubled in size and the royal coat of arms is back. Copyright Times Newspaper Ltd 1995

Figure 82 Modular yet strikingly bold – this features page from *The Observer* shows how a picture can be made to dominate. Copyright *The Observer* 1995

10 | 8 October 1995 the Observer

KENNETH ROY'S BRITAIN

6. Mrs Garbutt's greyhound.

'Jesus,' said a poster outside a church, 'is the winning ticket in life's lottery.' The streets were empty and it was not yet seven in the evening. There are lonely nights in life's lottery, sweet Jesus, when you would hug a lager lout.

On the right track: Going to the dogs in Castleford means you may first have to catch the bus that says Wakefield on the front. Photograph by Murdo MacLeod

High and dry on a dog day in Castleford

VOLVO IN DARLINGTON Mill CHESNUT STREET, DARLINGTON. Tel. 0325 53536

BRITAIN'S REGIONAL NEWSPAPER OF THE YEAR

The Northern Echo

No. 36,750 Founded 1870 WEDNESDAY, NOVEMBER 9, 1988 STILL ONLY 18p

NEWS BRIEF

Grange Hill actress found

TV actress Penny-Belle Fowler, 16, who plays schoolgirl Tracy Dean in Grange Hill, was found wandering on railway lines in Swindon last night after going missing for 24 hours.

Police had said they were "very concerned" about her safety. She was said to suffering emotional problems.

'Ogling' strike

MORE than 300 council workers struck in Liverpool after women workers complained an area housing officer was ogling them.

Escape scare

THE Princess Royal's son, Peter Phillips, was at the centre of a police alert when three teenagers on the run from a youth custody centre broke into his exclusive Dorset boarding school.

Pay up, please

THE RAC is fed-up with being summoned constantly by a minority of members, and in future they will be allowed ten emergency calls a year. More, and their membership will be doubled.

Marcos boost

AN American Supreme Court justice last night granted deposed Philippine President Ferdinand Marcos and his wife Imelda a temporary stay of an order requiring them to produce bank records and other evidence demanded by a federal grand jury.

Death charge

A FATHER and son were remanded in custody last night, accused in connection with the death of four-year-old Irene Morgan in a road accident in Glamorgan.

Walesa crisis

SOLIDARITY leader Lech Walesa faced a rebellion by hundreds of union militants who went ahead with a strike at a Gdansk shipyard, after he decided there should be no showdown with the Polish government.

Bomber crashes

A UNITED States B-1 bomber crashed last night in Texas. Four people were reported to have parachuted from the plane.

IN THE NORTH

Robbery charge

BROTHERS Brian and Ronald Robinson, aged 18 and 25, of Rothsay Terrace, Bedlington, were remanded in custody at Sedgefield yesterday, accused of the robbery of the North of England Building Society office in Church Street, Shildon, on October 28.

Crossbow horror

AN open verdict was recorded at an inquest yesterday on a Wheatley Hill man who shot himself through the head with a crossbow. Page 7.

Memorial row

PEOPLE in Chester-le-Street are enraged by a new, modern £5,500 war memorial, which the local council provided in time for Remembrance Sunday. Critics say it is an insult to the dead.

Pigeon eviction

THOUSANDS of pigeon fanciers in Easington may have to get rid of their birds if the council goes ahead with a plan to ban back-garden lofts. Page 10.

TODAY'S CHOICE: The Echo TV Editor makes these selections for the next 24 hours' viewing – Tyne Tees TV's series using private letters, diaries and photographs to tell the story of ordinary people during World War One ... *TV schedules — Page 2.

ECHO COMMENT ON... The next president of the United States, selling off the water board land and picking on the pigeons. Page 8

RING US

You can ring a Northern Echo reporter at head office DARLINGTON 381313; AYCLIFFE 311312; BARNARD CASTLE (Teesdale) 38626, or Teesdale 38584 (answering machine); BISHOP AUCKLAND 602232; DURHAM 384261; HARTLEPOOL 272578; MIDDLESBROUGH 247645, 247647; NEWCASTLE (Tyneside) 261 7743; NORTHALLERTON 3610; RICHMOND 5281; STANLEY 280613, 232929; STOCKTON 675678, 671375; SUNDERLAND (Wearside) 567 3422 or 567 9155; YORK 656476, 35860.

Fax Number: ☎ (0325) 380539

WEATHER

FORECAST: A few showers are likely, an odd one of which may be on the heavy side, and the day will be generally cloudy with nothing more than an occasional fairly brief spells of sunshine. It will be a mild day with highest temperatures 14C (57F) and winds will be light to moderate south or southwesterly.

OUTLOOK: Showers will die out by morning but rain will return later in the day.

WEATHER PLUS: Page 3.

Duke left reeling as Republican sweeps back

By George – it's President Bush

By PHILIP YOUNG

GEORGE Bush will be the 41st President of the United States of America.

But Democrat Michael Dukakis gave him a harder run than anyone expected. Mr Bush won a solid block of support throughout the conservative South and mid-West, but there was no sign of the whitewash some Democrats feared just a week ago.

Riding on a higher-than expected turn-out Mr Dukakis picked up significant popular support in many key states but at the end of the day came second in the winner-takes-all system.

★★★ Went down fighting

Late predictions showed Mr Bush's winning margin to be less than 10 percent of the popular vote. And the Democrats were set for victory in the prestige states of New York, Illinois and Washington DC.

An early result showed a Republican victory in Indiana, the home state of Dan Quayle who will become Vice-President when Mr Bush is inaugurated in January.

Mr Bush the winner to become the first sitting vice president since 1836 to be elected to the presidency.

But the Democrats went down fighting. Even as the polls were closing on the East Coast, Mr Dukakis was claimed: "It's a fight to the finish, a cliff-hanger all over the country. It's going to go right down to the wire.

★★★ Surprise a few people

"I think we're not only going to surprise a few people, but we're going to be doing the celebrating," he said.

Black civil rights leader Jesse Jackson, who had fought Mr Dukakis for the Democratic nomination, was among the first to recognise Mr Bush had won: "George Bush is to be congratulated. His basic projection of a kinder, gentler nation must be a commitment that both Republicans and Democrats rally to because the campaign is behind us now," he said in an interview with CBS.

1988 US PRESIDENTIAL ELECTION - RESULTS POSITION AT 4.00 am

KEY TO MAIN MAP: Republican won state; Democratic won state; P Projected result; VA. State abbreviation (See table below)

% VOTE — 1980: REPUBLICAN 50.7%, DEMOCRATIC 41.0%, OTHERS 8.2%; 1984: REPUBLICAN 58.8%, DEMOCRATIC 40.6%, OTHERS 0.7%

KEY TO STATES

KEY TO SMALL MAPS: Republican won state; Democratic won state

Many Americans told pollsters they were not greatly attracted by either candidate – raising questions about the strength of the popular mandate Mr Bush will be able to call upon in any battles with the Democratic Congress.

Surveys of voters leaving polls suggested Mr Dukakis was doing well in several Western states, including California, Colorado, Montana and Missouri. But losing running mate Lloyd Bentsen's home state of Texas was a major blow – no Democrat has ever made it the White House without the Lone Star state.

The Democrats were pinning their hopes on reports of much heavier than expected voting in many states. Nationally the party has the most supporters, but they are traditionally more reluctant to vote. A big turn-out by East Coast Jews and blacks could have rocked Mr Bush and Mr Dukakis a chance of victory.

And Mr Bush's celebrations were dampened by the prospect of both houses of Congress remaining under Democratic control. That would force him to compromise and bargain to win approval for legislation if he wins.

The Presidential candidates were chasing a 270-vote majority of the 538 "electoral votes" cast by each state. The winner of the popular vote in each state receives all of that state's electoral votes.

Profile of a President – Page 8

File on fans is favoured

ENGLISH and Welsh football fans will have to join a nationwide computerised membership scheme before they can go to matches if a Government working party has its way, Whitehall sources said last night.

Colin Moynihan: report urges national computer for fans

The working party's report, to be published later this week, also suggests setting up a football membership authority to administer the scheme as part of the Government's war against soccer hooliganism.

The Prime Minister has been urging such a scheme on Britain's soccer clubs for the past two years.

But the football authorities have, in the minds of the Government, been dragging their feet, claiming that such a scheme would be unworkable and too costly.

Mrs Thatcher has warned them that unless they adopt a scheme voluntarily, she is prepared to make it compulsory through legislation.

This is what will almost certainly happen following the report's recommendations.

The working party is chaired by Sports Minister Colin Moynihan and its membership comprises representatives of the police and the football authorities.

Those who have argued against membership schemes have been told by the Prime Minister that Luton Town has been successfully running one for several seasons, while other clubs operate 20 per cent membership schemes without any particular problems.

Most people expect the Government to introduce legislation in the next session of Parliament.

1,000-job hope from new units

UP TO 1,000 jobs may be created on a Spennymoor industrial estate following its purchase by London-based developers.

The scheme, claimed to be the first development of its type in the region, is on the former Courtaulds factory and is due to be launched on November 18.

Pavedelta Ltd, a subsidiary of London-based property company Smithfield Developments plc, recently purchased the former Sedgefield Enterprise Centre from Sedgefield District Council for more than £2m.

Pavedelta's sales and marketing manager Rory McMillan said last night that up to 60 companies would be housed in a variety of units at the centre and estimated 500-1,000 jobs would be created in the 350,000 sq ft facility.

Two units have already been let, to Boots the Chemist and Rothmans International.

Sedgefield District Council's deputy leader Fred Chaplin yesterday said he was delighted at the news.

Lords rebels lose eye vote

THE Government last night finally won its battle to introduce a £10 charge for eye checks.

Health Minister Kenneth Clarke: 'I'm extremely relieved'

The House of Lords voted 257 to 207, majority 50 to defeat a Tory rebel bid to scrap the charge, which, with the planned £3.15 fee for dental examination, has caused a major political storm and a serious Commons revolt.

Shortly after, ministers were victorious again when the House rejected a "fall-back" amendment by former Labour Health Secretary Lord Ennals, to exempt pensioners from the eye check charge.

But their majority was slightly trimmed to 41 – voting 237 to 196.

The Lords was packed, following frantic activity by Government whips to bring in "backwoods" supporters from the shires for one of the biggest roll calls this century.

Clear relief could be seen on the face of Health Secretary Kenneth Clarke as he sat on the steps of the Throne to hear the result announced by Government chief whip Lord Denham.

A revolt on dental checks was avoided after the Speaker of the Commons, Bernard Weatherill ruled that this was a financial measures, technically precluding the Lords from maintaining their opposition.

As the debate started, a concession was announced by Lord Privy Seal, Lord Belstead, who promised the Government would consider exempting poorer pensioners from the dental charge.

Health Secretary Mr Clarke said of the defeat of the rebel amendment: "I am extremely relieved.

Permissive price

Kenneth Baker

EDUCATION Secretary Kenneth Baker last night blamed the so-called Swinging Sixties for the "ambiguous" moral climate of the 1980s.

Backing criticism of the "do your own thing" decade, he said that "so-called liberalisation" culminated in killings by radical left-wing terrorist groups.

Meanwhile, the Church had become absorbed with social policies, and teachers were too reluctant to impart "traditional moral values" to pupils.

Speaking near Bolton, Mr Baker also declared: "I sometimes think the only four-letter word which trendy parents shrank from using, was the word 'don't'.

"The greatest offence of Sixties liberalism and permissiveness was its arrogance – the way it sought to sweep aside the accumulated wisdom and practices which had served society for centuries."

Charles the Silent

Charles

PRINCE Charles, a highly vocal critic of architecture at home, decided yesterday that the best course during his visit to France was to say nothing.

Yesterday, Charles was shown round a housing development by the Mayor of Paris, Jacques Chirac, and was aked by a reporter what he thought of it.

The Prince, who has recently been the scourge of architects and planners in Britain, diplomatically replied: "I am trying to keep my opinions private."

He kept his word by remaining silent about other developments that have attracted heavy local criticism.

Earlier, in a speech at a reception, Charles said he "wouldn't dream" of commenting on French architecture, although "one or two French people have already said they would be interested to hear my views".

Figure 83 *The Northern Echo*'s award-winning 1988 page one harnesses graphics to modular design

WORLD TONIGHT

Slimline Fergie told to ease up by trainer

Oyez! Cyril wins free prescriptions for men

By Geoff Meade

Get them out!

Loophole closed

The frog prince

Nato chief's plea

M-way poison peril

Patten landmark

Safe investment

Tidal wave warning

'Serial killer' held

Balloon horror

France to promise bomb test ban

First steps to relieve agony of war victims

By Eileen Murphy

Major and US clash on weapons claim

Two hurt in MoD blast

Police in second skunk drug raid

By Andrew Nott

Yeltsin block on Nato nuke arms

"My feet needed the treatment...

...I needed the treat."

FREE GIFT

Scholl

YOU'LL PAY LESS AT NATIONAL

...OR WE'LL REFUND THE DIFFERENCE!

REMOULDS £9.95

155 x 13: NEW STEEL RADIALS £19.95, SEMPERIT £25.95, UNIROYAL £29.95, MICHELIN CLASSIC £36.95

165 x 13: NEW STEEL RADIALS £22.95, SEMPERIT £29.95, UNIROYAL £35.95, MICHELIN CLASSIC £37.95

175/70 13: NEW STEEL RADIALS £27.95, SEMPERIT £33.95, UNIROYAL £36.95, MICHELIN CLASSIC £43.95

185/60 HR14: NEW STEEL RADIALS £33.95, SEMPERIT £49.95, UNIROYAL £57.95, MICHELIN CLASSIC £59.95

TYRES
EXHAUSTS
BATTERIES
BRAKES
MOT
SUSPENSION
CLUTCHES
LUBRICATION

BRAKE POINT SAFETY CENTRE

UNBEATABLE EXHAUST PRICES!

£18.50 £24.95 £21.95 £24.95

NATIONAL TYRES AND AUTOCARE

CALL YOUR LOCAL CENTRE FREE ON 0800 62 66 66

ALSO AT:

Figure 84 Combining a modular layout with their new sans type in no way detracts from this pair of busy *Manchester Evening News* pages

panelled-in segment resting on top of its down page selection of national and regional news. Its double column News in Brief display in columns one and two enables it to get nineteen items on the page, the 24 point sans headlines serving a useful foil to the otherwise all-lowercase serif format.

The election graphic, inputted from the Press Association picture service, and enhanced by the *Echo*'s graphics artist shows the state-by-state results and is a good example of how graphics can be substituted on big page one stories when pictures are either unsuitable or unavailable and, by so doing, give a modular page almost the freedom of freestyle design.

The page four and five examples from the *Manchester Evening News* (Figure 84) show a busy evening tabloid paper that is thriving on a sans lowercase type format and modular design. The discipline and requirements of both approaches in no way diminishes the *Manchester Evening News* traditional vigour and determination to pack its pages with stories and pictures. The modules are boldly put to work in the cause of readability and eye appeal.

New directions

The Independent, launched on October 7, 1986, is a classic example of a computer-designed creation that has benefited from electronic pagination. Editor Andreas Whittam Smith, uninhibited by existing print demarcations or any reader tradition, opted to utilize the full

INDEPENDENT 35p

WEDNESDAY 4 OCTOBER 1995

FIVE CLASSIC HOLIDAYS FOR TWO TO BE WON

See page 10

IN SECTION TWO

IS SHE THE TORIES' SECRET WEAPON?

The power of Cherie Booth

BROKEN HEARTS

The new technique to put them back together

MRS THEATRE

The woman who owns the West End

After one year, three months and 20 days, OJ Simpson walks free

Inside

Blair vows to 'change Britain for good'

Brain food.

Prospect

The magazine for a sceptical age.

IN BRIEF

COMMENT

section ONE

section TWO

...s Macedonian President

...nia deal

...n dinosaur footprints

...t on austerity diet

Chicago x 7

Seven daily nonstops from the U.K. to Chicago.

American Airlines

Something special in the air.

Figure 85 Bold picture use and increased readability are *The Independent's* style in the 1990s

THE GUARDIAN

Printed in London and Manchester Saturday June 8 1985 25p

Today

Open-air epic PAGE 13

Lord of the censors PAGE 17

Monday

FREEDOM FIGHTER The new ideological guerrillas: Guardian Women meets a terrorist from the Animal Liberation Front.

Tuesday

PLACE SEEKERS Where will 10,000 children end up? Education Guardian on the battle for parental choice of school.

Thursday

BOND AND BIRD Derek Malcolm reviews the new Bond and Alan Parker's Cannes prizewinner, Birdy. Movie Guardian.

News in Brief

Bank aids Sinclair

Threat to unions

Bitter switch

Bank charges

League gag

Whose gold?

Market moves

The weather

Inside

Supporters want legislation brought forward

Powell bill fails but embryo battle hots up

By Colin Brown, Political Reporter

Gerard Vaughan — Government should act

END OF QUEST: John Booker, Isle of Wight chief fire officer, and Tautrydas Girenas telling journalists that Tautrydas's brother Ramanus had been found dead

Militia takes Unifil troops hostage

From AP and Reuter in Sidon

Rescue workers find body in well

By Seumas Milne

Axe falls on pit that led strike

By Peter Hetherington

US fails to sway Europe

By Hella Pick

'Inside man' behind £6m security haul

By Paul Keel

Ramanus Girenas — trying to escape

Falklands war ferry holed

Honest Ed's Old Vic bit part

By Aileen Ballantyne

The Guardian

They've both got heart disease.

We want to know why.

We've already identified smoking and obesity as major causes of heart disease.

What the British Heart Foundation is trying to discover is why even fit, non-smokers can be affected.

That's just one of the many areas where we need your help to fund heart research.

To find out how you can give it, simply return the coupon to us today.

The more you help us, the more we'll find out.

Name

Address

Postcode

British Heart Foundation

The heart research charity.

Figure 86 *The Guardian*, June 8, 1985 – a low key type style shortly to be discarded

Figure 87 *The Guardian* of the 1990s – a bold title-piece and an emphasis on headline typography. Copyright *The Guardian* 1995

50p
Saturday
October 7
1995
Published in London
and Manchester

The Guardian

Multiple murder trial begins

Wests 'had bond of sex'

Duncan Campbell
Crime Correspondent

The party's over . . . Seats littered with publicity pamphlets after John Prescott's concluding knockabout success

Birt attacks 'crass' pressure from Labour

Andrew Culf
Patrick Wintour
and Rebecca Smithers

Next week
Virtually Normal

"This is the argument of my life and I have to win it"
Andrew Sullivan (above), the brilliant young English editor of the American magazine, the New Republic, tells how he came to terms with the fact he was gay in Virtually Normal — probably the most important book about homosexuality and society ever published.
Exclusive serialisation starts in the Guardian on Monday

Bribery investigation may force head of Nato to resign from post

John Palmer in Brussels

Rock and rhyme divide council

Owen Bowcott

Pop and poetry . . . Mick Jagger and Sydney Keyes

Figure 88 Justifying the unjustified – a daring approach to body setting in *The Guardian's* weekend Outlook section. Copyright *The Guardian* 1995

The Guardian Saturday October 7 1995 — INTERVIEW 29

'The visit to Nicole's grave is part of the deal so we know he's got to come. It looks worse each day he leaves it'

Sex, race, violence, a clash of monumental egos – the OJ Simpson trial had it all. IAN KATZ charts its final days

Likeable, funny, pretentious . . . Jake, left, and Dinos

Jake and Dinos Chapman are nasty and disgusting, or intelligent and inspired. Pay your money, take your choice. Photograph: DAVID SILLITOE

Brilliant bad boys of the galleries

The MEGAN TRESIDDER INTERVIEW

Beware bureaucrats blurring the Tory vision

The Conservative Party gathers in Blackpool next week, but where does the soul of the party now reside? ROGER SCRUTON on the quest for 'true' conservatism

Faulty meter leads to shocking electric bill

A MOTHER-OF-FIVE was hit with an electricity bill of more than £2,400 — for having a faulty meter.

Mrs. Katherine Bartlett, of Dulwich, was told her meter had been "under-registering" for two years.

The London Electricity Board (LEB) then demanded she pay £2,451.39 for power she was estimated to have used within that time.

But the determined mum refused and last month won a court declaration stating she was NOT liable to pay.

Her lawyer Nik Nicol said after the case, "My experience suggests that Mrs. Bartlett is not the only one who has suffered from such behaviour.

"I hope that other people will take heart from this case and not just give in if they disagree with their electricity bill.

"If you think your bill is wrong, seek advice to have it checked out," he added.

Mrs. Bartlett said LEB's estimate was totally unfair, especially since she had been paying regular quarterly bills of about £100.

LEB spokeswoman Anne Hewitt said the bill was totalled judging the types of appliances in Mrs. Bartlett's home, using past experience and advice from other electricity groups.

She added, "We try to follow up where we think there has been a history of electricity not being recorded correctly to make it fair on our other customers."

Council rapped over plight of mum

● *A HOMELESS mother-of-two had to wait THREE years to be re-housed after her marriage broke up.*

● *The desperate mum bedded down with family and her children shared cramped conditions while Southwark Council showed "a high degree of inefficiency," according to the local government Ombudsman.*

● *He has ruled the council should improve links with tenants and apologise to the unnamed mum concerned for unnecessarily delaying her application for a new home.*

● *Southwark housing officers incorrectly graded her application TWICE and allowed months to go by.*

WEST NORWOOD: A man armed with a revolver escaped with £600 after raiding the Woolwich Building Society in Norwood Road on Monday afternoon.

CLAPHAM: Firefighters battled for an hour to bring a blaze in a warehouse in Bromells Road, undr control on Monday. Most of the roof was destroyed.

Rubbish being burnt next to the building is believed to have started the fire

SOUTHWARK: A hospital chief has agreed to attend a meeting of health watchdogs to answer their questions on opting out.

Dr. Hugh Saxton, chairman of Guy's Hospital Unit Management Board, will attend a special meeting of Lewisham and North Southwark Community Health Council on Tuesday.

The meeting, which is open to the public, will be held in the Court Room, Guy's Hospital, St. Thomas Street, Southwark, at 7 p.m.

WATERLOO: Members of the London Cycling Campaign demonstrated by Waterloo Bridge on Wednesday, against the pollution caused by cars.

WANDSWORTH: Two secondhand dealers who sold dangerous electric fires have been prosecuted by the council.

The move follows a crackdown by Wandsworth trading standards officers which found a string of faults in fires being sold in local shops, including bare live parts, unsafe switches and cables with old colour coding.

The two traders were fined a total of £240 and ordered to pay more than £900 in costs.

(a)

● PLEA — Paul Daniel.

Setback for Aids helpers

A CASH lifeline for a voluntary group which helps support AIDS sufferers has been withheld.

London Lighthouse asked Wandsworth Council for a £7,120 grant to pay for this year's running costs in providing three teams of workers. The group has been in the borough for the last 18 months and has already given aid to 10 sufferers.

But Wandsworth has deferred a decision on the grant because of a lack of information about the group.

Labour councillor Paul Daniel, who is pushing for a quick result, said, "None of the councillors on the grants and IAP sub-committee had been given the proper information. No-one was aware that the scheme was up and running.

"London Lighthouse should be given the money now and not be made to wait for months."

The 24 Wandsworth volunteers come from all walks of life to comfort people with AIDS and help with everyday things like shopping, walking the dog and washing.

Communication manager Mary Pipes said without council funding the group may be forced to cut back service.

"We would find it difficult to operate the groups in that area and we would have to consider whether we could afford to run the service," she said.

(b)

● **BRITAIN'S 44 million voters go to the polls tomorrow to express THEIR views on Europe's future.**

● **THE BATTLE for the 81 seats in the European Parliament — 78 in England, Scotland and Wales and three in Northern Ireland — also gives them a chance to send Mrs Thatcher a message on her policies at home.**

● **THE SCORE at the moment is: 45 Tories, 32 Labour and 1 SNP. But polls predict Labour will return more MEPs than the Tories.**

● **MRS THATCHER'S hard-line anti-European stance has led to fears that the nation will not be prepared to meet the challenge of 1992 when the European Community becomes a single market.**

● **SWEEPING changes to be introduced this year by Education Secretary Kenneth Baker have raised fears over future standards in Britain's state schools.**

● **CAN Britain's education measure up to Europe? The Mirror investigates . . .**

ELECTION '89
USE YOUR
☆☆☆☆☆☆
EURO VOTE
☆☆☆☆☆☆
TOMORROW

'I may have to go abroad'

THE 16-year-olds at Holden Lane High School are sweating over their GCSE exams in long hot halls right now.

Each will take an

'We get on well with our teachers'

LIKE most French parents Mme Vitel Crepin, a 42 year-old laboratory analyst keeps a close watch on her son's education.

And she was delighted when teachers at Lille's Franklin Comprehensive school spotted 14-year old Xavier's computing skills.

Classes in technical training and two rooms filled with modern computer terminals at the school have helped him build up an advanced knowledge of complex programmes.

Mme Crepin hopes the knowledge will enhance her son's job prospects.

Extras

But if he loses interest, he still has TEN

Figure 89 Design faults: (a) the page rules here have tied the AIDS story firmly in with the advert instead of offsetting it; (b) blobs, stars, big quotes, drop letters, WOBs, halftone, underscore . . . the eye hardly knows where to turn; (c) a halftone picture instead of the WOB lead headline would have got the designer out of trouble; (d) a good picture that has turned its head firmly away from the story

(c)

Special survey

SERVICE

closest to a woman's heart

ESTEE LAUDER'S Re nutriv: Frosted Apricot: £7.50.

☐ Definitely kissproof and stayed put for

MARKS AND SPENCER'S Images Range: Golden Glow: 89p

☐ Perfumed and greasy. Stayed on for 1hr 15 mins. Not kissproof.

Detective died after freak car plunge

A DETECTIVE sergeant died when his unmarked police car inexplicably plunged off the road, demolishing a hedge and 85 feet of wooden fencing.

Neil Gibson, 32, was impaled on a fence post which smashed through the windscreen of his Vauxhall Astra car.

A police vehicle examiner told an inquest that Sergeant Gibson did not appear to have been

By Echo reporter

The inquest was told DS Gibson, a member of the Serious Crimes Squad, ran two colleagues back to Runcorn police station

FREE CAMERA

WHEN YOU BUY ANY ITEM £10 OR OVER!

TOMORROW ONLY (THURSDAY 9 MARCH)

ACTUAL SIZE!!

(d)

Eleven for for Imperial

THE William Hill Imperial Cup, Saturday's feature race, has attracted the smallest field for years.

Just eleven horses have stood their ground at the five-day stage with the weights set to rise at least a stone for the Sandown contest.

Toby Balding's Beech Road will carry top weight of 11st 13lb, which includes a 6lb penalty for his recent defeat of Vagador at Fontwell.

But the trainer, who also has Little Toro among the entries, was playing his cards close to his chest yesterday.

Asked whether his pair were likely runners he responded: "Maybe both, maybe neither."

Special Vintage, winner of the Tote Jackpot Handicap Hurdle at the Esher course last month, is set to carry 11st 10lb, but his participation depends on the going.

Trainer Jimmy Fitzgerald said: "As long as it's not heavy, he will run but we will leave a decision as late as possible. He wouldn't be happy carrying a big weight, but he's used to it."

Despite the poor turn-out the sponsors are not disappointed. Hills spokesman Don Payne said: "Considering that we have had such a mild winter, it is not all that surprising.

"There is such a glut of races for trainers to choose from with the Cheltenham Festival and the County Hurdle in particular offering valuable alternatives."

Tresidder can go one better

By MATT SEYMOUR

TRESIDDER, runner up at the last meeting, is fancied to go one better in the Robin and John Simpson Memorial Handicap Chase at Sedgefield this afternoon.

He looked a much improved performer landing the Hope Inn Handicap Chase over today's course in December and confirmed that impression when he followed up at Ayr, despite a last fence blunder.

Course specialist Jody's Boy, a prolific winner this season, having scored seven times in 18 outings, is closely handicapped with Tresidder on their running behind Castlevennon in the George Mulcaster Memorial Cup over today's course in January.

On that occasion, second placed Jody's Boy held a neck advantage over my selection at the line, but I am confident Tresidder will come out on top today.

Trailing Rose, prone to the odd jumping error, finally got her head in front at Edinburgh last time, and given a clear round could be concerned in the finish.

Tsar tackles today's trip for the first time and may prove best of the rest. Tresidder however looks a safe bet to outpace them all.

Extended

Arthur Stephenson, leading trainer at the course, saddles Polar Nomad in the McEwans Best Scotch Durham National. He has progressed quickly since tackling extended distances and has outstayed the opposition, taking the Pintail Handicap Chase at Newcastle and the Tote Eider Chase over the same course on his last two completed starts.

Sure to be well supported to win the race for the Leasingthorne handler, who was successful in the corresponding race last year with Sir Jest, I am, however, inclined toward the chance of Prince Metternich.

He showed a good turn of speed to defeat Cross Master at Southwell in December and divided Rausal and Little Polveir over today's trip at Bangor last month.

The Journal Racing Extra Handicap Hurdle has a very tricky look about it.

Peter Easterby's Sword Beach, caught on one of his better day's, came to head Lottie's Fury at the final flight of the Harry Lane Memorial Handicap Hurdle at the last meeting.

The latter may have held on with a more enterprising ride and is expected to come out on top in a very competetive race.

John Parkes, enjoying his best season since his move from Richmond to Malton, may have found a winning opportunity for his Beau Guest in the Crook

ARTHUR STEPHENSON — leading trainer at Sedgefield

GOING: Heavy.

Warwick

JANUS | TATTENHAM

BEST BETS

JANUS

DURBO (nap, 4.00 Warwick)

Phil's nine-dart game

DARTS player Phil Nixon of Middlestone Moor played the perfect nine-dart game at a competition to choose the Durham representative for the British gold cup singles. He hit two 180s, treble 17, treble 18 and double 18 but later lost to Laurence Pratt, his Spennymoor North Eastern Hotel team mate.

Pre-season friendly

ESSEX will play Middlesex in a 50-over-a side match at Chelmsford on April 17 as part of their preparations for the new season.

Tonight's football

EUROPEAN UNDER-21 CHAMPIONSHIP

Albania v England (2.00 Shkodra); Scotland v France (Dundee)

BARCLAYS LEAGUE — Division III

Huddersfield v Reading

Division IV

Rotherham v Grimsby, Wrexham v Hartlepool

B & Q SCOTTISH LEAGUE — Division One

Falkirk v Dunfermline.

Division Two

Albion v Montrose, Brechin v Alloa, Dumbarton v Berwick

SKOL NORTHERN LEAGUE

Easington v Billingham Synthonia; North Shields v Gretna; Stockton v Shildon.

Division Two

Cleveland Bridge v Prudhoe; Esh Winning v West Auckland.

HFS LOANS LEAGUE — Division One Cup

Droylsden v Whitley Bay.

Greyhound results

MIDDLESBROUGH — 7.30: 1 Whitton Farmer (2) 2m 7-2; 2 Cavan Harp (5) 12m 2-1f. 5½l. 28.84. £1.45. £4.76. 2.5.1. £9.10. 7.45: 1 Pickfords Tramp (3) 9-4jf; 2 Melly Park (1) 11-4. nk. 28.13. £1. £1.82. 3.1.6. £3.82. 8.05: Jubal (4) 4m 11-8f; 2 Sandy Park (3) scr 9-4. 6¼l. 28.77. £2.35. £1.72. 4.3.2. £3.10. 8.22. 1 Glengar Jewel (5) 5-2f. 2 Oakvale Breeze (1) 7-2. nk. 29.71. £1.80. £2.86. 5.1.2. £11.42. 8.39: Brave Fox (3) 6m 5-2;

capability of the Atex mainframe system and moved straight into direct input by reporters and writers, with electronic editing and page composition although at first pages still had to be printed out as bromides to take adverts and pictures by paste-up. The result is a model of modular design based upon computer inputs and so is a useful guide to what can be done.

The pages illustrated (Figure 85) show the development of its modular style after its change of ownership and editor in 1995. Headline sizes have increased and ten of its twenty-eight main section pages have full colour but the essential balanced elegance of its eight-column format remains. Also remaining is *The Independent's* noted use of original and well-cropped pictures as main focal points. Good use of white space gives easy readability and it is less grey than it used to be, although it carries fewer stories.

The pages are characteristically 'horizontal' and are well-flagged and sensibly by-lined; the page one In Brief column and down-page cross references are helpful to the reader. Gimmicks, in fact, are eschewed, the one exception in this edition being the girl's face that projects jauntily into the revised title-piece – an eye-catcher for the news-stands.

More revolutionary in modular design than *The Independent* was the drastic re-styling of *The Guardian* in 1987 from a format which successfully exploited text and pictures at the expense of headlines (Figure 86) to one dominated by eccentrically spaced headlines and wholly single-column body setting.

The Guardian, partly in the search for stylistic purity and partly to respond to the promise of on-screen page composition, imposed upon itself a rigid headline format of lowercase Geneva/Helvetica type alternating in modules in black and light across and down the page, with an occasional light italic.

Everything was changed from masthead to the television programmes. With rare exceptions, no headline exceeded four columns in width, and never more than three lines in depth, remaining mostly in two. The unusual visual innovation (Figure 87) has been to give each headline an eccentric amount of white space below it varying from page to page, from 2 to 5 picas. Bylines have been standardized to one or two lines of 10 point Helvetica black within fine rules, and subject labels to same size single-column WOBs, likewise standard through the paper. Features pages are differentiated by the use of an occasional small strapline or standfirst.

The effect of this rather austere type treatment is to give the paper a great deal of visual continuity and a distinct typographical character. Good sized and well-cropped pictures help to break up most pages, but the prevailing effect of the type use is to lessen the influence of the designer in helping the scanning eye through the pages. The stories, although some are set in bold type as a variant, lie curiously flat and unaccented in the pages and one is conscious of the obtrusiveness of the type style. A novelty is that drop letters, which appear on all stories apart from the smallest fillers, are chosen not to a house style but are of a size and type to match the headline above. There is a recognisable logic in this, but unfortunately the result is that two facing pages can have a variety of sorts and sizes of drop letters, some bold, some light, some roman, some italic.

It is a style that grows on you. It is helped by the fact that, the various parts of the paper – home, international, financial and sports news, comment and analysis, and the various features sections and listings are well tabbed across the top of the pages, and the paper is made easy to handle both from the editing and editioning point of view, and from the point of view of the reader, once its presentation becomes familiar.

The design theme is continued inside the paper's weekend sections, Outlook and *Weekend Guardian* (Figure 88). The same 18 point rules made up of eight fine rules are used to cut one feature away from the other. The headlines, though set to the same style, are lighter and bigger and look much softened and the longer texts make the type style less obtrusive. The doctrinaire design approach, however, has resulted in the entire reading text being set mainly unjustified ragged right.

Good and bad design

The variability of newspaper design, each within its market, nevertheless demands some sort of standards by which success or otherwise can be judged, and it is here that we return to general principles and common factors applicable to all styles. To give some examples of the sort of faults that count against newspapers in design contests, we can say that a page fails if it is:

Too fussy. Excessively busy leading to poor visual balance.

Too jazzy. Too many display devices crammed into one page resulting in a failure to establish effective highlights.

Overcrowded. Trying to get too much into the page.

Dull. Failure to exploit the page ingredients so as to attract the eye, that is, without clearly defined highlights and creative type and picture balance.

Has clashes. With headline type or picture clashing with the adverts or the adjoining page.

Muddled. Failure to properly sort out and order the page ingredients into a meaningful and attractive pattern, resulting in confusion to the eye.

Has poor type balance. Stories fighting each other, or all demanding the same attention. Headlines badly fitting.

Has poor page furniture. Unskilful use of typographical devices such as panels and special setting, bylines and standfirsts in wrong type; failure to utilize artwork; captions badly placed; poor use of crossheads and other breakers.

Has poor picture use. Badly sited pictures; poor cropping and sizing; failure to utilize picture potential.

The examples in Figure 89 illustrate the effect of some of these faults.

12
Creative use of typesetting

It will be clear from the last few chapters and the pages illustrated that each newspaper has a type style based upon a number of chosen faces and the use of a regular range of sizes, either in caps and lowercase, or just in lowercase. Within each style, however, will be found some special uses of type and distinctive sorts of setting that are resorted to by the designer. Their purpose is to solve particular problems that crop up in building the pages. We examine, under the headings that follow, the occasions when these are used and how they contribute to the creative process of design.

Breakers

Long texts, or pages with a number of longish items or that are weak in pictures, can be made easier to read by the use of typographical devices or variations in body setting which we call *eye breakers*.

Panels

Panels, or boxes, can be introduced into the display. A selected story is set, usually in bold body type to differentiate it from the rest of the page, inside a frame made up of type rules or borders, either plain or decorated, as can be seen from example pages in the last few chapters. This can be to create a focal point for the eye, or to give emphasis to the story enclosed in the panel – or to perform both functions at once (Figure 90).

The sort of border used depends on the newspaper's style. The more serious morning papers favour fine or 2 point or thick-and-thin borders, while the popular tabloids, with their bolder style, go for thicker black, tone or milled (Simplex) borders, or even 'shadow' boxes, which are made up of thick black rules on two adjoining sides, opposed to thin rules on the other two. House style should denote which side of the panel the shadow is on, preference for the thicker rule being usually on the right and bottom of the panel. Proportion is

Pickled bunions

FISHERMAN Chris Lowe was given his little toe pickled in a jar after it was bitten off by a 4ft-long conger eel.

Doctors in Truro, Cornwall, couldn't sew it on after the accident 10 miles out at sea.

Chris, 32, will keep it on his mantelpiece.

Raiders blow up firework factory

CRACKPOT crooks who tried to break into a £1million fireworks factory accidentally BLEW it up.

The building went up like a rocket after the clumsy raiders set their getaway van alight with sparks from a blow torch.

Police say the men ... trying to BURN ... door to the ... technics

Figure 90 Examples of panels in screen make-up: an elegant thick-and-thin border for a political logo, a fine rule panel layered on to black for shadow effect, and the dream of old-time comps – a panel with rounded corners

important. The thickness of the rules should not be the same on a chunky three-column panel as on a short single-column one. A shadow rule of wider than 6 point (opposed to, say, a 2 point) on a single-column story would leave the body setting at around 6 picas, which is too narrow for most body faces, whereas on a big display panel, a shadow rule on a bold tabloid can reach as much as 18 points in a display-orientated tabloid. There are endless permutations. The shadow can be constructed of multiples of rules; 12 point of screen tinted rule against 2 point on the other sides can give contrast in 'colour' and a real feeling of shadow compared to total black.

The liking for type panels noted in tabloid papers is said to have started in the hot metal days, when industrial troubles reduced the flow of 'blocks' to the pages and editors could not bear to have their carefully thought-out display patterns destroyed for lack of pictures. Instead, strategically placed panels and heavy white space were used to produce contrast and pull stories away from the surrounding material. These adventures in typography led to the increased use of type panels in tabloid and broadsheet design and proved that, on rare occasions, pictures could be dispensed with.

On features display a large area of a page or spread that requires cohesion can be effectively tied in with a panelled border (Figures 73 and 76), while headlines can be enhanced as part of the display by being let into the top rule of a panel, or 'winged in'. This device is useful where a headline is required to be centred for emphasis with white space on either side. In the case of multiple-column panels, care should be taken to balance the white space against the surrounding text and headlines so that it looks like 'creative' white and not a failure on the part of the headline writer to fill the allotted space. The placing of panels should also be checked against the adjoining adverts in case they are using the same device, while another caution is to check the backing page to ensure that show-through from a heavy black rule is not going to damage white areas. Advertisers on the backing page have been known to claim against this fault.

The use of panels, especially big ones, should always be checked in the day's dummy against the opposing page, either editorial or advertising, to see that there is no clash or mismatch.

Panels can be formatted without difficulty in computerized systems in a variety of borders, including shadow box, and delivered complete with text fitted, and are an important part of modular design in papers like *The Independent*. The computer even makes possible what always remained the dream of hot metal compositors the panel with rounded corners.

Cut-offs

A cut-off performs a similar function in isolating a story from its surroundings but is used more for highlighting an item that relates to a main story with which it is included. The setting, as with a panel, is often bold to differentiate it from the main text, but instead of having a border all round it is separated from the material above and below by a cut-off rule of perhaps 2 point or 4 point, and is often set indented to show white against the main text. It can be run either with or without its own headline.

Break-outs

A long text, especially a feature, can be eased for the eye and emphasis drawn to part of it by taking out a self-contained section and running it with a separate headline, though ruled within the area of the main story. This is often called a *break-out*. Similarly, a related story on a news or features page might be used as a *tie-on* and ruled in with the bigger story under its own headline. The headline type should be chosen to match the main headline, though smaller. Here again the text can be set bold to differentiate it, or in the same type to tie it in visually. Where a tied-on story appears directly under the main intro it is sometimes referred to as a *shoulder*.

Crossheads

Crossheads are the commonest devices used to break up a long text. On a news page they are usually set to a house style, sometimes in

12 point Metro black or similar, caps for a page lead and lowercase for other tops. A common practice is to place them above a paragraph every 4 or 5 inches of text, or sometimes to use them to break up consecutive runs of short legs. The idea is to select a significant word or pair of words from the text that follows. Some purists choose the words carefully so that the crossheads in a story link together lexically, or in idea. Another school of thought asserts that since a crosshead is an eye break it hardly matters what they say, and that readers would not notice if they were nonsense words. Some papers follow this through by using white spaces placed above a drop letter or a piece of decorative rule or graphic symbol (Figure 73) in place of crossheads, or simply do without them (Figures 77 and 78).

On features pages, with their generally longer texts, crossheads can serve a valuable display purpose, appearing often as two-liners in bigger type which deliberately matches the main headline in order to lend cohesion to the type dress. Sometimes the lines are underscored, and can be set right, centred or staggered. Here, more significant words are chosen with the purpose of giving a taste of the text to the reader. The placing of display crossheads in a feature projection is carefully chosen to help balance the page, and they are an integral part of the page design.

Highlight quotes

The projection of long features texts has called into use a number of devices which would rate of little importance on news pages with their busier contents and more numerous headlines. Highlight. quotes, in which notable quotations are lifted from the text and set in a special eye-catching type such as 18 or 24 point Ultra Bodoni or Ludlow black, can be found as visual breakers in projections of interview-based features (Figure 91). These look well in four lines of single- or double-column placed strategically around a page or spread. Unlike crossheads, these should be located in the middle of paragraphs so that the eye is not tempted, by their extra depth, to stop reading the text at that point. In the case of double-column quotes it is important to ensure that their display purpose does not interfere with the run of the text, and they should ideally be placed at the foot of the display or under an intro or picture where the text can conveniently thread round them. If the layout is heavy on headline, or if the layout style permits it, highlight quotes can give useful 'colour' by being used as WOBs or BOTs.

Bylines and standfirsts

Bylines, which on news pages are often simple lines of 8 point or 10 point Metro or Helvetica set to style above intros, can be harnessed for display purposes on a features spread. They can be panelled in

Figure 91 Quotes used: (a) as the main headline; (b) and (c) as text breakers

RONALD REAGAN YESTERDAY

'You can't massacre an idea . . . you cannot run tanks over hope . . . you cannot riddle a people's yearning with bullets'

(a)

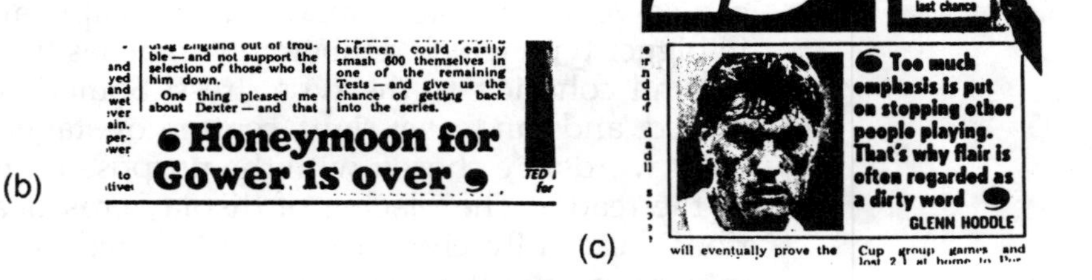

(b)

'Honeymoon for Gower is over'

(c)

'Too much emphasis is put on stopping other people playing. That's why flair is often regarded as a dirty word'

GLENN HODDLE

Figure 92 The standfirst – that is, special setting above the intro – can be used both as a useful explanation and as a design feature

Applaud the politicians, but don't raise the roof

The international ozone conference has been a step in the right direction, but there is a long way to go, says Jack O'Sullivan

IN A matter of months complacency has changed to platitude and thence to virtual panic over the destruction of the Earth's ozone layer. The international ozone conference in London closed yesterday amid general feeling that the world is seriously out of joint.

Quite a fright is required for the Earth's major polluters — Britain included — to advocate so strongly elimination of chemicals used in a multitude of household goods, from fridges to floor polish.

The fright over what we now

escapes, although scientists suggest it is used as an industrial cleaner in Eastern Europe. Each emission survives for up to 70 years.

The conference was a political triumph for Mrs Thatcher, whose "green" conversion was displayed to much congratulation on the world stage. "It makes a real change to see Britain leading from the front rather than following up at the rear," said one of the pioneers in ozone research, Prof Sherwood Rowland from the University of California.

Charities learn to be City slickers

● The long-awaited white paper on control of charities deals with a giving industry that has become big business. NICK FIELDING reports

SEVENTY years old this week, and about to mark the fact by a jamboree for 8,000 people at Alton Towers, Save The Children Fund is the very model of a modern major

and set in display type, perhaps with the word 'exclusive', and dropped into the middle of a page, or under an intro as a pivot point for the text to turn on. Or they can be incorporated in a *standfirst* run in special type next to the intro (Figures 78 and 92) to explain to the reader why this particular feature is significant, or even – where the name is important – be worked into a 24 point or 30 point strap line above the main heading to introduce the page. Where the writer's name is the most important part of the story, as with a big name columnist or a very special outside contributor, the name can be bigger than the headline, or be worked into a decorative *logo*, around which the display is constructed (Figure 51). Such a byline can be the most important offering of the day's edition.

Decoration

Every newspaper has its favourite pieces of type decoration and their use dates back to the flourishes and illuminated capitals of the monastic scribes, which were translated into typographical equivalents by the early printers. Take *drop letters* (which we described in Chapter 3). They have been in and out of favour with the most diverse variety of newspapers and just when they appear to be going out of fashion they reappear in some unexpected quarter, being favoured at the time of writing by such opposed papers as *The Guardian* and *The Times* (Figures 81 and 86) and rejected by *The Independent* and the *Daily Express*.

In many ways the drop letter in its normal position is the most expendable form of adornment since it is the body of the text and the centres of pages that need the benefit of decorative relief rather than the first paragraph of a story. Yet, used carefully, it can give a touch of style and elegance to a page. For instance, in long feature drops with generous white space above them, spaced out across a long read, can act as attractive eye rests while at the same time marking natural breaks in the text (Figure 17).

Blobs and squares

Blobs, either open or closed and ranging from 8 point to 12 point, were once associated with the popular tabloids but have now infiltrated all manner of newspapers. They turn up as text markers on columns of *nibs* (news in brief), summaries, list of things and at the beginnings of *tie-ons* (or *nuggets*) to stories. Black or open *squares* of similar size are an alternative to blobs for the same purposes (Figure 93). The one danger to watch in a permissive layout environment with devices like this is the spotted dog syndrome, which is to be diagnosed in pages in which a rash of blobs and squares has developed into a sinister disease. A side effect of the disease is that the emphasis intended to be given by their use is totally negated by their number (Figure 89).

Bold caps

For a less eye-boggling solution in such situations, setting the first word of a (limited) number of items in *bold caps* is effective. The device can also be used to give emphasis to a particular word or quality in a story, although *italic* is perhaps safer since the potential visual damage to the page is less.

Drop figures and quotes

Drop figures, used in the same way as drop letters – set against two or three lines of body type at the beginning of an item – can give a

■ DEL Monte Dried Fruits are a great healthy snack to give you an energy boost. And they're low in fat, too.

They are packed in re-sealable bags and canisters and are available in most major supermarkets, priced from 94p to £2.59.

■ SOLEMATES, a range of insoles that keep your feet cool and dry, are being launched by Sorbothane and DuPont.

■ SUNKIST C is a pack of chunky sweets in four refreshing flavours, orange, lemon, pineapple and citrus. Each pack contains the recommended daily intake of vitamin C. They are available at newsagents, supermarkets and chemists for 25p.

■ DURACELL have launched the flashing safety light for bikes, bags and clothes. The superbright light can be seen from up to 600 metres. It also has a sturdy belt clip so it can be fixed onto bags and clothes. It comes with two Duracell batteries, costs £10.99. Available from Halfords and bike shops.

FRUITY . . Del Monte's snacks

Are you at risk?

ONE in 12 women gets breast cancer but for women with the illness in their family, the risks are much higher.

Even if there's none in your family, you are slightly more at risk if:

● You've never had children or if you have children in your thirties.

● You have a late menopause.

● You bottle-fed your babies (breast feeding helps to protect against breast cancer).

● You eat a high-fat diet.

● You start taking the Pill in your teens and carry on taking it for more than eight years.

YESTERDAY, the town was bracing itself for an invasion as Mr McGuire claimed nearly 100 responses to his offer.

'I think the reaction will be of horror,' said town council chairman David Stebbens, an estate agent. 'This is a holiday resort which attracts people for peace and quiet.'

Zoe Hall, spokesman for local Tory MP Ralph Howell — who has promised to investigate Sheringham's own big issue — said: 'There are no jobs for anyone around here, it's a retirement town. People who respond to the advert will be like fish out of water. It just

Figure 93 Squares and blobs are useful in lists. A decorative stand-up drop letter, used with white space, makes long texts easier on the eye

distinctive touch to a bold feature layout in which a numbered list is important and needs highlighting (Figure 93). Drop quotes, another variation of the drop letter and set in a size equal in depth to two or three lines of body type, can enhance and give prominence to a special quoted section in a feature, but not on a number of such sections in the same story. Two drop quotes on a page, one to open the section and one to close it, can be safely accommodated, but half a dozen confuse the eye. When falling badly, such as along the bottoms of column legs of text, or fighting with blobs and squares on the same page, their use can bring on an extreme and terminal form of the spotted dog syndrome, a condition not unknown to some popular tabloid papers on their bad days (Figure 89).

Stars

Stars, black or open and from 12 point to 18 point, are used more to jolly up blurbs, horoscopes and competitions than to itemize columns of things and can look right on the right material. There is some justification for them on a column of show-business gossip, but little call for them elsewhere in the paper. Use with care.

White space

One of the simplest and most effective forms of decoration to the type area is white space used judiciously and artistically to separate the items and components of a page. It can stop a busy layout looking overcrowded and give great elegance when used in place of dividing rules on a features page (Figures 77 and 78). Used elsewhere, however, it needs to be planned into the page as an act of artistic judgement rather than fudged in as the spin-off from 'windy' headlines.

Special headline effects

There is little call on news pages for headlines outside the computer's normal type range, but on features pages most papers resort at some time to Letraset transfer type or drawn headlines to achieve a special display effect. Royal features look more seemly in older types such as Baskerville or Bembo lowercase, often blown above normal size, sometimes with instant art crowns woven into headline or strapline. An adventure series can be given great colour with a display heading in 'tea chest' caps from a Letraset sheet, or a period cameo feature in a Victorian cursive. A graceful seriffed lowercase Century light or Garamond (or, if in sans serif, then in the finest and lightest Univers or Futura) gives a feminine look to a women's page display (Figures 17 and 77).

In addition to the instant stick-on type available, the computerization of printing has made many new type tricks possible. The facility of laser printing of type bromides where paste-up is still used means that no newspaper need go without a wanted type effect.

Some manipulation can be carried out on the art desk on letters to improve display headings. On biggish descending lines in lowercase, knowledge of word shape can be used to enhance the visual effect by tucking the descenders into the ascender space of the line below. Sometimes a design purpose can be achieved by literally entwining the ascenders with the descenders. Kerning of the space between the letters, as we have seen (Chapter 3), can improve the visual effect of lines, especially those with a mixture of straight and rounded strokes.

Ligaturing

Ligaturing, used knowledgeably, can make an effective line in certain features. This is the method of connecting letters together as used commonly in the diphthongs *æ* and *œ*. Originally it was used only with lowercase letters, in particular double *ff*s or in connecting *f* and *t* together on their horizontal strokes, and the aim was to save space in consonants on wooden and metal type bodies. Early font creators were probably influenced by authors' preference for consonants. Dr Johnson Ball, in his biography of the type designer William Caslon, says: 'Charles Dickens will empty the vowel boxes long before those of the consonants, whereas the style of Lord Macaulay's will run heavily on the consonants.'

This purpose has long since gone, but, aided by the computer, ligaturing has come back into vogue, especially in book cover, magazine and advertising design, but also in newspaper features page layouts (Figure 94). It can draw the eye and in some cases replace the picture content with a stylized titling focal point.

Capital letters now play a greater part than in traditional ligaturing. Curved and otherwise distorted lines of type, in particular, demand that the corners of certain characters should overlap. Ascenders and descenders can be overlapped into the 'beard' spaces

Figure 94 Ligaturing – an old device – takes on a new design role in these examples from the drawing board

of the letters and juxtaposed to create a novel effect. An effective method is to use tints to define or highlight the overlap.

Drawn type

Some papers with predominantly seriffed type formats tried before the computer age to improve line counts by mechanically shrinking the width, but this never looked right in lowercase, though it worked in caps. At the *Daily Express*, in the 1970s, a layout artist produced a drawn version of the whole Century bold alphabet to 120 point, with all its furniture, as a mock Century extra condensed. This was used for a time on front page splash headings where it contrasted well with the masthead.

In its search for literate headlines which met with its requirements of heavy sans banner type, the *Daily Mirror*, in the 1950s, adopted a condensed sans face drawn in house by one of its young composing

Britain yesterday:

PULL YOUR FINGER OUT!

PRINCE Philip made an astonishingly frank remark yesterday to bosses and workers in British industry. He said "I think it is about time we pulled our fingers out."

Figure 95 A famous headline demonstrates the *Daily Mirror*'s distinctive splash type of the 1950s and 1960s, 'Kreiner Condensed'. Its hand-drawn origin is betrayed in the broad 'N' and odd exclamation mark

room deputies, Len Kreiner, who had heard that H. G. Bartholomew was having trouble finding the right type. Though the W, N and M were too full by normal sans standards, the type (Figure 95) struck the right chord for the *Mirror*'s poster style and Kreiner Condensed remained a stock *Mirror* headline face for nearly thirty years until editor Michael Christiansen replaced it with Placard in the 1970s.

Drawn type for several effects remains an option in page design since it can be transferred to the screen, but the computer can fulfil most requirements.

Special setting

In some areas of the paper visual familiarity is the paramount aim. Nowhere is this more so than on 'listings' features such as the television guide and race programmes. Not only should they appear in the same part of the paper in much the same shape so that they are instantly findable, but they should be set in such a way as to be easily readable for quick reference.

The formatting facility in computer systems has taken the hard work out of this sort of typesetting and it is now possible to bring together setting shapes and combinations of body type that will highlight times and names in bold caps for the reader at a keystroke, and identify special information entry by entry (Figure 96). Once the editor is satisfied with the body size, width and amount of indent required, and a workable overall page or spread has been devised to allow for pictures in stock positions, the listings subeditor has only to keyboard the daily or weekly details into the page.

Some newspapers have taken advantage of the ready-formatted listings material supplied daily on-line by the Press Association and other agencies. Such a format does not preclude corrections and updates being keyboarded in provided the overall setting depth (and

5.10 THE COUNTRY BOY By Bernard Ashley. Six part drama series about a Kent sheep-farming family who are directly affected when a local chemical company dumps pesticide in a river. Ben discovers that Duke is dead and realises that he must have swallowed the same poisoned water. (S)

5.35 NEIGHBOURS (Repeat of this afternoon's).

6.00 NEWS, WEATHER

6.30 WALES TODAY.

7.00 WOGAN Terry talks to Glenn Close and John Malkovich, who star in *Dangerous Liaisons.*

7.35 BEST OF BRITISH Sir John Mills introduces extracts from 50 years of Rank films, focusing on film portrayals of historical figures.

8.00 DALLAS Although Miss Ellie reluctantly agrees to the sale of Southford land and Casey Denault is jubilant, the business does not quite go according to plan. JR is surprised by a visit from his wife. (S)

8.50 POINTS OF VIEW With Anne Robinson.

9.00 NEWS, WELSH NEWS, WEATHER.

9.30 Q.E.D. The Mystery Of Tears. ***(See Starchoice)*** (S).

7.40 Every Second Counts. Six more spouses fight to win an African safari in Paul Daniels' hurry-up quiz.

8.15 Dynasty: Grimes and Punishment.★

9.0 News; regional news; weather.

9.30 Jumping the Queue.★ In this two-part love story dramatised by Ted Whitehead from Mary Wesley's novel, Sheila Hancock plays a middle-aged woman who has decided to end her life rather than slide into senility. But before she can swim to oblivion off her favourite beach, she meets Hugh (David Threlfall), a young fugitive on the run from the police for killing his mother.

10.50 The Mephisto Waltz. A dying pianist seeks to bequeath a journalist a ghastly legacy in this horror thriller starring Jacqueline Bisset, Alan Alda, Barbara Parkins and Curt Jurgens. Made in 1971.

12.35 The Rockford Files. Love Is the Word. James Garner renews his romance with a blind psychologist (R).

BBC1

8:30 am Roobarb. **8:35** The Raccoons. **9:00** On the Waterfront. **10:55** Cartoon Double Bill. **11:10** Film: "Song of Norway" starring Toralv Maurstad and Florence Henderson with Harry Secombe, Robert Morley and Edward G Robinson. **1:25 pm** News. **1:30** Grandstand including FA Cup Final between Everton and Liverpool from Wembley Stadium (kick-off at 3:00). **5:15** The Pink Panther Show.

5:35 News. **5:45** Regional News and Sport. **5:50** MacGyver. **6:40** That's Show Business. **7:10** Bob Says... Opportunity Knocks. **8:00** Columbo. **9:15** News and Sport. **9:30** Midnight Caller. **10:20** Match of the Day: The Road to Wembley. Highlights of today's FA Cup Final between Everton and Liverpool. **11:20** The Odd Couple. **11:45** Film: "The Anniversary" with Bette Davis, Jack Hedley and Sheila Hancock.

BBC2

2:45 pm Network East. **3:25** Film: "Thousands Cheer" starring Gene Kelly with Kathryn Grayson. **5:25** The Week in the Lords. **6:05** Civilisation. **7:00** Ways of Seeing. **7:40** Newsview. **8:25** The Shock of the New. **9:30** Film: "The Go-Between" starring Julie Christie, Alan Bates and Dominic Guard. **11:20** Film: "The Quiller Memorandum" starring George Segal, Max von Sydow and Senta Berger. **1:00-1:35 am** Rapido.

LONDON

6:00 am TV-am Breakfast Programme. **11:00** The Chart Show. **12:00** ITN News followed by ITV National Weather. **12:05** Local News and Weather. **12:10 pm** The Cup Alternative: "Carry On Again, Doctor" starring Kenneth Williams, Sidney James, Charles Hawtrey and Barbara Windsor. **1:50** The Cup Alterna-

Figure 96 Listings setting should combine readability with the successful highlighting of names and times. In these examples the reverse indent of the first two is more helpful to the eye than the run-on style of the third

in effect the number of words) of the format is adhered to. The programmes can then either be printed out in blocks for paste-up on page cards, or dropped into position on screen for full page composition.

To achieve speed in page production using this sort of setting it is necessary to stick to a format once adopted and to ensure that the page layout for which it is designed is not jeopardized by differences in advert shapes day by day and week by week. A drastic change of setting or page display should be necessary only where a listings feature is felt after long use to have grown stale visually, where it has proved to be faulty in its purpose, or where a radical relaunching of the listings part of the paper is being undertaken to fit in with a new editorial policy.

13
Mastheads

We have said that the aim of the designer is to give a newspaper a recognizable character as well as to make the contents of the pages readable and attractive to the eye. Nowhere is the recognition factor more important than in the title piece or name of the paper which occupies the area at the top of page one usually referred to as the masthead. Strictly the masthead includes the information on date, price and serial number and sometimes the edition marker. It can also include a pictorial motif, or badge, such as the eagle on *The Independent* or the crusader on the *Daily Express*, and so when referring solely to the name of the paper we will call it the *title piece*, although the term *logo* (word) will also be used.

Newspapers traditionally look to continuity in their title piece. Page one has an inevitable shift and flux in contents as the main news of the day is presented to the best advantage; there might, during the course of a few years, be the occasional typographical revamp to coincide with a drive for readership; there might also, though less frequently, be a change in size from broadsheet to tabloid, or vice versa. During all this, great store is placed on the function and typographical appearance of the title piece in assuring the reader that it is the same paper. Consequently, change in the shape and size of the title piece needs to be carefully thought out.

This does not mean that it should never change. An important relaunch of a paper that had gone into decline can be signalled by giving the logo an eye-catching new design. Such a title piece can be an integral part of the publicity surrounding the relaunch, although it will not, of its own, convince readers that things have improved unless the fundamental causes of the paper's decline have been attended to. In the same way a new paper, having carefully thought out its masthead to suit its market, must also ensure that the contents have been properly thought out, too.

The title piece, in short, is the brand name that helps sell the product. Through its use in advertising, street bills, slogans and in other forms of publicity it can push the sales of a paper by planting the notion of its expected character or excellence in the mind of the recipient. It is thus a valuable promotional aid to readership and circulation.

A title piece can also overstay its welcome. Designer and typographer Alan Hutt, writing in 1960, said: 'Papers which redress themselves with agreeable headline and text types, effectively made up, retain hanging over their now smart shop window – the front page – the ancient and grotesque sign which accompanied the spindly news titling heads and the muddy minion text of their founding fathers. It is a position that no ordinary self-interested shopkeeper would tolerate for a moment, and for obvious reasons. That these reasons seem often not to be obvious to newspaper proprietors and managements is a curiosity of trade conservatism quite out of place in the second half of the twentieth century.'

Updates and spot colour

Nowadays one would want to qualify Hutt's general condemnation on the ground that a change of title piece is a serious undertaking akin to changing a brand name of a product. However, in the early 1960s there was some truth in what he said. Worries over the rivalry of television led to newspapers of the period making a close examination of their image and the decade, to some extent, took note of Hutt's strictures by witnessing a revamping, and even the extinction, of some familiar, if dated, mastheads.

The *News of the World*'s famous scroll, showing Britannia presiding over the paper's name in an elaborate Victorian face that would have looked more at home on the side of a canal barge, was jettisoned after more than 100 years in favour of a Rockwell slab serif logo in caps across the top of the then broadsheet. With it went the advertising ear pieces on either side which had earned handsomely but had added to the dated look of the masthead. The new title piece was broken into a two-liner when the paper was relaunched as a tabloid in 1983 (Figure 97). It became a white on red colour seal printing off a separate cylinder in the style that had become the practice with the popular tabloids. An interesting postscript to the change occurred during the 'tabloid or not' debate in the 1980s when it was suggested to the proprietor, Rupert Murdoch, that the original logo might be brought out of the cupboard, reduced and used as a nameplate on the new tabloid. 'I'm not in the antique business,' was Murdoch's comment.

The Sun also underwent a traumatic change of masthead in the 1960s. Launched by Mirror Group Newspapers in 1964 as a successor to the ailing *Daily Herald*, which the group had bought from Odhams, it appeared at first as a 7.5 cm wide panel containing a tightly spaced, and blown up, single word SUN in Franklin extra condensed with an orange coloured disc standing alongside it (Figure 98). The idea was that the panelled masthead could be moved about the top of the page from day to day, and even from

Figure 97 *News of the World*: 1910, 1982, 1995

edition to edition, to suit the size and shape of the splash headline – not a concept that has found many imitators since. One problem was that the orange disc, inked and printed off its separate cylinder under the letterpress system, would wander away from the title piece which printed black and so remained static in its hot metal forme. A more serious defect was that the type of the logo did not differ enough from the surrounding headline type and tended to merge into the page. Thus, in trying to be new and different the masthead gave up the one great advantage of the old black letter or hand-drawn title pieces – their visual differentiation from the type on the page.

Differentiation in appearance from the rest of the page has to be the key to an effective title piece. True, there are successful mastheads designed in the same typeface as the rest of the front page and printed in black as part of the page forme. *The Times* (Figure 81) is one, and there are others, but they succeed by being placed in a static position clear of the page contents and surrounded by protective white space. Ideally they are in caps against lowercase splash headlines, or lowercase against caps splash headlines.

A total disregard for *The Sun*'s broadsheet title-piece was shown by its new proprietor, Rupert Murdoch, and its editor Larry Lamb, in 1969 when they sat down to dinner to discuss relaunching the paper as a tabloid. As the meal drew to a close the new editor doodled the words THE SUN with a red Biro on a table napkin showing white words on a red background. 'That's it!' said Murdoch.

Figure 98 *The Sun*: 1968, 1969, 1995

The Sun's new title piece which has since influenced other tabloids, was successful for two reasons. One, the colour red attracts the eye more than any other, and two, the typography of the carefully hand-drawn sans letters was eye-catching in its own right. Two decades later the only change has been to widen the letters of the word SUN to produce a wider panel of red (Figure 98).

One of the reasons for the more decorative mastheads found before the 1960s was the reluctance by editors to choose colour as a means of differentiation due to the unreliability of spot colour as used in the letterpress process. It had long been realized that with the method of printing stop press or edition markers or title motifs in red, blue or green from cylinders separately inked from their own reservoirs it was impossible to place spot colour accurately.

Only by printing the entire title piece in colour, as in *The Sun* example above and as adopted by other popular tabloids, could spot colour be safely used. Even then, in creating a position for it to print on page one, a spare pica of space had to be left vertically and horizontally around it in the page forme to allow the whole panel to 'wander' as the colour cylinder operated independently against the printing plate while the page went through the press.

Old ways and oddities

Up to about 1890 the convention in mastheads had been fairly simple. The Victorians loved the decorative quality of the Old English type as the most effective form of differentiation. Often

(a)

(b)

(c)

(d)

Figure 99 (a) How *The Times* started life as the *Daily Universal Register* on Saturday, January 1, 1785; (b), (c) and (d) some nineteenth century mastheads – a variety of approaches

Figure 100 *Daily Telegraph*: 1872, 1995

(Figures 99 and 101), the title piece consisted of the name of the paper across the top of the page bisected by the Royal coat of arms – to which they were not entitled – in the manner used by *The Times* today. In a modified form, and shorn of some of its embellishments, the Old English has retained its adherents since, as can be seen in the *Daily Telegraph* (Figure 100) and the *Daily Mail*. Invariably, the type was thickened in stroke so that it would not break up or fill with ink, especially where a brass title plate was exposed to frequent re-use in the hot metal process, while the letters, at first in-lined, were adapted to simpler solid strokes.

ASK FOR CHRISTMAS HARMSWORTH. **Daily Mail.** ASK FOR CHRISTMAS HARMSWORTH.

NO. 1,117. LONDON, MONDAY, NOVEMBER 20, 1899. ONE HALFPENNY.

Figure 101 *Daily Mail*: 1899, 1955, 1962, 1995

With the growth of the popular press, starting with the nineteenth century popular Sundays, of which the *News of the World* is an example, elaborate drawn title pieces came into their own. The 1890s *Daily Graphic* (Figure 102) displays a highly stylized masthead the full width of the tabloid-sized page and filling nearly a quarter of its depth in which cupids whisper into the ears of goddesses flanking the words *Daily Graphic*. The symbolism of motifs was to endure in masthead design.

Oddities abound in the world of drawn title pieces. The Nazi populist illustrated paper *Illustrierter Beobachter*, published in Munich at the time of Hitler's installation as Chancellor of Germany in 1933, produced an amazing logo which featured its initial letters in an opposed shadow format, making a name plate that covered 20 per cent of the page depth with the initial letters flanked by an enormous scrolled WOB design on which the full name is projected

HOW I FLEW ACROSS THE CHANNEL: BY M. BLERIOT. (See Page 7.)

Figure 102 *Daily Graphic*: 1909

Figure 103 *Illustrierter Beobachter*: 1933

Figure 104 *Hoy*: 1982

in bold upper and lowercase script. The entire masthead is surrounded by the price and date of the publication, together with the Nazi emblem and eagle. The result (Figure 103) combines dignity with a feeling of revolutionary fervour. The tints laid against the black of the shadows are daring in view of the crudity of printing presses of the period.

The new printing systems of the 1980s have spurred more ingenuity in mastheads. The American designer Rolfe E. Rehe, in his 1980 revamp of the masthead of the Ecuador daily, *Hoy*, which means 'today', produced a brilliant design (Figure 104). He turned the letter O in Hoy into a rising sun with the use of graduated white rules across the perfect disc. Small bleach-out motifs alongside the title piece and referring to coverage of the day, and the date lodged around the descender of the 'y', complete the masthead furniture.

Reversals and direct colour

A simple device for differentiation that became popular in the 1950s was the reversal of the title of the paper as white type on black or on tint, or black type on tint, as shown in the example from the *Daily Mail* (Figure 101). It was also used by the *Sunday Graphic* and *News Chronicle*. Though few of this genre have survived (the *Daily Mail* quickly went back to a simplified Old English logo) reversal was to become the accepted way with spot colour logos as in the example of *The Sun*, described above. The movement into web offset printing in the 1970s and 1980s gave further scope for these papers preferring a colour logo since it provides the facility of direct colour printing with

South London Press

South London Press

Figure 105 *South London Press: 1960, 1989*

its fine accuracy of register. The morning paper *Today*, launched by Eddie Shah in 1986, broke new ground by being the first national daily to have a directly printed colour title piece (Plate 3 in the colour plate section) designed into the page. The initial failure of *Today* under its first management to take the share of the national market it anticipated added fuel, however, to the arguments of the detractors of direct colour. The other web offset launching of 1986, *The Independent*, noticeably rejected colour for its pages, except on special occasions, and opted for a traditional black title piece across the top of the page. In this way it aligned itself in style with The *Daily Telegraph* and *The Times* whose readership market was closest. A number of provincial papers (Figures 105 and 106) printing on web offset presses have taken the opportunity to use the direct colour facility to bring in new masthead designs, some of them displaying great ingenuity. Such mastheads have become a characteristic feature of the many free-sheet papers that have been

Hastings & St. Leonards Observer

Hastings & St. Leonards Observer

No. 7161 SATURDAY, AUGUST 9, 1958 Price FOURPENCE

HASTINGS & ST. LEONARDS OBSERVER

SATURDAY, APRIL 18, 1959 No. 7197 PRICE FOURPENCE

Telephone 51351 No 8734 THURSDAY, MAY 25, 1989

Figure 106 *Hastings and St Leonard's Observer:* from black letter to the latest web offset title piece

launched in the last two decades, which have not had to worry about preserving the reader identity of old title pieces.

Designing a title piece

Hand drawing of title pieces, colour and otherwise, has occurred not only in the elaborate cupids-and-scrolls genre but on more recent occasions where a standard titling face has proved inadequate for the effect wanted, or an update has been called for. Northcliffe's title piece for the *Daily Mirror*, the 'women's' paper he launched in 1903, was designed appropriately in a nineteenth century roman Old Face, used at first across the top of the page and, from 1939, broken into two lines in a left-hand corner position. Here it continued unchanged until the 1950s, when it was redrawn in-house slightly bolder. In this form, in an astonishing example of continuity, it carried on into the 1970s, undergoing reversal into white type on a red panel before, under Robert Maxwell's ownership, the paper was given a new hand-drawn italic sans title piece, which was later changed again to its current one (Figure 107).

The launching of the *Daily Star* on to the national market in 1978 required a logo that would enable it to compete in the popular tabloid field against the *Daily Mirror* and *The Sun*. With the briefing that the design should be something of a 'spoiler' against the other two titles while at the same time having an independent character of its own, Vic Giles designed a drawn logo, using capitals as against the lowercase of *The Sun*. The development of the logo is shown in the illustrated example (Figure 108).

Figure 107 *Daily Mirror*: 1918, 1939, 1959, 1972, 1995

The crucial part was the emphasis given to the letter S to match that of *The Sun*. The intention was to make the design roman but the way in which the S developed, with more emphasis across the bottom of the letter against the slope, determined that italic would give a greater legibility at a distance. Compressing the T, A and R, while keeping the boldness of the basically Futura or Antique Olive letter was more difficult until it became obvious that the only way was to ligature the ST and the AR, while pushing the bottom of the A character against the upright of the letter T. The result enabled space to be left for the word DAILY centred on the visual space between the extreme right of the S and the far right of the R, in the same way as with the word THE in *The Sun*. In a subsequent change the style of the logo was preserved but given greater distinctiveness by cutting the red away from around the letters, leaving a thick red

Figure 108 *Daily Star.* 1978, 1986, 1988, 1995

Weather
Cold and windy, with threat of sleet and snow. Details page 2.

THE IRISH PRESS

Vol. LVII. No. 15 TUESDAY, JANUARY 19, 1988 The Truth in the News C PRICE 50p

Figure 109 *Irish Press*: January 1988 to April 1988 – a dramatic change in market and style

outline. Subsequently, the editor changed to a red type title-piece following closely the original white on red.

The revamp and relaunch of the broadsheet *Irish Press* as a tabloid in 1988 presented another example of a logo drawn to suit a particular purpose – in this case the broadening and popularizing of an old-fashioned paper's content aimed at expanding the sales. Again, boldness was called for in a market against existing tabloids. One of the requirements, however, was to incorporate the newspaper's phoenix motif, known as the Gaelic eagle, which had graced its broadsheet masthead and had survived several previous revamps.

Vic Giles, the designer on this occasion too, produced a bold two-line title piece with the lettering in red, and incorporating on the left a silhouette in blue of an alighting eagle overprinted with the paper's motto THE TRUTH IN NEWS – also a wanted part of the design (Figure 109). Another requirement was that the whole masthead could be easily reduced or enlarged for use on labels, stationery, posters, vans and television advertising. Due to the paper's imminent replanting, there was more scope for design in that the masthead would be colour printed by web offset as part of the page.

Giles decided, for the sake of continuity, to use the Times bold of the old design in upper and lowercase for the word Press, which would give space in the ascender area for a small version of the word IRISH in extended caps. A difficulty occurred immediately because of the enormous width of the capital P in Times bold when increased in size. This was resolved by using Letraset stick-on type and cutting away the rondel of the P and replacing it with a lower-case O, which worked after retouching.

A good flow was achieved on the word Press by ligaturing every letter. At 167 mm wide the depth proved to be 67 mm. With the eagle touching the P on the left the whole design stretched to 200 mm, leaving enough space to the right on the tabloid page for blurbs, captions, narrow pictures, etc. to suit the needs of page layout. The word IRISH in Antique Olive Nord 18 mm deep was centred on the

Figure 110 Launching of a new daily: three of the many title pieces designed for *The Independent*; fourth, the design that was chosen; fifth, a further change in 1995. Reproduced with permission

THE INDEPENDENT

Published in London 25p THURSDAY 28 AUGUST 1986

FRIDAY 13 JUNE 1986

THE INDEPENDENT 25p

The Independent

PRINTED IN LONDON MANCHESTER AND PORTSMOUTH

visual white above the word Press. Finally, the masthead was separated from the page contents below by an 8 mm deep WOB strip containing date and price.

The deliberate contrast between serif and sans serif in the title piece reflected the mix of serif and sans serif in the type dress of the pages, while the caps of the logo meant that splash headlines could be in lowercase, as wanted, with the foil of smaller sans headings in caps.

Thus was created a tabloid page of distinctive appearance.

Totally different in its birth was the masthead designed for the new broadsheet *The Independent* for its launch on October 7, 1986, which was the product of a committee working relentlessly at a variety of prototypes by many hands, some hand-drawn, some typeset. The final version, an elegant chiselled serif line in black across the top of the page with an alighting (or taking off?) eagle on the left was, in fact, a re-hash of an earlier version produced close to launching day in almost a mood of desperation (Figure 110).

Michael Crozier wrote in *The Making of The Independent*: 'The masthead was still giving us problems. We had all seen so many that it was extremely difficult to make an absolute choice.... In the end we returned to the beginning and the original idea of Hitchens for a chiselled masthead and an idea subsequently repeated by Thirkell. However, all the previous versions seemed to go wrong in some way and I asked Michael McGuinness if he knew of someone who could hand-draw a new version. He did and the present masthead drawn by Ken Dyster of the Mike Reid Studios in in-line Bodoni is unique.'

Figure 111 *The Guardian*: 1951, 1952, 1985, 1995

More daring and controversial was *The Guardian*'s new masthead (Figure 111) which appeared with the revamped paper in 1987 displaying a row of cut-out head and shoulder pictures of inside page personalities standing above the title piece. The faces each carried alongside a cross-reference to the appropriate page. They may subsequently be replaced by more conventional blurbs. The title-piece itself broke new ground by having the word Guardian in 120 point lowercase of the Geneva black lowercase splash headline, with the The lodged closely next to it in a Garamond italic serif face of the same depth, and the whole set right to leave eccentric white space on the left of the page.

Giving the pages identity

Not only is a newspaper *The Guardian, The Sun* or the *Yorkshire Post* because the title piece tells you it is. It must also, as the pages are turned, feel like what it is supposed to be if the reader is to be satisfied and comfortable with it. This is not just a question of content or attitude, which is what fixes a newspaper in its readership market, but of typographical style. We have considered the relationship between design and market and how this shapes the typographical style of a newspaper. We will now extend this to show how design, while grabbing the eye with the juxtaposition of headlines, text and pictures, also gives the reader this feeling of comfort and continuity by the use of what we call typographical 'signatures'.

Figure 112 Signatures: the title piece as an identity motif

Sun TRUE CONFESSIONS

ALL of us have a guilty secret we try to hide, but are desperate to share. So get it off your chest in True Confessions.

Did you NICK £10 from mum's purse and keep quiet? Did you BUMP into the boss's car and never admit it? Did you BREAK your neighbour's window and blame lager louts?

Are you burdened with remorse for something even more serious? Make a clean breast of it today. Send the details, in a letter, to TRUE CONFESSIONS, The Sun, 1 Virginia Street, London E1 9XP.

Please include a daytime phone number so we can talk to you about your story. We will publish the best, but in a way which will NOT identify you.

TELL US YOUR GUILTY SECRETS

The Northern Echo

Tuesday, March 7, 1989

Time for rail safety answers

YESTERDAY'S second train crash in 48 hours tragically underlines the point we made yesterday about the

DAILY STAR

Madness on the railways

RAILMEN are rushing headlong into the buffers. The national strike called for next Wednesday may do exactly what the union

The title piece, which we have just discussed, is the first and most important of these – the piece of typography that embodies the brand name of the paper – and its use on stationery, labels and various forms of advertising is vital to the image the paper projects. It can also be usefully deployed inside the paper to stress those areas where the paper's attitude or service to the reader are paramount (Figure 112). The leader page, on which is found the paper's daily or weekly opinions, is an example of this image projection.

To illustrate this we turn to the leader page of the *Daily Mail* (Figure 113). Over the word Comment in white on black type above the regular 10 point setting of the leader, or editorial opinion, is a reduced replica of the *Daily Mail* Old English logo. It is a logo that, with minor changes, derives directly from the very first *Daily Mail* in 1896.

The pica black vertical rule alongside the leader, the well-spaced Rockwell Bold lowercase of the main headline, the 4 and 6 point black rule breakers, the Century Bold crossheads . . . all are likewise unchanging stylistic signatures on a page on which high regard is paid to continuity and familiarity.

The byline logos of the *Mail*'s regular columnists Keith Waterhouse, George Gordon's Letter From America, with its stars-and-stripes breakers, and Nigel Dempster, plus the adaptation of the paper's title to Femail and TV Mail and Sportsmail are all signatures planted to draw the reader to regular offerings in the paper, and stamp the paper's identity on the pages (Figure 114).

The Independent, more restrained in style, deploys its name on its leader page and its Section Two supplement and on a 'readers' offer', while its eagle signature turns up on an advertising page panel.

Page 8 EXS1

Daily Mail, Monday, October 9, 1995

Daily Mail

COMMENT

Defector's flight from reality

NO sooner has the curtain come down on Tony Blair's triumph at Brighton than a Tory MP defects to Labour just as the Conservative show is scheduled to open in Blackpool.

If the agony of Alan Howarth's conscience is not in question, nor can there be any doubt about the damage he has inflicted on his erstwhile colleagues desperately battling against the electoral clock to regain public trust.

For the Tories, the denouement to Mr Howarth's long night of the soul could hardly be more untimely. For Mr Blair, it is a coup de theatre.

What, then, has driven this troubled man to change allegiance in so traumatic a fashion?

He fears for the welfare state. He cannot abide what he stigmatises as the 'harsh' and 'authoritarian' tone he now claims the Government is adopting towards the poor.

While most certainly that is not the Daily Mail's perception, we do recognise that in the parliamentary party, he is not entirely alone in his worries, as the article by Edwina Currie on this page testifies.

Sincere as may be Mr Howarth's human sympathies, in our view he betrays a woeful disregard both of the abuse of benefits and the need to free people from dependency.

He appears profoundly unaware of the threat to national prosperity posed by the ever mounting cost of an unreconstructed social security system.

Belatedly, Ministers are nerving themselves to do something about this.

They are bringing in the jobseekers' allowance to wean men and women off welfare and into work. They are insisting that invalidity benefit goes to those who deserve it and not to skivers. They are attempting to ensure that absent fathers contribute towards the upkeep of their children.

Such changes are bound to cause some controversy. But they are moves in the right direction. They do demonstrate an awareness of what is wrong.

Nor are these changes as out of tune with the way most people feel as the high-minded Mr Howarth appears to have convinced himself.

Although he is much concerned to espouse 'one nation' policies, which do not divide the poor from the rich, he shows scant understanding for the millions of decent folk, who as they struggle to make ends meet on low pay deeply resent having to subsidise the feckless and the workshy.

The truth is that whichever party wins the coming General Election — Labour or Tory — will have to tackle the crisis engendered by paying for a welfare state that has grown out of all proportion to the noble principles on which Beveridge founded it more than half a century ago.

There are signs that Mr Blair recognises this, even if he chooses not to trumpet it before polling day. Indeed, on his own side there are already articulate and impressive welfare reformers such as Frank Field, whose fiercely radical analysis makes the current thinking of Mr Major's administration sound tame and tentative.

Too late, anguished party-hopper Alan Howarth could find that he has jumped off the Tory hot plate into the New Labour melting pot.

Meanwhile, should he not resign his parliamentary seat rather than cross the floor of the Commons?

In all conscience, that would be the democratic thing for this conscience-stricken MP to do.

OUT OF TUNE HERSELF WITH MUCH OF CURRENT TORY THINKING, A FORMER MINISTER EXPLAINS WHY SHE WON'T BE JUMPING SHIP

Worried but loyal: Edwina Currie, and inset, Alan Howarth, who has defected to Labour

I admire him, but Alan was wrong to quit

A PERSONAL VIEW

by EDWINA CURRIE

THE journalist's voice on the phone was silky. 'It's rumoured that a high-ranking Tory MP, an ex-Minister, is about to defect to the Labour Party. Any ideas?'

'Well, it's not me,' I retorted. 'But it could be Alan Howarth.'

I was right, but that's no comfort. For some months Howarth has wandered around Westminster like a man under torture. By contrast, yesterday's photographs suggest a new-found serenity: that of a man at peace with himself. He is a decent and intelligent man and for his own sake, I hope he is content.

Why did he do it?

Like many of us, Howarth warmed to John Major's early declarations about a 'classless society' and 'a nation at ease with itself'. When I was fighting to lower the homosexual age of consent from 21 to 16 last year, Alan was a great support. He was pro-European, and he was deeply concerned about education.

He and I worked together recently over sex education in schools — unlike some, we both believed good sex education was essential, not as a way of fomenting trouble but as a way of avoiding it in advance. He is a liberal-minded person who has changed sides out of spotless motives. I cannot but feel a twinge of sympathy.

The problem is that many of those splendid early ideals that John Major pronounced so bravely have not been followed by his current Government. Instead of a classless society, the Cabinet sometimes seems hellbent on the very opposite — 'the politics of divisiveness'.

Thatcherism used to encourage everyone, including the poorest, to aspire to the best. Thatcher's aim, in promoting the sale of council houses and wider share ownership, for example, was to give ordinary families more control over their lives.

The men Margaret Thatcher most admired were self-made — like her, they had risen from humble origins. Along with millions of new Tory voters, I passionately supported her in this view.

Nauseating

More recently, the tone has changed, and not for the better. In that, Alan Howarth is spot-on. When the Back To Basics campaign was in full swing two years ago, a member of my staff, a single parent who has made a great success of raising her child alone, asked: 'Do you want me to go on working for you?'

Of course I did. But the sneering tone of some of the remarks made by certain leading Tories was nauseating.

Gossip has it that there may be other MPs who are tempted to defect. Some — less honourable men than Howarth — believe the writing is on the wall for the Tories after what, by the next election, will be an 18-year stint in government. Many of them have marginal seats and dislike the idea of years in opposition. Many of them have been bamboozled by the cosmetic changes wrought in the Labour Party by young, inspirational Mr Blair.

But my warning to them is this: if they go over to the other side they will find that a defector's lot is not a happy one.

Most of Labour's safe seats have picked their candidates. Nobody will make way, certainly not for an upstart newcomer. In the marginal seats, with the polls looking so good for the socialists, hopeful new boys and girls will dig in their heels. And former Tories, especially those like the elegant, patrician Mr Howarth, may find themselves made only superficially welcome.

In those smoke-filled rooms replete with trade unionists, low-paid workers and the unemployed, they'll be able to cut the atmosphere of suspicion and hostility with a knife. I wish Alan luck: he'll need it.

Tory Conference delegates arriving at Blackpool are stunned and demoralised. Howarth was well known and widely liked. He gained the CBE for political service at Central Office even before he entered Parliament.

They will feel his criticisms have hit home. No wonder John Major is said to be very upset.

The battle is on for the soul of the Conservative Party. The outcome of the General Election will depend on who gains the upper hand. Are we a bunch of money-grabbing opportunists, out only to enrich our friends, sans principle, sans moral scruple, sans everything?

Or do we believe in a modern, responsible society, in which talents can flourish, the less fortunate are cared for, and the nation looks to its neighbours and the outside world with confidence and pride?

In our electoral system, the party which wins the centre wins the country. Pandering to the extreme is electoral disaster. Labour knows that from its 1983 and 1987 failures. Blair's answer is to woo the middle ground: Middle England, the middle-class. Howarth is exactly the sort of person he's trying to win over and his staff must be crowing with joy.

Here is the real danger. If in their efforts to create 'clear blue water' between themselves and Labour, the Cabinet decides this week to shift further to the right, they will be righting themselves out of office with a vengeance.

I joined the Tory Party 30 years ago and I will stand at the next election as a Tory. No one who has stood, as I have done, face to face with the contorted rage of the miners and other striking trades unionists who held us to ransom during the early years of Mrs Thatcher's premiership, could ever contemplate doing otherwise.

Natural

The Labour Party is still the natural home for people of this ilk — I came up against one of them at Staffordshire University only the other day. For nearly an hour I stood up to him, a New Labour union official who tried to throw me out because he did not agree with my views. That man proudly sported his Labour Party badge.

I will stay in the Tory Party and do my best, along with like-minded colleagues, to keep it firmly at the centre right of politics.

I just wish Alan Howarth was still at my side to help.

The infamous grouse

THE Queen was given the bird the other day when a grouse dive-bombed into her chest at Balmoral. She was briefly shaken, but soon laughing the incident off.

Let's hope Her Majesty was right to laugh. In the Ancient World, birds were believed to be messengers of the gods. It was a bird omen that marked out Tarquinius Priscus as the next king of Rome in the seventh century BC. He was entering Rome when an eagle plucked off his cap, then replaced it and flew off. The reigning king then died after 24 years on the throne and Tarquinius took over.

Not all visitations from the sky were so agreeable. One killed the Greek tragic playwright Aeschylus. He was out walking when an eagle, looking for something on which to smash the tortoise it had plucked from the ground, mistook Aeschylus's bald pate for a rock and scored a resounding, and deadly, bullseye.

PETER JONES

Figure 113 Heart of a paper's identity: the *Daily Mail* leader page

The regular format of the Andy Capp strip in the *Daily Mirror*, the Page Three girl in *The Sun*, the various 'agony aunt' columns of the popular tabloids, even the repetitive typographical shape of the television programmes, plus the fixed position of these items in the paper, perform a similar function. They give easy access to favourite features while providing oases of familiarity amid the flux of the day's news. Such devices are planted discreetly in all manner of

Figure 114 Signatures: an inside page logos map out the *Daily Mail*

Figure 115 Signatures: a lower key approach in *The Independent*

newspapers to help give a continuity of image a newspaper cannot do without.

Typographical signatures also perform a useful service for the designer. They enable the visual identity of the paper to be secured in strategic areas, leaving the designer to display the editorial contents without feeling so tied to the pursuit of continuity that design relapses into formula.

14
Special markets

In examining newspaper design we have seen that there are common factors that apply across the board and that there are techniques that apply to particular types of newspaper and styles of design. Differences in market and purpose can influence design style even with general newspapers, whether national or regional; in fact, geography of distribution is of least importance in arriving at an acceptable format.

In this chapter we look at some of the special requirements and special markets outside the main run of newspapers and at how these affect the approach to design, and how the methods we have discussed might need to be adapted.

Sectional newspapers

Some newspapers have discovered that their content and market can be best served by dividing their product into a number of separately folded sections, each with its special logo. The practice originated in the US as a means of coping with the large editions that grew in response to heavy advertising placement in newspapers serving big city conurbations. It enabled publishers to offer advertisers space in particular editorial environments – world news pages, local news, sport, women's, weekend magazine, financial, etc. – and also of offering readers a bulky product in which areas of interest could be found more easily, and which could be split up for reading within family and works groups.

The advertising-conscious retail trades and services and extensive mail order businesses thrived on the penetration achieved by the dailies published in the cities, and by the relatively cheap advertising rates that concentrated circulation made possible. The result was the growth of monster papers with ever increasing numbers of sections, culminating in the celebrated occasion on October 17, 1965, when the *New York Times* brought out fifty-six sections totalling 956 textsize pages.

Sectional newspapers – and the ultimate section, the colour magazine – first appeared in Britain in the 1960s when Sir Roy, later Lord Thomson, bought the *Sunday Times* and other Kemsley titles and set about building up their advertising revenue by high-pressure space-selling techniques. The method suited the wide news coverage and lengthy magazine content of a quality Sunday paper and was eventually taken up by *The Observer*, and to a lesser extent by *The Times* (when it also became a Thomson paper) and the *Financial Times* and some of the bigger provincial papers. It did not, however, become widespread in town and city papers as in America.

Identity is the important thing with sections. In design terms they need to have enough individuality for them to be distinctive one from the other, and to give the pages of each a particular character, while sharing sufficiently in the overall design style to be recognizably part of the same newspaper. This can be done by introducing into each section a distinctive 'signature' type which will blend in with the types used elsewhere in the paper and common to all sections; or a condensed or lighter version of the paper's stock type can be used as a dominant headline type for the section. Another way is to keep the same type range but to introduce design characteristics such as indented setting or pages set within panel rules (using fine or 2 point rules). Creative use of rules and white space will give opportunities of making the pages look special without making them look too different.

Figure 116 The sectional newspaper: elegance and variety for *The Times*, but the logo is always there

For easy identification a distinctive logo or title piece is necessary, which should incorporate a miniature of the main masthead (Figure 116).

A decision has to be taken about pagination – whether to page the sections through from the carrier paper or to page them separately. A useful ploy, if the pressroom folders allow it, is to fold the odd section into half size as in the books section in the *Sunday Times*.

Pull-outs

Now common, and especially so in the popular tabloid dailies and many provincial papers, are pull-out sections consisting usually of the four or eight centre (tabloid) pages. They can be devoted to special news or features content, often under a separate logo such as 'four-page supplement on . . . etc.', or to such things as holiday, shopping or property guides with heavy advertising sometimes accompanied by editorial back-up.

Most are paged through with the rest of the paper, a practice used originally to circumvent print union restrictions which insisted that separate supplements to an edition should be charged extra on the wage bill. The notion was that a separately designed but paged through section in the middle fulfilled the purpose of a supplement though the use of the word 'supplement' was carefully avoided without incurring extra print charges. It was also less of a problem for the press-room folders than separate sections.

Pull-outs offer two main options to the designer. They can either be designed as an extension in style of the main paper, with perhaps a simple motif (which can be numbered) flagging each page; or they can be given their own special design characteristics within an overall format in the run of the paper or with separately folded sections. The latter way is adopted by papers running regular labelled supplements such as the *Daily Mail* (Figure 117) the *Financial Times* and the *Evening Standard* (Figure 118).

A particular problem arises with pull-outs in the case of popular tabloid papers as a result of the practice of carrying 'under-spreads' across what is in fact the centre spread of the paper proper. What would normally be a spread on pages 24 and 25 of a 48-page paper thus becomes a spread on pages 22 and 27 where there is a 4-page pull-out. The psychology of readership assumed by editors is that people will detach the pull-out and read across the under-spread as if the other pages were not there. This works with readers who expect to have to do it, but can annoy a percentage of readers who start reading the first half of a spread to find their way unexpectedly blocked in mid-headline by an unrelated page, and that they have to turn from page 22 to page 27 to find what happens next.

One answer is to make the under-spread two separate pages so that the interruption of the pull-out does not upset the sequence of page numbers. Another is to keep the under-spread but have material on it that can be divided so that neither headlines or pictures cross the gutter. The problem for the page designer who has to

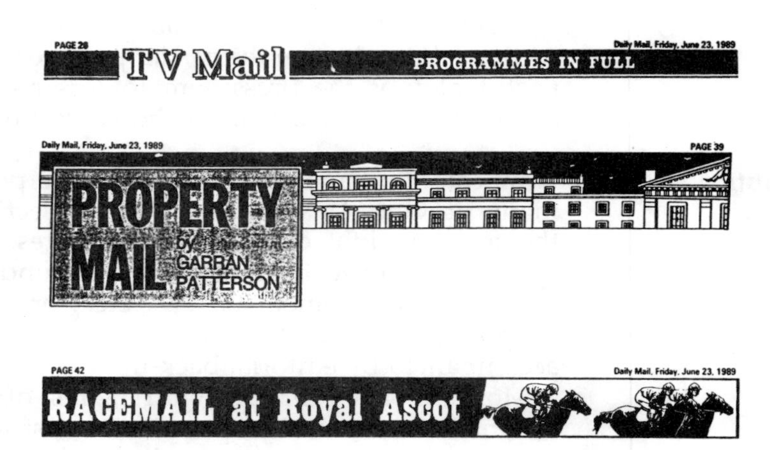

Figure 117 Pull-outs and other devices: the *Daily Mail*'s way

Figure 118 Title and motif mark the *Evening Standard* sections

accommodate a central pull-out is one of reader psychology as much as design – how, by means of typography and pictures, to make the pull-out seem an additional benefit to the reader rather than an intrusion. If the pullout embodies the main editorial offering, such as a four-page special feature or series instalment, it becomes the vital part of the paper and to try to maintain an under-spread would seem to be in danger of distracting attention from it.

A solution to the page sequence problem caused by a centre pull-out (though now little used) is to have a *wrapround*, in which a folded sheet forming four pages, separately paged and with a suitable recognition title piece, is attached to the outside of the newspaper. This can be useful for a commemorative edition in which a paper is, say, celebrating a centenary, or wants to display special material with perhaps flashbacks to earlier papers; or for a Royal or similar occasion in which a pre-printed colour souvenir is prepared on special quality paper to be detached for keeping.

Unlike separate sections, neither pull-outs or wraprounds offer any problem in printing and folding, and the pagination adopted either paged separately or paged in – is a matter for editorial choice.

Desktop design

The use of desktop publishing (DTP) has greatly increased the number of organization, civic and similar tied circulation newspapers which exist in forms as varied as the systems upon which they are produced. It is difficult to lay down design techniques that would apply to all DTP systems; but a good yardstick would be to apply the factors common to all newspaper design as described in Chapters 2 and 8 of this book and then to formulate styles that take into account the market of the product and the facilities built into the system.

Programs used with the Apple-Mac computer can offer a wide range of typefaces and graphics facilities. It would be pointless, however, to try to imitate newspapers produced with great resources. The result might simply be an inferior looking newspaper. The style would have to relate, for instance, to:

1 The sizes and ranges of types available for body setting and headlines.
2 The sort of pictures available.
3 Required text lengths.
4 The method of page make-up, that is, paste-up or electronic composition.
5 The paper quality.
6 The size of page and column format adopted.

The last two items are important in DTP publications. A small format of, say, five 7 pica columns would offer scope for some varied layout ideas and headline patterns, and a variety of picture sizes, provided the items were not too long, whereas fewer but wider columns – a three or four column format – would suit longer texts, especially where the emphasis is more on words with just occasional pictures. A few pictures of good size would be the perfect foil here.

If the paper to be used is of reasonable quality and the right pictures are available it would enable the design to utilize magazine techniques with perhaps bleed-offs and the creative use of controlled white space, or special setting round cut-out picture shapes, producing a result which is a hybrid between a newspaper and a magazine. Fine paper means that a finer screen can be used for halftones so that with correct inking subtle picture tones can be rendered and an effect produced which approaches that of gravure.

The best sort of effect is that which arises naturally out of the materials used, both editorial and software. Thus a factory newspaper on fine coated paper might make an attractive feature out of a series of linked pictures of an industrial process accompanied by a descriptive text printed in the knowledge that it would be meaningful to the paper's captive readership. Such pictures would be less effective on ordinary newsprint and less interesting to more general readers.

Pictures of social occasions can give a repetitious feel to a company newspaper writing mainly about its workforce, but variety can be injected and useful layout ploys set in motion by, for instance, inviting readers to submit their best snaps or best holiday pictures or pictures taken of novel or unusual things.

As far as design and page make-up go, with the facilities and software now available with the Apple-Mac computer it is possible for a working editor plus a couple of reporter-subeditors to produce a publication of a good professional standard (Figure 119).

The machine can be programmed at the outset to the size of the publication, and the pagination style established by a tool called the Show Document Layout on the View menu. Formats are established in the computer's memory for retrieval and use in answer to machine-posed questions on the appropriate pop-up menus – for instance the size of your page and its required margins, the number of columns, the font and size of body type to be used and the spacing and character kerning (tracking).

Where two Apple-Macs are in use a story can be put together and input into its own document box on a slave screen and transferred when ready to the companion screen for building into the page. In a one-computer system, document and page can be displayed simultaneously on split screen for the one to be layered into the other.

Headlines are written into their page boxes to the required size; pictures retrieved by the Get Picture command from the memory to fit their boxes, with four-colour images (using Quark XPress) coming together with perfect registration. Another tool enables the picture to be enlarged and moved about for editing inside its box. Reduction of the picture image can also be achieved inside the box or a frame added to the picture. Headlines can be reversed into black, tint or colour, or given outlines or shadow effects.

Stories chosen for use from stock are highlighted by command key and the Edit menu used to copy and paste them by mouse control into their text boxes. The machine will produce the story in the type size and face designated by the text box commands you have ordered. If you are truly a one-person outfit you can write your copy

Barbican
RESIDENT

The Bi-Monthly Magazine of the Barbican Association ● October/November 1995 ● Issue No. 150

At 150 issues old we thought your magazine deserved an easier-to-read format. Hope you like it!

Care housing project for City residents gets underway

This is the Barbican Association's magazine of views and information.

Published six times a year, its objective is to keep residents in touch with events throughout their estate and to record the evolving relationships between the City Corporation

Coming in your next issue

LORD MAYOR'S SHOW-

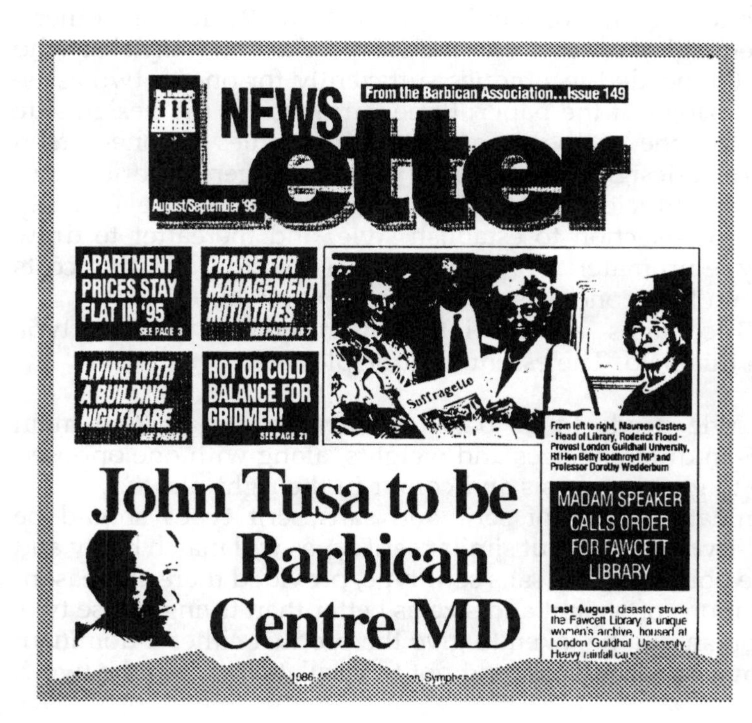

From the Barbican Association...Issue 149

NEWS Letter

August/September '95

APARTMENT PRICES STAY FLAT IN '95
SEE PAGE 3

PRAISE FOR MANAGEMENT INITIATIVES
SEE PAGES 6 & 7

LIVING WITH A BUILDING NIGHTMARE
SEE PAGES 9

HOT OR COLD BALANCE FOR GRIDMEN!
SEE PAGE 21

From left to right, Maureen Castens - Head of Library, Roderick Floud - Provost London Guildhall University, Rt Hon Betty Boothroyd MP and Professor Dorothy Wedderburn

John Tusa to be Barbican Centre MD

MADAM SPEAKER CALLS ORDER FOR FAWCETT LIBRARY

Last August disaster struck the Fawcett Library a unique women's archive, housed at London Guildhall University. Heavy rainfall ca

Figure 119 Desktop design: two ways of doing a locality newsletter. A magazine of either sort, varying in size from 24 to 48 A4 pages and keystroking 11,000 to 17,000 words, is well-suited to two people and an Apple-Mac. Ideally it needs a fast keystroker, a good input of copy-filled disks, readily available picture sources, plus a page designer able to write headlines and assess copy. These versions show an upmarket and a downmarket display approach to a typical news content, the difference being typified by the logo designs

into already designed page shapes to fit text boxes without the need of raw copy.

Type is no problem. Thousands of fonts are available on disk or CD which can be copied on to hard disk. For easy viewing, text can be pulled up on screen to three times its size. Conversely, pages can be viewed together for checking at 50 per cent and 75 per cent under actual size. For the innovative, free shapes for pictures such as circle or ellipse can be created to any size by use of the Polygon tool in the Quark XPress Item menu (Figure 65, on page 136).

When complete, the publication can be downloaded from copied disks or hard disk on to your printer, either laser or inkjet, or the page disks taken to a print shop to be turned into printing plates for offset printing. The printer there will feed your disk information into the equipment directly to plate or will take 1400 dots to the inch laser proofs to create negatives from which to produce plates.

There are still some print shops who will take story and picture proofs from your disks to marry them together on paste-up sheets in order to produce page negatives for plate-making but they are already a dying breed.

Most DTP products fall down on poor type use, lack of layout ideas or a general failure to make proper use of the system and paper available. Thus a product produced on a shoestring, but making full use of what there is can score over an amateurish job turned out on good coated paper with full colour facilities.

Poor type use and lack of layout ideas are the inevitable result of the absence of trained staff. This can be solved in three ways: one, by making a serious study of similar products to learn from others' ideas; two, engaging at least one trained production journalist who can pass on the needed techniques sufficiently for one or two assistants to help bring out the paper; three, farming out the design side of the work to one of the many small companies or one-person outfits who do a design and pre-press package. A general briefing of requirements is sufficient for a skilled designer to produce a variety of dummies for selection to establish style, and thereafter to draw pages and type up material on a regular basis so that printing costs are kept within the economics of the desktop system.

Chapter 3 contains all the information needed to use type creatively in this field. The essential principles are:

- A type style needs to be based ideally on one or two main typefaces in different sizes and weights, along with one opposed typeface to give emphasis and colour in the right places.
- Indiscriminate mixing of serif and sans serif types should be avoided. Two different but similar serif types will match badly and would be pointless. One sans serif type, provided there are reasonable variations of weight and size, is better than trying to use two. A decision should be taken to give the product either a dominant serif format or a sans serif, and not to vacillate between the two.

- To refine the type format further a decision should be taken whether or not to use a wholly lowercase format, or perhaps to exclude italic type as a matter of style. A small product needs fewer types and variations than a general newspaper or big magazine. The greater the mixture of faces, sizes and variations the more difficult it is to achieve a consistent design approach.
- Variations in choice and setting of body type should likewise be limited to avoid pages becoming hotch-potches. Panelled items should be used as structural points in the design in the same way as pictures.
- White space should be used creatively as a design element and not to point the failure of type to fill its allotted space. Space within and around headlines should be consistent. If type is indented or set 'ragged' there should be a design reason for it. On good paper with a sharp register white space is a boon to the designer. A study of gravure fashion magazines is a useful source of ideas in the creative use of white.

Chapter 5 will help with picture use. Points to note are:

- Pictures should be cropped for their relevance to the text. What they need to show is enhanced if the image can be made bigger by the exclusion of what they do not need to show.
- Similar pictures, whether of work processes or of people doing similar things, can usually look better if grouped together on a page or pages.
- Good tonal quality of photograph is essential to avoid disappointment in reproduction. In choosing a picture any greyness or indistinctness will look worse, not better, when the picture prints, however good the paper.
- Avoid big enlargements of small snaps unless the negative is available. The detail disperses rapidly in proportion to the increase in enlargement.

Free sheets

Free sheets – not the favourite term of the Association of Free Newspapers – is the name given to newspapers in which advertising income is pitched at a level which enables the proprietor to give them away. To bring in the advertising they usually go for house-to-house distribution or pick-up at retailers in heavily populated areas so they can offer saturation coverage to potential advertisers. Circulation can thus be pegged to a potential target figure, unlike with paid-for papers, although distribution cannot be said to guarantee readership.

To win readership a free sheet has to make itself a viable alternative to a paid-for paper serving the same market, and therefore it has to take seriously the presentation of its editorial content, even if it does occupy a smaller percentage of the paper. In fact, the impact of frees has resulted in paid-for papers in some parts of the country increasing the proportion of advertising to boost falling revenues and the local free sheet increasing the proportion of editorial matter to win readers, to a point where the two have drawn closer together in their overall mix.

In design terms, in order to succeed, frees need to pay at least as much attention to layout as paid-for papers, which means that the editorial content should be used to the paper's advantage. A type style should be evolved as part of an identity. Too many frees fail not through insufficient editorial material but through insufficient care being taken in presenting what there is. Whether they are produced through desktop systems or on contract by printers (the more usual way) the typesetting and graphics facilities should be studied to get the best possible utilization from them and design principles as examined in this book applied.

This does not mean a rash of 'hype' or self-publicity but rather a measured use of type and illustration to create attractive page patterns that are helpful to the eye and right for the sort of readers receiving the paper. Identity and continuity of format allied to properly planned editorial content and services can win over readers who might otherwise equate frees with mail-shot advertising. It is not a market in which readership should be taken for granted.

Contract printing

Many specialist and tied circulation publications, including free, and some bigger papers, are printed by contract printers who specialize in this sort of work. Some regional groups who have invested in modern computer systems and new web-offset presses have found it profitable to make their facilities available to other newspaper publishers. This should not alter the approach to design. The contractor will print whatever you prepare for the press and if need be make up the pages as well. An editorial department and an art desk can operate just as effectively without the printer being just round the corner, though the separation has to be taken into account in arranging time scales for production. The geography of such an arrangement, with reproduction plant sometimes situated hundreds of miles away, does slow down the production cycle.

The simplest form of contract printing is where the entire editorial operation up to, and including, page make-up is carried out at the newspaper or magazine office and pages sent to the production centre as negatives to be made into printing plates or perhaps, in the case of electronic pagination, direct from screen to plate.

There are endless variations. Some small publications provide typed text, pictures and adverts to a freelance designer/editor who designs and prepares the whole package for the press and either sends it back complete to the publication's office, or deals directly with the printer. An improvement on this is where the designer/editor prepares the package on Apple-Mac and submits it on disk.

Other publications have their own art desk and send drawn pages and edited copy by fast delivery for input, by what ever system, at the contract printers. This can be cumbersome where the printer is some distance away since it entails sending back galley proofs or page proofs for cuts and adjustment, or dispatching an editorial person to the printers on press day to make sure everything fits.

An improvement on this is the system where a publication is edited and designed at source and the material entered electronically into the system in use at the contract printers (where ever it may be) for make-up by paste-up or on screen. Even here an editorial presence is desirable at the printers on press day.

Whatever the method, and at whatever level, the design parameters are those dictated by the publication's market and its agreed type format. A disadvantage (except where the contract is simply for a press room operation) can be a loss of the fine control over the product by editorial and art desk that exists with in-house page production and printing.

15
The international scene

The development of a strong mass circulation national press, arising out of high population density and short lines of communication, has given Britain a special position in newspaper readership. The fact that national titles aim at a variety of social and income groups means that they exhibit, on the news-stands, a diversity of content as well as of presentation which leaves every taste catered for.

At the same time the pervasiveness and high readership of the national press means that most provincial papers, their main topics having been hijacked by their national rivals, are pushed towards a predominantly regional, or even local content in the effort to woo readers. This content colours their presentation both in the pictures that are used and in the topics and words, and even the attitudes, that leap from the pages.

One of the consequences of this is that regional morning papers published in provincial cities have found it hard to be accepted by readers as purveyors of a spectrum of national news and have, during this century, suffered a decline.

Thus, while the design and the quality of the writing of many provincial titles is of a high standard, the feel of these papers is usually markedly different from that of a big national paper. The effect of this – as becomes clear to anyone studying the subject – is to divide the British press into two camps: the national press and the provincial press.

The first noticeable thing about looking at newspapers abroad is that this dichotomy does not seem to apply. There is a vigorous growth of city and region-based papers in most countries that compete successfully with the few mainly political titles that go for national distribution.

This situation is justified in big countries such as the USA and Canada by the simple logistics of time and distance and, to a lesser extent, by the rich pickings of conurbation advertising. Out of this

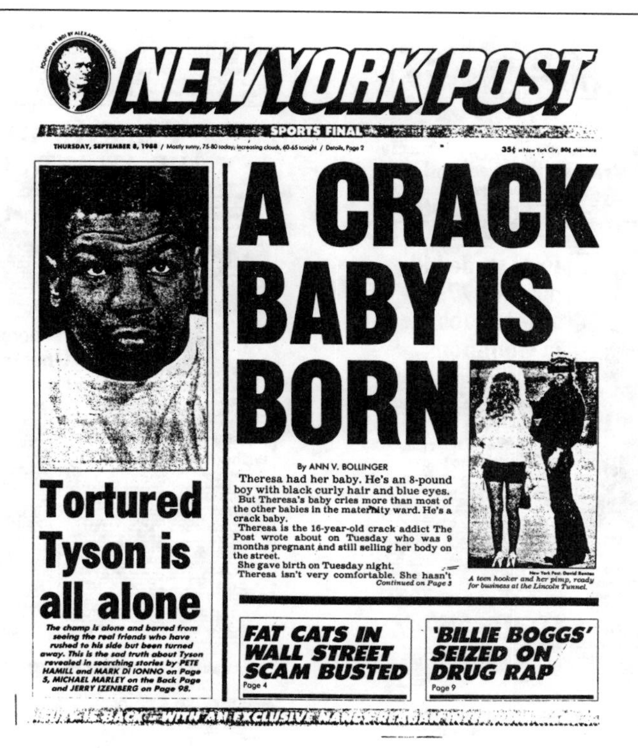

NEW YORK POST

SPORTS FINAL

THURSDAY, SEPTEMBER 8, 1988 / Mostly sunny, 75-80 today; increasing clouds, 60-65 tonight / Details, Page 2

35¢ in New York City 80¢ elsewhere

A CRACK BABY IS BORN

By ANN V. BOLLINGER

Theresa had her baby. He's an 8-pound boy with black curly hair and blue eyes.

But Theresa's baby cries more than most of the other babies in the maternity ward. He's a crack baby.

Theresa is the 16-year-old crack addict The Post wrote about on Tuesday who was 9 months pregnant and still selling her body on the street.

She gave birth on Tuesday night.

Theresa isn't very comfortable. She hasn't *Continued on Page 3*

A teen hooker and her pimp, ready for business at the Lincoln Tunnel.

Tortured Tyson is all alone

The champ is alone and barred from seeing the real friends who have rushed to his side but been turned away. This is the sad truth about Tyson revealed in searching stories by PETE HAMILL and MARK Di IONNO on Page 5, MICHAEL MARLEY on the Back Page and JERRY IZENBERG on Page 98.

FAT CATS IN WALL STREET SCAM BUSTED

Page 4

'BILLIE BOGGS' SEIZED ON DRUG RAP

Page 9

Figure 120 New York Post

can grow a strong regional loyalty to titles. The punchy *New York Post* with its blockbuster sans headlines (Figure 120) is in a different geographical and cultural as well as typographical world to the sedate *Los Angeles Times*, while both are light years away from the *Detroit Examiner*, in whose city neither titles would find favour. The *Toronto Star* (Figure 121) which presents the whole world to Ontario, gives no thought to the *Vancouver Sun*, at the other end of Canada. Likewise Sydney, Brisbane and Perth, Western Australia, happily go it alone.

Yet in old world Europe the same situation is found to apply. There is nothing about *La Gazzetta del Mezzogiorno*, with its bold serif dress and confident use of pictures, spacing and rules, and front page news choice (Figure 136) to suggest that it is published and printed far away from the centre of the universe down at Bari, at the heel of Italy. The raucous poster layout of *Abendpost* (Plate 3 in the colour plate section) might encapsulate the world in Frankfurt, but not in Hamburg or Munich where the world and Germany are looked at from a different axis in the pages of their own newspapers. The *Tribune de Geneve*, with its air of final authority (Figure 132) might trickle into airport bookstalls in Athens and Rome, but it speaks not for Zurich or Berne.

What we are looking at in the old and new worlds, from our vantage point in the home of mega-circulations, is a rich and varied

BUSH VOWS MAJOR CUTS IN ACID RAIN Page A24

BOLD FASHIONS FOR FALL Page J1

REFUGEES FLEE CHAOS IN PANAMA FOR TORONTO Page A3

SMEARS AND LOATHING ON THE CAMPAIGN TRAIL Page A32

MAN CHARGED AFTER 2 DIE IN HIT-RUN Page A20

Metro Weather

THE TORONTO STAR

Thursday, September 1, 1988 30 CENTS

METRO EDITION

Deal could end 9-day-old postal strike by this weekend

Up to 112 lose jobs as CFTO ends lockout

HATS OFF TO THE GOLDEN BEAR

Free trade bill passes 177-64 amid chaos in Commons

It will be flag versus the buck in election

Analysis

New chief executives appointed by Torstar

Workers vote to end strike at shipyard in Polish city

Recycle trash or else, Midland residents told

Lotto 6/49 winners

Admits defeat

"Everything I own I leave to my ventroloquist dummy."

Figure 121 *Toronto Star*

How Braves are doing down on the farm 1D

Software glitches that can be deadly 1E

Risen Star runs away with Belmont Stakes 3D

HIGH STYLE

The gowns of renown: Priscilla takes the cake

The Atlanta Journal

THE ATLANTA CONSTITUTION

SUNDAY, JUNE 12, 1988

$1.25

"I've called and called the police. I've told them what happened. But they won't ever believe me."

2 get new hearts at St. Joseph's

Georgians well aware of AIDS risk, but knowledge has gaps, poll shows

Family waits for justice without a unified court

A tear at UGA's commencement

Drought that comes on heels of '86 crisis leaves many high, dry

INSIDE TODAY

Sunny, warmer

ARTS & ENTERTAINMENT

BUSINESS

DIXIE LIVING

HIGH STYLE

HOME & GARDEN

METRO & STATE

NATION & WORLD

PERSPECTIVE

SPORTS

TRAVEL GUIDE

Democrats draw battle lines on taxes

Youths flee as Miss. officials take over school

Figure 122 *Atlanta Journal*

growth of – by our standards – small- and medium-sized titles that combine within their pages the national and regional function of a newspaper. They are bold and vigorous in opinion and record, while staying alive through sensible regional distribution arrangements and the enjoyment of national advertising income unmolested by mass circulation national predators.

It would be natural, looking at a cross-section of foreign and Commonwealth papers, to expect design styles to be as varied as the geography of the titles, and indeed there is variety. Yet the more we examine the world's newspapers – those printed in the European alphabets at least – the more a number of consistent traits can be identified.

America

The US has been a leader in print technology and type design for nearly two centuries and American designers claim for it a powerful influence on the world's press. By British standards their wide

FRIDAY

South County

It's fair time
Event listings in Weekend section

Pistons in lead

THE ECONOMY D-1

$2 billion deal
Texaco proxy showdown today

The San Diego Union

Friday, June 17, 1988 — A Copley Newspaper — 25 Cents

Senate passes welfare reform

Emphasis on work is first big change in laws since '35

Drought may send up food prices

Bribery probe said to target 3 in Congress

Arizona jury acquits former Gov. Mecham

Impeached chief, brother cleared of concealing loan

Gang cops carrying beefed-up arsenals

FAA curtails tracking plan

Altitude device rule for small planes narrower than proposal

Drug testing of federal prison workers blocked

Trade
President Reagan's sanctions against Japanese electronic exports are being compared to the "coming of the black ships" more than a century ago, when the American gunships of Commodore Perry forced Japan to open its ports to the outside world. See story, Page B5.

State.
Bill Waller, governor of Mississippi from 1972 until 1976, will try again for the mansion. He started his trek this morning in Southaven. See story, Page A6.

Sports
The Milwaukee Brewers extended their winning streak to 12 games with a 6-4 win over the Texas Rangers Sunday. Details are in today's Sports section, which starts on Page B1.

Vicksburg Evening Post.

MONDAY, APRIL 20, 1987 — 35¢ — VICKSBURG, MISSISSIPPI — VOL. 105, NO. 110

No Connection To Candidacy Seen In Murder

By BOB YARBROUGH JR.

NOPSI Loses Rate Appeal

Postmaster Says 25 Cents Sounds Good

Three Firefighters Hurt During Main Street Blaze

Today

A Sampler Of Statistics For Use By The Candidates

Figure 123 *San Diego Union*

Figure 124 *Vicksburg Evening Post*

columns, generally 9 point body type and heavily spaced headlines and body setting lead to pages that lack impact and waste space. And yet broadsheet papers such as the *Atlanta Journal* (Figure 122), and the *San Diego Union* (Figure 123), with their open six-column format and simple horizontal layouts typical of American state journals, are easy to read, as is the Mississippi town evening, the *Vicksburg Evening Post* (Figure 124). The interminable turns from the front page to inside pages in the *Union* and the *Journal* would annoy European readers, and excessive advertising on many pages makes design as we know it impossible, yet their long expanded lowercase headlines, wide fat paragraphs and well-flagged pages have a dignity and a comfortable feel that grows on you – and with the number of pages there is, in the end, plenty of room for the words.

The American style, if it might be called that, in its simplicity is the product of thick, many-sectioned papers crammed with advertising in which technique has to take second place to the sheer logistics of

FOCUS
Licenced to kill:
Britain's SAS
is on a rampage . . . D1

TRAVEL
The last call for paradise
on Mexico's changeing
Pacific coast . . . A15

THE ARTS/BOOKS
Cultural broker David Silcox
has a swashbuckling
approach to the arts ... C1

DRAKE EXECUTIVE
The right people to find the right people.

The Globe and Mail

The Hundred Antiques

REPORT ON BUSINESS

QUOTE OF THE DAY

NEWS

SPORT

INDEX

'They're torturing people'

Processing refugee claims shut down due to backlog

PLANE CRASH KILLS 75

Tuition fees have doubled in a decade

FEDERAL PARTY SUPPORT 1984—1988

Tories take lead in voter popularity as Liberals falter

Few PCB traces found, residents of St-Basile told they can go home

Setting rate on CSBs to force tough choice between votes, deficit

Prices for most services, commodities adjusted monthly

Brazil's inflation destroys capital, dashes hopes

Figure 125 *Toronto Globe and Mail*

getting the hundreds of pages of each edition together. It is an unlikely model for the very different societies outside North America. It is perhaps natural, therefore, that its nearest clones can be found in Canada, where the *Toronto Globe* and *Mail* (Figure 125) and the *Toronto Star* (Figure 121), though with less weight of advertising, exemplify the American approach with the same six-column format (seldom found in Europe), long headlines and stories in horizontal segments. It is noteworthy that the *Union*, the *Journal*, the Vicksburg paper and the *Globe and Mail* all have traditional lowercase Old English title pieces.

With the prestigious titles commonly seen in Europe, such as the *International Herald Tribune, Washington Post* and *Wall Street Journal*, following the traditional American pattern it comes as a novelty to find that New York also supports a bold sans type tabloid in the *New York Post* (Figure 120). One is tempted to see in it the influence of the British popular tabloids through its long spell under the Rupert

THE AUSTRALIAN

Nationals facing crisis on Sinclair

The Iranian who has gambled all on bringing peace to the Gulf . . .

N-plant bombed in air battles

INSIDE

Secret deal and oil price

Countdown to ceasefire

War with no winner

Cynicism of history

Your dollar

PM defends multiculturalism

'Aboriginal jobs' to be cut by third

HAIR PROBLEMS? HAIR LOSS?

Top Chinese dissident to visit

The bleyer clinic

NATIONAL WEATHER

Figure 126 *The Australian*

Daily Mirror

U.S. sailor AWOL in Aust

MY 18 MONTHS ON THE RUN

EXCLUSIVE

Introducing . . .

Marcia Hines' girl Denni

STAR PHOBIAS

Secret fears of the rich and famous

LATE FINAL EXTRA

Figure 127 *Sydney Daily Mirror*

Murdoch ownership. Its bold caps headlines BORN A CRACK ADDICT, WIFE-MURDER TRIAL DAD WEDS WITNESS, and TORTURED TYSON IS ALL ALONE have the taut whiplash effect of authentic low-count tabloid blockbusters. On the occasion of the Pope's visit to New York, twelve TV stations broadcast football games and only two the Pope's arrival. The *New York Post* splash headline was FOOTBALL 12, POPE 2. Yet after the first dozen eye-catching layouts it settles down to being a staider newspaper than it first appears and gets on with the job of packing text and adverts into its 100-plus pages in a way unlike a British popular tabloid.

The Commonwealth

With the formative years of the Commonwealth press coinciding with the growth of empire and political institutions in the UK, British newspaper practice was well placed to have global influence. It is not surprising, therefore to find in India, Australia, parts of Africa, and even Hong Kong, British language papers that have a familiar look.

In Australia there is the same division into eight-column broadsheets (occasionally nine) at the serious end of the market, and seven-column city tabloids at the popular end. Rupert Murdoch's *The Australian*, the country's only true national paper (Figure 126) achieves eye comfort and readibility for its up-market readers with an elegant modular design based on Century bold and light lowercase well broken by bold pictures. There is a feel of the British quality Sundays of a couple of years ago here, though without the preoccupation with artistic rulery, daring picture crops and utilization of white space, or self-conscious artiness.

Down-market, the *Sydney Daily Mirror* (Figure 127) and the *Brisbane Truth* and its city clones proclaim, though again with a slightly dated feel, their kinship with *The Sun* and Britain's *Daily Mirror* through their use of heavy sans headlines, page leads mainly in capitals, comic strips, glamour pictures and liberal use of WOBs and BOTs. Only the fatter size – 72 pages and more – betray the influence of American-scale advertising. The 78-page broadsheet *Sydney Morning Herald* (Figure 128), while sharing the modular design approach of *The Australian*, moves further towards the American model with a three-section Wednesday edition crammed with advertising, with strips of news horizontalized above. It also has the Old English title piece much favoured in American papers in comparison with the *Times* style logo of *The Australian*.

The *South China Morning Post*, of Hong Kong (Figure 129), and its stablemate the *Sunday Morning Post*, both many-sectioned and crammed with news and advertising, could have been produced by the same art desk and on the same presses as the Australian broadsheets, though displaying more original and imaginative feature page ideas inside. Both are presented with great style and a newsy feel. Unaccountably, the daily has a nine-column format and the Sunday an eight-column.

European

European newspapers look to a different lineage. Here exist such anomalies as the broadsheet-sized poster-layout and the tabloid sized quality sheet, thus contradicting the old notion that design philosophy is somehow related to page size.

Le Monde (50 mm x 34 mm) with its small wordy seriffed headlines and 10½ pica columns of readable 7 point (Figure 130), is by no means the exception that proves the rule that a quality newspaper opts for broadsheet. Spain's *El Pais* (41 mm x 29 mm) and the Swiss *Tribune de Geneve* (49 mm x 33 mm), have a similar headline dress and words-orientated design, and appeal to a similar readership market. Having in recent years finally permitted the use of pictures (cartoons were always a feature), *Le Monde* is relaxing sufficiently these days to allow such departures as the occasional drop letter, the occasional cut-out picture, WOB page flags and the odd bit of bold setting and spot red. It remains, however, a distinctive, well-signposted, comprehensive journal of comment giving emphasis to

The Sydney Morning Herald

HEALTH COVER

PRE-SCHOOL Why it's a must

NIGHT OF KNIVES THE INSIDE STORY

Battle in the sky as Iran prepares to make peace

COLUMN 8

Child care places up by 26,000

So this is what they call a snowfield

Bitter Aust takes on US over sugar quotas

In ACT a black hero may oust a white

Sinclair defiant as Nationals split on leadership

INSIDE

WEATHER

Figure 128 *Sydney Morning Herald*

South China Morning Post

140 PAGES TODAY with 102

Bid for Hotels rejected

Hostile overture not welcome – Kadoories

Saturday Review

THE CATHAY STORY

AMNESTY IN AFRICA

JUNGLE FOR CENTRAL

Call to end defendants' right to remain silent

Pull-out of troops possible

Governor to focus on repatriation in talks

Osman warned in bid to avoid extradition

US jury indicts Marcos, wife

Price row grounds tycoon's satellite program

AP Audemars Piguet

Figure 129 *South China Morning Post*

Le Monde

VENDREDI 22 SEPTEMBRE 1995

Les marchés financiers redoutent une nouvelle crise monétaire

La chancellerie a fait pression par écrit sur un magistrat pour protéger M. Juppé

Un éventuel délit d'ingérence pourrait être reproché au premier ministre

Un niveau historique en 1996 pour les prélèvements obligatoires

Une polémique oppose Matignon et le CNPF

Les promesses de « Laloo » aux damnés du Bihar indien

L'OTAN suspend ses frappes en Bosnie

Le football européen dans l'embarras

Les interrogations des catholiques français

Les guerres perdues de l'ONU

Un dissident cubain de l'intérieur

Les arbres poussent plus vite qu'autrefois

Le premier Salon des beaux-arts

Les éditoriaux du « Monde »

DANIEL MAXIMIN

Éditions du Seuil

Figure 130 *Le Monde*

EL PAIS

DIARIO INDEPENDIENTE DE LA MAÑANA

Los chilenos deciden hoy la suerte de la dictadura de Pinochet

La policía había perdido la pista a los 'grapo' que mataron ayer a un agente

HOY, EN EL PAIS

Comienza el curso universitario.

No haga números, Iberdealer se lo resuelve con letras...

BONO LOTO

LETRAS DEL TESORO

Telefónica enmascara la subida de tarifas, según los sindicatos

Figure 131 *El Pais*

the words. Its small headlines work through being attractively whited out, while the geography of its items, clearly identified in strap-line or label, persuades the reader by the time page 48 has been reached that the world has been well and truly reviewed. It gives generous listings to 'arts and spectacles', but has no place for sport.

The Madrid daily *El Pais* (Figure 131), with similar small size seriffed heads (no more than 42 point) all in lowercase and well whited adopts the same pattern of flagging its coverage to help the reader. The pages are uniformly of five-column, as opposed to *Le Monde*'s six, and the type a particularly elegant and readable 8 point set to advantage on a 9 point body. It uses thick and thin rules effectively across the tops of its pages, often incorporating subject labels. There are rather more pictures, a good business coverage but still no sport. It could be almost a Spanish edition of *Le Monde*.

The *Tribune de Geneve* (Figure 132), like *El Pais*, is a five-column tabloid, though slightly bigger, with a 13 pica column width in 8 point, seriffed headings of up to 60 point page lead size, and with capitals used on the page one splash head and on strap lines and identification flags. The typographical treatment of the pages and the feel of the paper is similar to *Le Monde* and *El Pais*, but underscored standfirsts and spot colour are used as eye-breaks and there is slightly

Mercredi 31 août 1988

JA No 203

Dans ce numéro

Evénement	3	Entracte	13
Monde	4, 5	Télévision	15
Bourses	4	Rendez-vous	16
Economie	6, 7	Genève	17 à 21
Suisse	9, 11	Décès	22
Mémento	12	Régions	23
Urgences	12	Sports	31, 32

TRIBUNE DE GENEVE

Quotidien fondé en 1879

Fr. 1.20

Prix du numéro: Suisse Fr. 1.20, France FF 4.20, Italie Lit. 2100.–, Espagne Ptas 180.–, Grèce Drs 110.–

L'USINE DE BIÈRE VA FERMER

SKIS: ROSSIGNOL SUICIDE AUTHIER

La Suisse va perdre sa dernière fabrique industrielle de skis. Le groupe français *Rossignol* a décidé de cesser la production dans l'usine *Authier* de Bière, au pied du Jura vaudois, qui fermera ses portes à la fin de l'année. La fabrication sera transférée dans les autres unités du groupe et notamment à Voiron, dans l'Isère.

Le Syndicat du bois et du bâtiment se dit révolté de voir ainsi cesser toute production industrielle de skis dans un pays qui en consomme 360.000 paires par année. La FOBB met de même en cause la Fédération suisse de ski (FSS) qui a fait de nos champions les porte-drapeaux des firmes étrangères ainsi que l'inertie des pouvoirs publics: «On a sauvé le creteau suisse, pourquoi ne pas sauver le ski suisse?»

page 7

Brigitte Gadient, sur Authier, quatre podiums, vingt-cinq fois des points marqués en Coupe du monde. Elle se classe trois fois deuxième à Waterville et Madonna en 1984, et à Savognin en 1985. (Photo ASL)

CATASTROPHES

La Chaîne se déchaîne

Une fois de plus, la Chaîne du bonheur se déchaîne, aujourd'hui sur nos ondes. Quatre semaines après les inondations au Soudan, dix jours après le séisme en Inde et au Népal, les secours sont insuffisants: impossible d'éviter le grand show médiatique. Mais d'où vient le matériel d'urgence Swiss made, tant apprécié à l'étranger? Daniel Wermus a visité les cinq étages du «supermarché de l'humanitaire», à Berne. De la brosse à dents à la tente géante, un étalage saisissant.

page 3

brico loisirs MIGROS

SAHARA

La paix des sables

André NAEF

MEETING DU WEEK-END PROCHAIN

Ecuvillens ne veut pas des «Frecce Tricolori»

La patrouille italienne des «Frecce Tricolori» ne se produira pas au Meeting d'Ecuvillens, à Fribourg, le week-end prochain. Ce sont les organisateurs qui ont demandé, par l'intermédiaire de l'ambassade d'Italie à Berne, à la patrouille italienne de renoncer. Les organisateurs ont annoncé qu'ils avaient pris cette décision d'une part par respect pour les victimes et les trois pilotes décédés dimanche dernier d'autre part parce que les conclusions de l'enquête sur les causes de l'accident ne sont pas encore connues. L'ambassade d'Italie à Berne devait préciser que les pilotes de la patrouille avaient aussi demandé d'être libérés de la participation au meeting fribourgeois.

page 9

Jürg Stäubli veut relancer *Donnay* page 6

Mostra de Venise: audace, ambition et ratage! page 13

Ex-mari escroqué page 17

Le justiciable indigent choisira son avocat page 17

emploi TRIBUNE pages 33 à 50

Figure 132 *Tribune de Geneve*

more variety in page design. It has an unusual masthead in blue and there are colour pictures on page three and also an adventure strip cartoon in colour. The paper is in three sections, one of which does include sports pages.

The poster broadsheet

At the opposite poll to these sedate tabloids comes the genre of poster-style popular broadsheets in colour, illustrated on Plates 2 and 3 in the colour plate section, which are perhaps the most distinctive aspects of Europe's contribution to newspaper design. It is a trend that has drawn no takers in Britain. Yet it is a style that can be found on bookstalls all over Europe. The fact that the examples shown come from Germany, Greece and two from France is coincidental.

The poster broadsheet is for the newspaper reader who wants everything on page one. It is a contents lists, a taster, a vast, colourful, concentrated distillation of the day's news; it is a grandiose buy-me, read-me-now feast for the eye; it is the ultimate use of display, not in wild or 'circus' fashion but in a controlled assault upon the impulse buyer who is the mainstay of the popular market. It is saying: 'This is it! This you must know! We have the lot – look no further!' It turns upside down the traditional notion that a popular paper should be a small sheet containing, on the front, blockbuster type on the best story of the day with one big eye-catching picture, offering the reader instead a cornucopia of goodies that dazzle the eye.

The *Abendpost*, of Frankfurt (Plate 3(b)) is an extreme manifestation of this genre. Page one manages to carry three lines of 144 point heavy sans lowercase type on a splash that consists of a one paragraph summary of a jet crash with cross-reference to page 2; a 29 cm wide picture of a broken bridge with headline in 84 point, one paragraph taster and cross-reference to page 4; a sports summary with 60 point headline, main results, picture and cross-reference; a self-contained down-page story 15 cm long with an 84 point headline; a picture with caption of a couple in fancy dress on a tandem; seven short self-contained stories, competition winners, weather report, TV programme cross-reference and, 12 cm down the page on the left-hand side, the *Abendpost* title piece used as a fulcrum for the stories swirling around it. The main items are clearly flagged with WOB labels, and the stories are kept apart by a mixture of 6 and 10 point red and black rules so that in the vast mosaic of type no one piece can be confused with another.

It is a page for the reader looking for instant easy-to-read news, who knows the rules of the typographical game by which the paper makes its name. Yet note the careful slotting of text in relation to headline and halftone, the strength derived from the long horizontal picture of the bridge, the studied intrusion above the fold of the sports summary and vital big type results – an important item for a large body of a popular paper's readers. The picture of the jets let into the WOB above the splash sends the eye to the centrally sited text. The bridge caption is arrowed, and so is the tandem one. The weather and competition winners are instantly findable in what, on closer inspection, is by no means a jumble of type.

The Athens daily, *To Fos* (The Light), in a similar market (Plate 2(b)) relies on the same technique with the added facility of highlighting in blue and yellow as well as red. Again, the title piece (in a red and black panel) is lowered from the top of the page for the stories to swirl around it. Many of the twelve items consist only of headlines and cross-references, an important football result taking the entire top of the page in yellow on black. The top half is virtually an extended contents bill for the news-stands, with the main pictures used as focal points round which to assemble the stories below the fold. It is this form of presentation that makes *To Fos* distinctive to its down-market readers. It is bold and instantly recognizable, although the ruling and separation of items is less helpful to the eye of the uninitiated than in the more disciplined German product.

The Paris evening *France-Soir* (Plate 3(b)) and the *Nice-Matin* a big seller on the Riviera, are less strident examples of the broadsheet poster technique. *France-Soir* follows the German pattern of using sans lowercase in a wide range of sizes, though introducing italic and even a serif variant for some down-page items. The splash story consists only of a headline and cross reference, as do three other items, while the dominant colour picture also funnels down into a caption and cross-reference. The title piece again drops below the top

Plate 1(a) The quill work on the right is from Einhard's *Life of Charlemagne*, written in the early ninth century. It demonstrates the style of illuminated letters known as the Caroline, upon which modern Black letter and Old English typefaces are based.

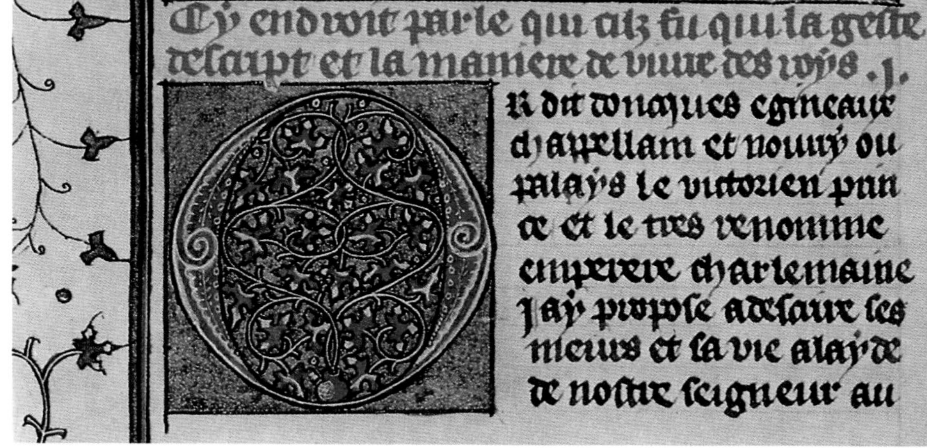
Cy endroit parle qui cilz fu qui la geste
descript et la maniere de viure des roys .j.
R dit donques eginaux
d'apellam et nourry ou
palays le victorien prin
ce et le tres renomme
empereur charlemaine
j'ay propose a descrire les
meurs et sa vie a la rede
de nostre seigneur au

De officio psalmistatus

mationepontifex sedēs accepta mittra patrinis et matrinis annunciat quod instruātur ⁊ informent filios suos bonis moribus ⁊ operibus: quod fugiant mala ⁊ faciant bona ⁊ doceāt eos **Credo in deum.** et **Pater noster.** et **Aue maria** quoniā ad hoc sunt obligati.

¶ De officio psalmistatus.

Psalmista idest cantor pōt absq licentia Episcopi sola iussione presbyteri officium suscipere cantandi: dicente sibi presbytero.

Vide vt quod ore cantas: corde credas: ⁊ quod corde credis operibus comprobes.

Et si episcopus clericū ordinās hec faciat benefacit.

¶ De ordinibus conferendis.

Quoniam collatio ordinum ecclesiasticorum maxime est importantie. pontifex multa prius considerare debet ⁊ videre ne cui cito man⁹ imponat; iuxta traditionem apostoli: Ordinandorum conditiones et qualitates: ne vel indignis conferantur: vel

De ordinibus conferendis

vitio suscipientium ordinum dignitas minuatur. ¶ Committat igitur prius examinationem ordinandorum Archidiacono vel Archipresbytero: pro more ecclesie: ⁊ illis addat sacerdotes timoratos: doctos ⁊ prudētes: prout sibi videbit: vel si placet ipse examinationi intersit. ¶ Examinentur primo de etate legitima: sufficienti litteratura: ⁊ morum honestate. Non enim debet q̄s ad subdiaconatum ordinari: nisi decimumoctauum: neq ad diaconatum nisi vigesimum: nec ad presbyteratum nisi vigesimumquintum sue etatis annum attigerit. ¶ Prima tonsura vero et minores ordines ante septimum annum dari non debent. ¶ Litteratura in subdiaconatu ⁊ diaconatu mediocris sufficit vt saltē sciat bene ⁊ expedite legere: ⁊ competenter cantare: det operam grāmaticalibus: sciat ordinare officium: ⁊ habeat modū ac commoditatem illud dicendi, ad quod obligatur. Neq debet ordinari nisi videatur q officium commode dicere possit. ¶ In sacerdote autem promouendo sufficiēs litteratura requiritur. Sciat grāmaticā: presertim si sit secularis q illis cura animarum committi solet. que sine litteris exerceri non potest. ¶ Examinentur etiā de natalibus quia illegitimi sine dispēsatiōe apostolica ad sacros ordines promoueri non possunt.

(b) This example of early sixteenth-century printing shows how closely the formation of the letters follow the quill work of the monks as in the above example.

DAILY Mirror

TWO GAMES FOR INSTANT WINNERS

4 CHANCES TO WIN £50,000

THE MIRROR'S LOTTERY

HONESTY, QUALITY, EXCELLENCE

27p

MIRROR INSTANT SCRATCH

TODAY'S BONUS CASH FIGURE: PAGE 26

Bomb Bosnia plan puts Brits in peril

MAJOR: Called summit

GAZZA BABY WILL HEAL FAMILY RIFT

FULL STORY: PAGE 3

BISHOP WANTED ME OUT

Cleared dean: It was Church witch-hunt

Turn to Page 7

Turn to Page 5

Plate 2(a) This electronically made-up *Daily Mirror* page one demonstrates the range of colours and effects being utilized on screen by the designer. The paint-box is exploited principally to activate the blurb, yet it is the bold white-on-black splash headline that grabs attention first, leaving the eye to move on to the pretty girl in colour and then upwards to the package of goodies.

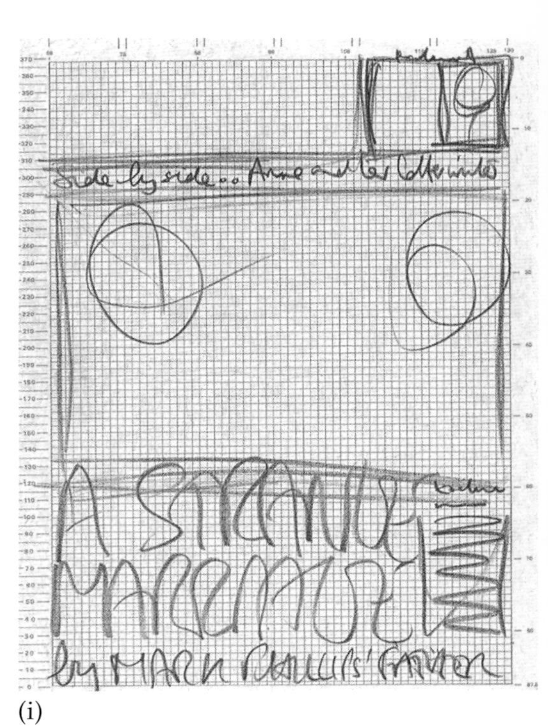

(i)

Η ΑΕΚ κανόνι 4-1

Εχασαν 1-0 οι Ελπίδες

«Με τον Σαραβάκο θα μιλήσουμε όταν έρθει η ώρα»

ΦΩΣ

Δ/ντής: Θ. ΝΙΚΟΛΑΪΔΗΣ

«Εγώ φταίω για Ντέταρι»

παραδέχεται ο Γιάτσεκ Γκμοχ!

Γκμοχ: Ο Ολυμπιακός είναι πρώτο φαβορί για το πρωτάθλημα

ΑΓΩΝΙΑ ΓΙΑ ΑΝΤΩΝΙΟΥ

Η Λάρισα κόντρα με τον Ηρακλή

Το σκάνδαλο του αιώνα απόψε στην Ολομέλεια

Μια Ολυμπιακή βραδιά στην Κέρκυρα

Η Εθνική σήμερα με την Αν. Γερμανία

Εννέα νέα μέλη από σήμερα στο ΑΣΕΑΔ

Οι ενοχλήσεις του Μπατίστα προβληματίζουν την Εθνική

Συμπτώματα απειθαρχίας στον Απόλλωνα

CENTRE COMMERCIAL NICE ETOILE

LES SOLDES QUI EMBALLENT

DU 15 JUILLET AU 15 AOUT

Cannes-Grasse

nice-matin

LE PREMIER QUOTIDIEN D'INFORMATIONS DU SUD-EST ET DE LA CORSE

MOGHADAM

HIPPODROME

MITTERRAND : VISITE IMPROMPTUE AUX "POMPIERS DU CIEL"

puis à la Sécurité civile de Brignoles

CAGNES

Hold-up avec prise d'otage

Transports aériens

Doublement des fréquences entre Nice et la Corse l'hiver prochain

Banque Worms

L'otage était complice

JULIA MIGENES-JOHNSON :

"L'Unique" à Monte-Carlo, ce soir pour le gala de la Croix-Rouge

Sécheresse : pluie de mesures

Nice devant Rennes :

La victoire à tout prix

PLANCHE

En vente chez tous les marchands de journaux

SPECIAL PLANCHE

Gymnastique

Un super show à Antibes ce soir et demain

JOURNEE SOIREE NON-STOP

Force 10. Direction "Alizé"

FRED

(b) The Athens broadsheet on the left, *To Fos* (The Light), is almost surrealist in its approach with a very European saturation blurb technique. *Nice-Matin* uses a more reserved typography to project similar ideas.

Side by side .. Anne and her letter writer

EXCLUSIVE

A STRANGE MARRIAGE

by MARK PHILLIPS' FATHER

(ii)

Today

NEWSPAPER OF THE YEAR

EXCLUSIVE

FALDO

Torment he hid on road to victory

CENTRE PAGES

Side by side .. Anne and her letters writer

A STRANGE MARRIAGE

by MARK PHILLIPS' FATHER

(iii)

Plate 3(a) A page one from *Today* in its three stages of production - first the editor's rough, which is interpreted by the art desk in the second version, and third, the finished page.

Katastrophe von Ramstein

Eine Mutter fleht: Hoffentlich stirbt mein Kind!

Es ist von Brandwunden für immer entstellt

Frankfurt • Mittwoch, 31. August 1988

Abendpost

Nachtausgabe

Beckenbauer: Mit 3 Spitzen gegen Finnen

Aschaffenburg

Autobahnbrücke stürzte in den Main

Offenbach 4:1 Hertha BSC

BW Berlin 2:1 Darmstadt

Detari kommt unters Messer

Wetter-Hugo meint:

Wie lange ist Brot knackig?

Gericht muß es klären ● Kunde verklagt Bäckerei

Scheidung mit 40: Frauen einsam

Doch Probleme mit dem Airbus A320

PROGRAMM

FOOT Bordeaux n'a rien pu contre Maradona

«France-Soir» vous révèle

Une cathédrale pour Evry-Ville-Nouvelle

France-Soir

QUARTÉ dernière

Le juge Boulouque gagne la 2e manche contre les terroristes

Il n'est pas dessaisi du dossier des attentats de septembre 1986 malgré son inculpation

Delphine (9 ans) appelle au secours

• Née d'un viol, violentée à son tour, l'enfant de la D.D.A.S.S. veut vivre chez ses parents nourriciers • Elle attend, isolée du monde, une décision qui tarde

LA CHRONIQUE de JEAN DUTOURD

Ne prenons rien au tragique

Hier, les larmes étaient vraies

UNION SOVIÉTIQUE

3 morts, 126 blessés au cours de manifs en Azerbaïdjan

BOUVARD ET toC!

Prudence par omission

TV VOTRE MAGAZINE TÉLÉ DU SAMEDI

(b) *France-Soir* page ones are noted for the strategic positioning of colour. *Abēndpost*, left, suffers, as do all German broadsheets, with 'portmanteau' words which demand plenty of Extra Condensed type. However the page - helped by the effective use of spot colour - cries out to be read.

NO. 1 IN THE USA — AVAILABLE AROUND THE WORLD — VIA SATELLITE

Vol. 14, No. 23

SPORTS FINAL

BRONCOS' COACH GETS REVENGE ON RAIDERS

DENVER WINS 27-0, 1,6B
▶ 49ERS' YOUNG HURT, 1B
▶ NFL NOTES, 6B

Bryan Kelsen, AP
JOHN ELWAY: Two TD passes for Denver, 1B

NFL REFS HAVE NO ROOM FOR ERROR

PLAYERS, FANS, LEAGUE SEE EVERY MISTAKE, 1,3B

WEDNESDAY

QUALITY: IS IT DEAD IDEA OR SOP?

BUSINESSES PUSH ASIDE TREND OF TOTAL QUALITY MANAGEMENT, BUT SOME TENETS ARE INGRAINED, 10B

BRUCE SPRINGSTEEN EVOKES AMERICANA IN NEW 'TOM JOAD'

SOLO, ACOUSTIC ALBUM IS 'INTIMATE, PERSONAL' 9A

By E.G. Gallagher
SPRINGSTEEN: Album due Nov. 21, 9A

WEDNESDAY, OCTOBER 18, 1995

NEWSLINE

A QUICK READ ON THE NEWS

DOLLAR: Closing quotes in London for key currencies: Pound at $1.5740 from $1.5755; U.S. dollar at 1.4134 marks from 1.4206; 100.28 yen from 100.42. Table. 9B.

▶ Gold closes in London at $383.20, from $383.95.

CONTRACT WITH AMERICA: Colin Powell says he supports "most of the elements of the Contract with America," while calling some parts "too harsh." 4A.

CAL-ATOLLAH: Willie Brown is making waves in quest to become mayor of San Francisco. 3A.

BETSY'S HOTEL: Undaunted by war and want in Georgia, Betsy Haskell of Baltimore proves you don't have to be a corporate giant to do business in one of the former Soviet Union's most violent trouble spots — Tbilisi. USA Abroad column, 5A.

CHILD SUPPORT: Increased enforcement of child support is running into political and practical barriers in the USA; 10% tax on all child support and nationwide computerized enforcement system are criticized. 4A.

KOREAN KILLING: South Korean troops shoot and kill a suspected North Korean spy. 4A.

VALENTINE FIRED: The Chiba Lotte Marines fire

Blast rocks Paris subway

Smoke and cries for help

By Eric Gaillard, Reuters
HELPING THE HURT: Firefighters and paramedics evacuate one of the injured from Tuesday's bomb blast in a subway train near the Orsay Museum, at least the eighth blast since July.

Police focus once again on militants

By Rose Marie Burke
Special for USA TODAY

PARIS — France's vulnerability to terrorism was again exposed dramatically Tuesday when a bomb tore through a Paris underground train during rush hour, injuring 28 people.

Despite the latest act of violence, "France won't let itself be intimidated," Prime Minister Alain Juppe told the French National Assembly just hours after the attack. He said France will go through with controversial talks with Algeria.

No group has yet claimed responsibility for the latest attack. The Armed Islamic Group, composed of Islamic militants fighting the military-backed government in Algiers, has claimed credit for most of the previous ones.

The bomb went off at 7:05 a.m. as a regional express train was going 35 mph through a tunnel between the Saint-Michel and Orsay Museum stations.

By Agence France Presse
CHRISTOPHER: Says sending troops now avoids more war

Push made for troops to Bosnia

By Steve Komarow
USA TODAY

The Clinton administration on Tuesday promised a skepti-

(b) Electronic pagination at work in colour: a *News of the World* page one in its later stages on an Apple-Macintosh terminal and, below it, the edition as it came off the press. Reproduced with permission.

Plate 4(a) The combination of a striking blue masthead, dominant full colour pictures and use of spot colour create a unique blend of tabloid and broadsheet techniques in *USA Today.* Below: the Snapshots feature produces a down-page eye-catcher using colour drawn from the halftones in the page. Copyright 1995 and 1989, *USA Today.* Reprinted with permission.

CORRIERE DELLA SERA

Una difficile scelta per il Likud dopo la mezza vittoria elettorale

Le due strade di Shamir

Il «ricatto» dei religiosi potrebbe resuscitare la coalizione

UN VOTO, NON UNA SCELTA

Fu un missile italiano ad abbattere il DC9 Itavia?

Il governo chiede all'Aeronautica la verità sul disastro di Ustica

Il leader dell'Olp a Roma

Arafat è pessimista: la rivolta continua

S'inasprisce la battaglia degli spot tra la Fininvest e l'ente pubblico

Berlusconi attacca la vigilanza Rai

«Troppi legami tra i commissari controllori e i controllati»

Quel tetto che scotta per Craxi e De Mita

Sacharov, Gorbaciov ha troppo potere

Contro i trafficanti di stupefacenti il ministro Gava mobilita i servizi segreti

Occhetto: guerra alla droga ma niente schedature

Le madri di Siracusa: meglio schedati che tossicomani

PLATINI SI CONFESSA «LA FRANCIA SONO IO»

Aumentano i prezzi delle sigarette

Figure 134 *Corriere della Sera*

GASOLINA PASSA A 438 E ÁLCOOL A 321

Bom dia, leitor!

Projeto de esperança

BANCOS NÃO ABREM 2ª FEIRA

Sarney joga tudo contra a inflação

MEDIDAS PROVÁVEIS:

Ultima Hora

Habitação é prioridade de Mandarino

Presidente anuncia hoje reformas no Governo com novos nomes no Ministério

Atendimento médico está garantido na passarela do samba

Assalto tumultua prédio no Centro

Criminalista garante que ninguém vai preso no caso do Bateau

Trens de subúrbio já custam Cz$ 50

uhTurfe Hipódromo vira circo

Ninguém sabe por onde andam os juniores

Feema interdita Leme e Leblon para banhistas

Fuga de metrô após assalto em Botafogo

uh Revista

uh MODA

OITO CHACINADOS NA GUERRA DO TÓXICO

Figure 133 *Ultima Hora*

LA STAMPA

ANNO 123, N. 145 — GIOVEDÌ 29 GIUGNO 1989 — L. 1000

Intervista con De Michelis: ecco le condizioni del psi

«Governo breve»

«Realistico il traguardo di febbraio, prima delle amministrative» «De Mita? Non poniamo veti. La confusione è nella dc»

ROMA. Un governo a termine. Che «realisticamente» può arrivare fino a febbraio, cioè alla vigilia delle prossime elezioni amministrative. Gianni De Michelis, nell'intervista a *La Stampa*, detta così le condizioni del psi per uscire dall'attuale crisi. «Occorre un programma con poche cose e chiare», spiega il vicepresidente del Consiglio. Nessun veto su De Mita, aggiunge De Michelis, ma «constatiamo che anche su chi debba guidare il governo c'è confusione nella dc».

Ma non basta. Nella formazione della nuova maggioranza «c'è il grosso problema del polo laico». «E' difficile fare un governo — dice De Michelis — con dei partiti che sono in rapporto a un terzo, il cui leader (Pannella, ndr), antisocialismo a parte, è portatore di un'altra coalizione». Restano difficili anche i rapporti con il pci: «La posizione dei comunisti, che è indubbiamente in evoluzione — spiega De Michelis —, rimane fortemente polemica e ostile verso i socialisti». Quanto al psi, secondo il vicepresidente del Consiglio «il risultato elettorale delle europee è stato buono». «Non c'è ragione di modificare la posizione tattica e strategica di Milano, cioè una costruttiva disponibilità per la governabilità e l'azione di costruzione dell'unità socialista».

Ieri, intanto, De Mita ha avviato il suo secondo giro di consultazioni, incontrando i rappresentanti della piccola industria e dei commercianti. Le trattative per il nuovo [illegible] entreranno nel vivo solo [illegible]

SCALA MOBILE

Confindustria da De Mita

ROMA. La Confindustria, dopo il fallimento delle trattative [illegible] martedì con Cgil-Cisl-Uil, ha deciso di rinviare di 48 ore una [illegible] sione sulla disdetta dell'accordo sulla scala mobile. Il presidente Pininfarina ha ricevuto un mandato a recarsi oggi dal presidente del Consiglio per un incontro nel quale sarà esaminata [illegible] questione della fiscalizzazione degli oneri sociali. Appare [illegible] che De Mita inviti gli imprenditori a non procedere con la d[illegible] ta. Proseguiranno intanto anche i contatti con i sindacati. [illegible] dovessero verificarsi fatti nuovi la Confindustria potrebbe [illegible] ciare l'accordo entro domani sera.

Ieri il presidente della Confcommercio, Colucci, ha confermato invece al presidente del Consiglio che la sua organizzazione [illegible] intende disdire l'accordo.

(Gian Carlo Fossi a pag. 2)

L'ESULE DELLA TIENANMEN

«Il mio popolo abbatterà il sole nero di Deng»

GENERALI, LE PAROLE E LE ARMI

Delitto Calabresi: il pm chiede il rinvio a giudizio per l'ex leader di Lc

Sofri, omicidio premeditato

«La testimonianza di Marino è credibile»

OGGI — LA STAMPA

Di nuovo crisi in Comune

Zanone: «I militari delusi ma fedeli»

Presagi la Rivoluzione francese

Partendo per l'Italia il nuovo ambasciatore Usa spiega come «si è preparato»

Secchia: io sono di stomaco forte

«Sarà difficile giostrare tra i capi del pentapartito»

La mappa dell'inquinamento

Figure 135 *La Stampa*

LA GAZZETTA DEL MEZZOGIORNO

La Gazzetta di Puglia - Corriere delle Puglie

Polemiche dopo la tragedia di Ramstein. Il capo di S.M. dell'Aeronautica Pisano dice...

«Ma le Frecce voleranno»

Saliti a 62 i morti, circa 350 i feriti

Sabato la Pattuglia acrobatica nazionale si esibirà a Friburgo, in Svizzera. Sulla strage di Ramstein nominate tre commissioni d'inchiesta (una italiana, una tedesca e l'altra americana)

ROMA — Le «Frecce tricolori» continueranno a volare, a volteggiare tra le polemiche e gli interrogativi. Non è arrivato dal ministero della Difesa il contrordine che molti si aspettavano. Al capo di stato maggiore dell'Aeronautica, gen. Franco Pisano (che ha incontrato ieri mattina), Zanone ha consegnato solo un lungo messaggio di «dolore e solidale partecipazione» al lutto dei familiari delle vittime, a nome suo e delle più alte cariche dello Stato. Anche al collega della Repubblica Federale tedesca, Rupert Scholz, ha fatto giungere parole di cordoglio.

Stanotte, al quartier generale dell'Aeronautica militare nessuno riesce a sorridere, il giorno dopo la tragedia di Ramstein. Tre di loro, tre dei migliori piloti della Pattuglia acrobatica nazionale sono caduti seminando morte su un campo gremito di folla. Perché e come un simile disastro sia potuto accadere lo stabiliranno tre commissioni d'inchiesta: una italiana (presieduta dal gen. Olivero, capo di stato maggiore della Terza Regione aerea, di Bari), una statunitense e una tedesca.

Saranno le carcasse delle tre "Frecce Tricolori", quei rottami anneriti piombati come proiettili sulla gente, a rivelare agli esperti le cause del disastro.

La verità. Il capo di stato maggiore dell'Aeronautica, gen. Franco Pisano, ha detto ieri ai giornalisti di non poter escludere, ma neppure privilegiare, una delle tante ipotesi finora emerse: «Deleghiamo la ripresa del discorso a quando le commissioni d'inchiesta avranno dato i loro risultati».

Di aver disposto l'immediata costituzione di un gruppo di esperti militari, Pisano aveva dato notizia in mattinata al ministro Zanone. Con lui si era accordato circa l'opportunità che la Pattuglia acrobatica non venisse meno agli «impegni già programmati per la stagione 1988», nonostante dei dieci velivoli solitamente impiegati nelle manifestazioni, oggi, dopo l'incidente di Ramstein, ne siano rimasti "in piedi" solamente cinque.

Durante la «cardioide»

Proprio quando c'era rischio zero

L'aereo esplode in volo

Ma veniamo a quanto il generale Pisano ci ha riferito ieri.

La manovra e l'errore. «La pattuglia stava eseguendo la figura cosiddetta "cardioide" (a forma di cuore): forse la più semplice, nei venti minuti di esibizione. Una manovra a rischio zero per coloro che l'effettuano. Il "solista" è giunto nel momento in cui le due formazioni si stavano incrociando, ha impattato contro il gregario di sinistra che, esplodendo, a sua volta ha colpito il capo formazione, tagliandogli la coda». Mentre il gregario precipitava a terra e l'aereo del capo formazione finiva incontrollato in un boschetto, quello del "solista" piombava come una palla di fuoco sulla gente. E il resto è cronaca di una tragedia.

Per l'aviazione il bilancio delle perdite ammonta a cinque velivoli: oltre ai tre "MB 339-A" andati distrutti, altri due sono rimasti seriamente danneggiati dalle schegge e non potranno volare. Gli altri cinque aerei torneranno questa mattina alla base di Rivolto. «Vedremo come la pattuglia è stata penalizzata — ha proseguito il gen. Pisano — Dovremo valutare in un momento più sereno le conseguenze della tragedia».

I programmi per il futuro. «Almeno per quest'anno, per far fronte ad impegni già presi, operare con cinque o sei velivoli sarà difficile. L'operatività sarà limitata». Ma non interrotta, a meno che le manifestazioni in calendario da qui a fine ottobre non vengano nel frattempo revocate (è dell'ultima ora la notizia che l'*Air Show* di Aviano del 2 ottobre è stato annullato). Il prossimo appuntamento dovrebbe essere a Friburgo, in *Svizzera*, sabato 3 settembre.

La sicurezza. E' l'argomento che più farà discutere. «Tutto quello che potremo trarre dall'incidente sarà utilizzato per rivedere i nostri programmi, ma contrariamente a quanto si afferma, la pattuglia ha sempre lavorato su fattori di sicurezza estremamente approfonditi, accurati. Soprattutto riferiti alla sicurezza di coloro che sono a terra». Nelle manifestazioni «i parametri vengono fissati di volta in volta, in misura diversa» — ha spiegato ancora il gen. Pisano — a seconda dell'orografia del luogo. «In questo senso abbiamo la coscienza tranquilla, anche se la dinamica dell'incidente è stata tale che comunque non si sarebbero potuti evitare guai».

Barbara Modesti

Ultim'ora

Fitto muore in un incidente a Francavilla

Il presidente della Regione Puglia falciato da un camion. Carbonizzato l'autista

Chiesta la testa del ministro della Difesa

Un'ondata di emozione scuote la Germania

San Pancrazio Salentino

Uccide la moglie a colpi di forbici

Ministro Amato conosce Galileo?

di Natale Gilio

Gelo russo a Venezia

NELL'INTERNO

Alto Adige

Oggi si vincono due milioni

Jaruzelski fa marcia indietro

Ci riprova Glemp col rebus polacco

Figure 136 *La Gazzetta Mezzogiorno*

of the page so that it is surrounded by items with once again a football story taking top position although, as a result of the large picture and the longish story on the right, the overall pattern is less complex than the German or Greek examples.

Nice-Matin (Plate 2(b)), rich in colour pictures, and with smaller, lighter sans type, carries nine items with the minimum of text, while slotting in half a dozen advertisements at the top and bottom of the page. Every item is cross-referred to inside pages with the exception of one picture with a caption story. The page, with its clearly ruled off items and spread-around pictures, gives the impression of an attractive news menu containing a little something for everyone.

Two New World examples of this genre are to be found in the worldwide circulating *USA Today* (Plate 4(a)) and the exotic *Ultima Hora* (Figure 133), of Brazil. In a front page remarkable by any standards, the title piece of *Ultima Hora* comes 43 picas down the page. It is difficult to identify a columnar format since on all pages there is an eccentricity of measure which, as on page one, manages to convince the eye by the sheer confidence of the presentation. The page illustrated is immaculately put together with the minimum of body matter, giving the reader sixteen headlines in assorted sizes and dwelling in poster form on serious issues arising out of Brazil's mega-inflation. Not sport but the price of petrol is at the top of the page.

Varieties of style

There is, of course, a wide middle ground between the two extremes discussed above. While there is little influence to be seen of the mass market British tabloids, there is a variety of styles to be found in the broadsheet newspapers published in the main European cities. Milan's *Corriere della Sera* (Figure 134) with its unvarying nine-column format and sparse use of pictures is typical of the more serious city-based Italian papers, strong on comment as well as news, and with excellent world news coverage, as is Rome's *La Stampa* (Figure 135). Headlines tend to line at the top with stories running across the page in a series of legs resting on similar horizontal segments lining below. Type is a mixture of serif and sans serif, often condensed and invariably in lowercase, with multidecking column and variation provided by light and bold headline versions. Pictures tend to be mainly small and used for relevance to the text more than for design purposes.

La Gazzetta del Mezzogiorno, of Bari (Figure 136), is typical of the middle market, both visually and in readership, using a version of Century bold and italic, all lowercase for headlines, but up to 84 point. Here a more creative use of type, pictures and typographical devices results in pages that allow a good read to the stories while still providing bold focal points to move the eye about.

In the more austere class of product, even by the withdrawn standards of *Le Monde*, comes Axel Springer's broadsheet *Die Welt* (Figure 137), published in Bonn which, on page one, has only recently begun to use halftone pictures regularly for its celebrated

LE CONCOURS DU FIGARO

L'INVITATION AU VOYAGE

QUESTIONS 32-33
(en dernière page du journal)

LE FIGARO

premier quotidien national français

Israël : poussée de la droite

Enfants : les dangers de la maison

Football : Platini au secours de l'équipe de France

Société générale : vers un nouvel affrontement

OFFREZ UN VISON

MAURICE KOTLER

Séisme : la prévision progresse

Super Portfolio

Figure 138 *Le Figaro*

Der Fall Rupert Lay: Ein Streit um Dogmen, Macht und ketzerische Gedanken / Seite 3

DIE WELT

UNABHÄNGIGE TAGESZEITUNG FÜR DEUTSCHLAND

Die schönsten Ziele im WELT-Report Reise

Frauenquoten verstoßen gegen EU-Recht

WELT NACHRICHTEN

Schäuble wirft Rot-Grün „Fortschritts-Skepsis" vor

Viele Verletzte bei neuem Anschlag in Pariser Métro

Scharpings und andere Schultern

Chopard

Figure 137 *Die Welt*

De Telegraaf

Burgemeester Groningen: Agent moet de straat op!

POLITIE PRAAT TE VEEL

Vader Stef: „Mijn dochter gaat opnieuw door een hel"

Claim van 12.5 miljoen in zaak au-pair Corina Peeze

Middagje winkelen in Wenen met prinses Margriet

Musical vertelt Surinaamse geschiedenis

Legendarische MG is terug

Kluivert reed in auto die niet was verzekerd

Toren van Pisa voorlopig gered

Zegelactie gaat concurrentie aan met Air Miles

Jan Lammers terug in de Formule 1

ONTHULLEND

Dollar krijgt zware dreun

Olvarit bevatte 3 jaar lang schoonmaakmiddel

De Telegraaf 12x gratis

Figure 139 *De Telegraaf*

αλήθεια

16 ΣΕΛΙΔΕΣ

ΣΥΖΗΤΗΣΗ ΕΦ' ΟΛΗΣ ΤΗΣ ΥΛΗΣ

Παπούλιας με πολιτική ηγεσία και κυβέρνηση

Ο Παπανδρέου αντιμετωπίζει τελικά πρόβλημα ήπατος

Παπούλιας: Χρήσιμη και ενδιαφέρουσα η συζήτηση με Κληρίδη

Με πολυκατοικίες στην καρδιά της Λευκωσίας θα «λύσουν» το θέμα των κυβερνητικών κτιρίων

Οι κατηγορίες εναντίον του κ. Παπούλια

Το Εφετείο αποφασίζει σήμερα για τη σύνθεση του δικαστηρίου στην υπόθεση του τριπλού φόνου

Στα χέρια του Βενιαμίν το θέμα της απελευθέρωσης του Κοϊρδού

Καταδικασμένες εποχές νοσταλγεί ο Μιχάλης Δούντας

Όχι να βρυκολακιάσει το παρελθόν

Figure 140 *Alethia (Truth)*

Пролетарии всех стран, соединяйтесь!

Коммунистическая партия Советского Союза

ПРАВДА

Газета основана 5 мая 1912 года В. И. ЛЕНИНЫМ

Орган Центрального Комитета КПСС

№ 241 (25593) • Воскресенье, 28 августа 1988 года • Цена 4 коп.

За рубежом

Успехи нефтяников

ВОСКРЕСНЫЙ ТЕЛЕТАЙП

Сегодня—День шахтера

Там, на шахте угольной

Одним абзацем

ПОДРОБНОСТИ

Бойцам и детям

КОЛОНКА СОБКОРА

Закон что дышло?

Выполнять женевские договоренности

Преследуют борцов за мир

Для морских разведчиков

24 часа планеты

Великобритания

Афганистан

Сингапур

УРОЖАЙ-88

А ЕСЛИ ПОСЧИТАТЬ...

ОСТРЫЙ СИГНАЛ

Опять без горючего

«ВАЛУНЫ» НА ДОРОГЕ

Figure 141 *Pravda*

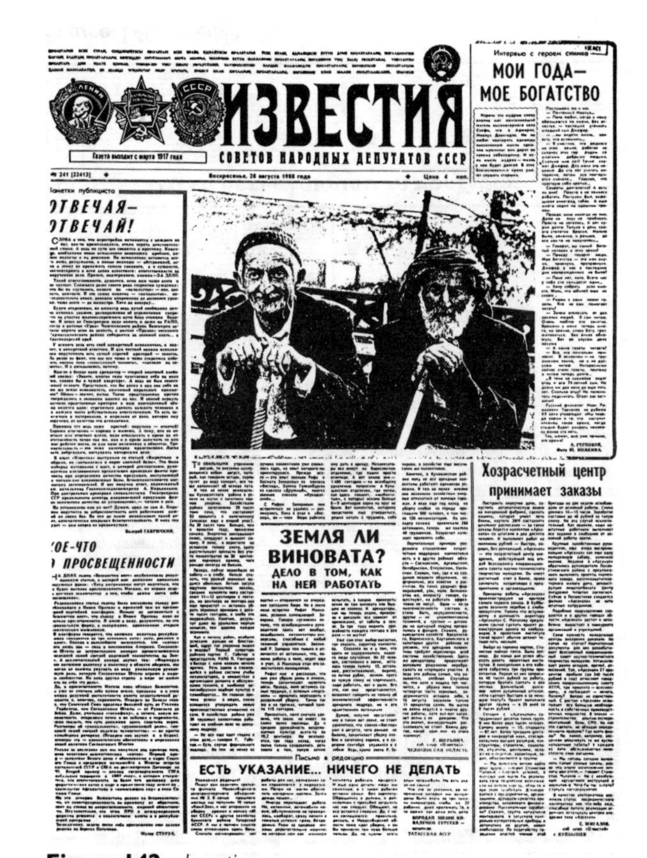

ИЗВЕСТИЯ

СОВЕТОВ НАРОДНЫХ ДЕПУТАТОВ СССР

Газета выходит с марта 1917 года

№ 241 (22413) • Воскресенье, 28 августа 1988 года • Цена 4 коп.

Интервью с героем снимка

МОИ ГОДА—МОЕ БОГАТСТВО

Заметки публициста

ОТВЕЧАЯ—ОТВЕЧАЙ!

...ОЕ-ЧТО О ПРОСВЕЩЕННОСТИ

ЗЕМЛЯ ЛИ ВИНОВАТА?

ДЕЛО В ТОМ, КАК НА НЕЙ РАБОТАТЬ

Хозрасчетный центр принимает заказы

Письмо в редакцию

ЕСТЬ УКАЗАНИЕ... НИЧЕГО НЕ ДЕЛАТЬ

Figure 142 *Izvestia*

world coverage. Meanwhile, *Le Figaro* (Figure 138), as national and conservative a paper as can be found in France, is daringly moving away from its stiff format of old, and adopting, though without their boldness, the shop window techniques of the poster broadsheets. All stories in our example, apart from the column one comment piece, are run at short length and then turned to inside pages, or are simply headlines with cross-references, thus enabling fifteen items to be aired on page one.

Another well-stocked page one relying on turns and cross-references is that of the Dutch daily, *De Telegraaf* (Figure 139), which manages to combine the elements of a mixed type design, with traditional masthead and splash position with a flavour of the poster broadsheets. The result is a page one which has great vigour above the fold, while degenerating into a hotch-potch below. The Greek version of this halfway house is the prestigious daily *Alethia* (Truth) which injects boldness and business into an old vessel with more dignity, marred only by the discovery and overuse of reverse video straplines (Figure 140).

An oddity in any review of Continental styles has to be the slim Russian newspapers. Our examples of *Pravda* (Figure 141) and *Izvestia* (Figure 142), show how daunting can be solid 6 point unleaded type in the Cyrillic reading text to editors who have to pack a lot of words into a few. In the face of this, the designers of both papers manage to inject some variety into the layouts by the creative use of rules and panels and the disposal of such white space as can be spared.

Both opt for a good sized picture on page one, though in *Pravda* the halftone would have looked better moved two columns to the right. *Pravda* is nevertheless the more readable. Its columnar white is effective and its two serif faces include what looks like a Russian version of Ultra Bodoni which, though an oddity on the front page, prints well both in large and small sizes. Also, its columns are mercifully free of the miniscule two-column setting on the two signed opinion pieces in *Izvestia*.

Appendix 1
Colour and the modern printing press

When to and when not to use colour depends on the type of newspaper and even on the part of a newspaper. While features offer more scope for the mood and atmosphere of a good colour picture and news seems to demand black and white, certain vivid news subjects can produce a shock effect in colour. A general point to note is that research has shown a colour picture is at its best when surrounded by black and white. There is general agreement that coloured does nothing to enhance a headline.

Colour can be effective in columnist or feature logos or motifs, especially those where the typography does not follow house style. Cartoons or sketches drawn by artists should nowadays be submitted with colour. Where the artist has not identified the area where colour should be applied the art editor will usually decide. A common usage is to apply colour in the areas to which mono tints would have been applied. Rarely should the main outlines of the drawing be in colour. Care must be taken in artwork with flesh tones since it is easy for such colour to give the impression of crude comic strips.

Using colour on body type is not to be recommended. It is difficult even with modern sophisticated methods to prevent the register from fractionally slipping on printing, thus giving an out-of-focus appearance to the type. To take colour; body type should be set at least above 10 point and be ideally in a sans serif face, that is, Metro or Helvetica. A point to remember is that newsprint is inferior to magazine paper and has an inescapable blotting factor which makes the fine work achieved on magazines impossible.

Spot colour for certain rules and borders can be effective in newspapers, if not used indiscriminately – and not on the news pages. It can also illuminate charts and graphics, as in the example *USA Today* in the colour plate section.

Where run-of-press colour can go in a newspaper depends, of course, not on the editor's whim but on the arrangements in the press room. *Colour imposition*, as it is called, is based on the concept that priority for colour will be given to the 'title press', that is the press that prints page one and the back page of a publication, together with the middle pages. What additional colour can be made available is a local decision since machine planning in the press room will have established where the colour units stand in the press line. Editorial choice is determined by this fact. Allowing for this there are more single page possibilities in a broadsheet paper than in a tabloid since the tabloid pages print in pairs (although new generation presses will print every page in colour if required – see below).

Pre-press

The computer has greatly reduced the time between page make-up (the old off-stone time) and the printing plate reaching the press. Under cut-and-paste it could be done in just under half an hour; with electronic pagination it is even quicker.

After being passed by the editor and lawyer or legal reader (usually reading in print-out) the page is sent electronically from the make-up screen to the pre-press department. Here, a machine called a raster image processor (RIP) turns it into a high resolution negative for feeding into the platemaker, or alternatively sends it in electronic form for platemaking at a satellite printing site.

Since with full electronic pagination there is no longer a composing room, paginating the edition, the pairing of pages in the case of tabloid production, and the sending of pages to the RIP is the responsibility of the editorial department. A production editor sees that the edition times are met and that the pages go in correct order to the plate-maker and thence to the presses.

With cut-and-paste production the composing room overseer remains responsible for the press times and the camera room where the completed pages are photographed and prepared as negatives for the plate.

New generation presses

Press development has kept pace with advances in electronic pagination and pre-press arrangements by providing good run-of-press colour and shortening even further the time between page make-up and press start-up.

The latest machines and their software and plate-making facilities do this by accepting pages direct from screen to plate, once the editorial and advertising elements have been assembled electronically, dispensing not only with camera room but also with RIP imaging. The laser engraving system takes down-the-line page data generated by journalists on screen using first generation dots that ensure the finest reproduction and make double scanning necessary.

Provincial newspaper companies centralizing their production on new sites and looking to recoup the costs are moving rapidly into this hi-tech area, installing state of the art presses to print not only their own newspapers and magazines but those of other companies on contract.

Typical is the development of Southern Newspapers' new headquarters at Redbridge on the outskirts of Southampton where advertisers are offered a full colour facility on every page of all the company's publications, including evening papers, with direct screen-to-plate production. A Cavalcade software system collates and stores condensed pages of thirty publications, oversees the input and retrieval of each item, and selects pages in correct plate-making order to match the command of the press controller.

The Southern Newspapers installation is based on the Man Roland Geoman press (Figure 143) which can deliver up to 70,000 copies an hour and has a variable web width to suit different publication sizes.

Even the maintenance is electronic. The German-made press is connected by modem to a plant at Milton Keynes which monitors the control system, and with the manufacturers in Augsburg, so that faults while running can be detected instantly on screen and be put right by operatives on site.

Southern Newspapers is setting up special training programmes to go with the new machines. 'It is difficult for people to harness technology without a basic understanding of its possibilities,' says Peter Henderson, the company's project manager. 'The right training is vital.'

From the editorial point of view the new press and plate-making developments mean more time in which to accommodate late news and pictures, hone page designs, and carry out easier and faster edition and slip changes.

Spot colour

The use of a particular colour – spot colour – has been possible for many years under the letterpress system through the provision of a colour duct on the presses, with its own ink reservoir, printing separately via a rubber cylinder. This enabled such basic things as the stop press (fudge) seal, the edition marker and – in more recent years – the title piece to be printed in a chosen colour, though not to a fine accuracy. Usually the colour was one base colour of cyan, magenta or yellow.

On some of the older high-speed rotaries this 'fudge box' colour could be used in tabloid newspapers only across a horizontal line at the top of the front and back spread. Two fudge boxes would run simultaneously on the press, one carrying the page one title piece, and the other the back page sports logo, often using red.

Although greater use of spot colour was possible by this method the heavy long print runs of national papers did not allow printing time, though some regional papers pioneered the greater use of letterpress spot colour. Advertisers, in particular, were keen to buy such limited colour as could be made available where newspaper companies were happy to offset the slower running of the presses against the extra income.

The switch to web-offset presses, with their easy use for accurate run-of-press colour, has revolutionized the provision of spot colour

Figure 143 Full colour facility on every page: the 70,000-copies-an-hour Man Roland Geoman press. It accepts direct page-to-plate editorial output

for editorial and advertising purposes and made possible a more ambitious use of a wider range of mixed colours in the same publication. The production by the Letraset organization of a British Standards Institution colour guide for the industry some years ago, called the *Pantone Guide*, has made things easier both for the ink manufacturers and art editors. The guide lays out the primary colours with a standard series of lighter and darker versions in easily comparable 15 mm squares of colour. Each one is given a Pantone number, and this can be used on a layout to indicate to the printer the exact colour required. The system is now universal, and is used to call up colour on a computer screen.

Each of the swatches in the *Pantone Guide* refers to solid flat colours. Further dilution, however, can be made. Instructions can be given to break down a chosen number by percentage in which the printer or litho house lets 'white' into a colour by screening it.

The technique is used effectively by the newspaper *USA Today*, the undisputed leader in this sort of colour (Plate 4(a)). Its four-colour run-of-press reproduction, good though it is, is outshone by the clever editorial and graphic application of a range of spot colour both in rules and panels, used to enhance its characteristic charts and graphics. A broadsheet paper, *USA Today* has evolved a type of 'tabloid' journalism in which colour is used instead of big headlines to increase readability and convey information. This quantum leap has been achieved by thinking colour where in the days of hot metal the designer achieved variation through the use of tints and 'shades' of black.

The freedom resulting from the easy resource to spot colour has proved invaluable in the enhancement of information graphics in many British newspapers and in the range of options available to advertisers to highlight parts of an advert.

Appendix 2
Some problems occurring in the designing of a newspaper

by Vic Giles

New and unusual tasks come to light for the designer engaged in launching or revamping a newspaper. Laying down a format for a new newspaper, as was the case of my work for the *Daily Star* in 1978 involves, for instance, not only the creation of a masthead and a type style to suit the size and market of a newspaper but to decide such things as the printed area of the pages, and with it the column width. Here, a false judgement can land a paper with an obstacle for the rest of its life.

In the case of the *Daily Star* the figures provided by the production director for the size of the new paper, that is, based on newsprint width – were 600 mm wide by 380 mm deep for a double-page spread. From this the column measures could be established together with the width of the column gutters and the gutter space of the fold.

To work this out I needed to establish the accuracy of paper tension through the presses, which were at this time rather old rotary presses. On older presses there will always be a tendency for the printed area to 'wander' on the paper surface, and therefore an awareness of the extent of the wander is necessary.

If the presses are already printing a newspaper, as was the case with the *Daily Star* presses, then the evidence of the eye should reveal the movement from page to page in the existing publication. Some new newspapers 'take fright' at this stage and create type areas at the outset that are wasteful of space and produce oversized margins. Looking at the *Daily Star*'s existing sister paper, the *Daily Express* it was obvious that the press executives had little confidence in their machines since the leading edges of the paper were excessively wide. The aim was to achieve 2 cm each side of the page. There were, in fact, variations of up to 3 and sometimes 4 mm divided between each outer edge.

My decision was to take the pages very close to the centre fold in order to achieve the maximum type area overall. The added advantage of the narrower fold gutter – a gain across the spread of 2 picas – was that when splitting a picture across the fold there was a visual improvement. With the metal chases in use at the time this meant that a narrower tabloid 'bar' had to be put into the chase to separate the pair of pages. More accurate and modern rotary presses such as those used for *The Times* allow a much greater saving on the leading edges, while the problem is eased even more with modern web-offset presses. The effect from the design point of view is that a well-filled page gives a more compact look to the design. It also puts expensive newsprint to better use.

The resulting setting for the *Daily Star* worked out at 8.3 picas to give a pica of gutter between the columns. This made a type page width of 64 picas (380 mm), and with a double-page spread taking 132 picas. To give a margin top and bottom that balanced with the leading edge I made the depth of the type area 81 picas with the folio space standing clear on top.

The change in the *Irish Press* from broadsheet to tabloid in 1988, which I was commissioned to undertake, was intended to broaden the readership of an old established paper by moving moderately downmarket. The journalists producing the paper found this something of a shock. They had become used to the grey columns of their paper and subediting attitudes were relaxed towards the wordy writing styles of reporters and feature writers. Pictures were not being used to advantage even on the huge broadsheet canvas and my first job in reforming the design was to talk to the staff and reveal, against examples of rival papers, how their product had become unwieldy to all but its most devoted readers. The breakthrough needed was to wrap up the facts in fewer words and it was going to take some practice if the wanted change in design format was to work.

The change would also have to be sold to readers unused to the tabloid format. There was a need to label pages with 'puff' strap lines to help the eye accept the contents and at the same time emphasize the confidence that can emanate from good tabloid style layout. I used the repeat title piece of the paper (as redesigned) whenever it was appropriate thus, in an almost subliminal way, involving the readers with the paper's name. The word PRESS of the title piece was used in Press Briefing on columns of news briefs, and appeared again in the sports logo on the back page . . . I placed the detailed weather information on page three where it could be suitably anchored on a page where spot colour was expected to become available for the maps and weather graphics.

When it came to the Gossip page, as in the style of all classic gossip columnists, I used the writer's name as the headline in a font different from those used throughout the paper. I put it in Antique

Olive capitals, very bold, with the *Irish Press* eagle motif, which had always been part of the paper's masthead, as an integral part of the graphics, thus giving a useful touch of continuity. At the time the name was Michael Sheridan and since the surname was longer it made a very good perch for the eagle to alight upon when the two words were set left. I allowed small stories to run above the byline logo to combat any visual interruption from the full page advertisement on the left. This also had the effect of pushing the main focal point down the page and thus got away from the news page style with the main headlines at the top.

A problem with the revamping of the Newcastle-based *Sunday Sun* (1989) was with the masthead, which suffered from show-through and from poor printing quality. The Pantone colour chosen for the title was an indifferent orange. Even a full cardinal red would have lost something to show-through due to the way the masthead was printing. The washed-out appearance was aggravated by the poor ink impression, which gave more ground for show-through from page two.

A cause was in the styling. The word 'Sunday' was slipped into the space left by an upper and lowercase 'Sun' and in both cases the characters were kerned with too much space. This visually reduced the impact of the words because the orange was not packed tightly enough across the area. The weak colour could have been improved by using a reversal (white on orange) in a squared-up shape. Even more apposite, instead of using the old method of the fudge box to print the masthead, the paper should have printed it using its run-of-press facility, which would have produced a denser result to combat show-through.

There was a problem with the adverts. The holiday feature were unconnected with the holiday adverts which should have been part of the attraction to both readers and advertisers. There seemed to have been no attempt to proportion the weight of advertising against editorial requirements on most pages. The use of page seven, a premier right-hand page, to a full page advert robbed the paper of valuable space for editorial projection and would not have been allowed by the paper's competitors, especially the national Sundays.

The use of reverse, or hanging, indent for the columns of small adverts on a seven-column page was wasteful. The area taken up by the indent could have provided an eighth column of adverts in the same type area, thus increasing the revenue while at the same time economizing on the space available.

We have said in this book that familiarity is an important factor in reader loyalty. Yet there are times in a newspaper's life when changes have to be made. A dramatic example occurred in 1953 when, as one of my early tasks, I was despatched by the Kemsley Organisation's Editorial Director, Dennis Hamilton, to oversee the removal of the Classified adverts that filled the front page of the *Western Mail* at Cardiff. The Kemsley directors in London, who had made the decision against the wishes of the morning paper's then editor, had instructed me to change the appearance of the *Western Mail* into one 'worthy of the second half of the twentieth century.'

My first move was to increase the size of the paper's Times Roman masthead by 50 per cent. I then had to plan a virgin broadsheet news page beneath it. I chose Century Bold, Italic and Roman at the unheard size of 72 points for the splash typography. The real break-through, however, was to up all the other headings in size; including those used on the inside pages and to give boldness to one page by using a big picture. In fact, apart from briefs with two lines of 18 point, there wasn't a news heading of less than 48 point.

The editor was horrified that I had acted against the traditions of this great and ancient Welsh national newspaper, and showed it. Angry memos were exchanged between Cardiff and London during the run-up to the re-launch, but Dennis Hamilton was insistent that the enterprise should go ahead.

So it did – with the result that on the morning of its debut in its new clothes the circulation of the *Western Mail* fell by more than 30,000 copies.

A man of few words, Dennis Hamilton's reply to Giles's timid telephone call on the dreadful day was, 'Stick with it man! The readers will come back.'

The opposite occurred. As day followed day, the readers, not recognizing the new version of their own newspaper, fell away by the thousand.

If Giles had lost his nerve, Brigadier Hamilton was not going to lose his. 'Stick with it, Giles, damn it!' he repeated.

Three weeks later, the circulation manager began to smile again, and the smile grew broader day by day as he was able to report the return of the faithful readers to the fold. Two months later the figures had recovered to their pre-launch level and had begun climbing above it. The battle was won!

Thirteen years later, in 1966, Sir William Haley, in his last year of editorship, followed suit and did the same for *The Times* – the last daily paper in Britain to lose its front page adverts and replace them with news.

Glossary of journalistic, computer and printing terms

AA Author's alterations. Used on reader's proof or output.

Aldus Pagemaker A trade name. One of the pioneer software programmes used in on-screen setting and page design.

Accents Stress and acute marks above appropriate letters, e.g. acute, breve, cedilla, circumflex, grave, macron, tilde, umlaut.

Absolute leading Line spacing of text in points.

Activate Computer term for making a typesetting or picture box live by clicking mouse arrow on them.

Additive primary colours Red, green and blue. These create all other colours when a computer monitor is being viewed and will reproduce on a print-out. The term additive is used because when each colour is superimposed on top of the other they will create white.

Adobe A software trade name (Adobe Systems Inc.). Many applications that represent tools for the printing industry, e.g. Adobe Typemaster, Adobe Photoshop, Adobe Postscript, Adobe Illustrator.

Alert A dialogue box warning of a computer problem, e.g. when you try to perform a command that cannot be undone.

Alphabet length The comparison of different font character lengths by line-ups of A to Z one under the other.

Alignment Under the style head, left, centred, right, justified, unjustified.

Alpha channel An 8-bit greyscale representation of an image. Often used for making masks to isolate a part of an image.

Ampersand A one character version (&) of the word and.

Anchor To paste a text or picture box inside the text so that it acts like a character and flows on with the text. On the style menu.

Anti-aliasing Rough edges smoothed, e.g. when using painting selection or type tool.

Application A software program performing a specified task, e.g. page composition, illustration and word processing.

Archive Off-line storage of date in the computer memory.

Artwork Original copy and/or pictures made ready for scanning.

ASCII Acronym showing software will import and save text in the ASCII format.

Ascender Part of the type character in lowercase that rises above the x-height, as in b, f, h, k, l and t.

Asymmetric setting Type line set with no attempt at pattern.

Asterisk Reference mark alongside a word that takes the reader's eye to the foot of the page for explanation.

Auto leading Line spacing value entered in the auto leading title in typographic preferences in the dialogue box.

Autologic Inc. Responsible for original digital construction systems in software form. They provide Autologic Image Setting machines with a range of font building programs.

Automatic hyphenation A feature dividing words at syllable junctures. Known as H&J (hyphen and justification) at the end of text lines.

Automatic imprint and page heading Placed on consecutive pages by a command before the design process begins.

Azerty European keyboard that incorporates accents as opposed to the English qwerty keyboard.

B-series paper size International ISO range of sizes falling between the A-sizes for use in posters, wallcharts and larger printed areas.

Background colour Background of a text or picture box. Also referred to as a box tint.

Background processing The computer system's ability to deal with lower priority tasks when you perform other work on screen.

Backnumbers Older copies of a publication still available.

Backspace key To back up over and erase a previously keystroked character. The delete key on a Macintosh keyboard.

Barcode Unique symbolized code in vertical bar format readable by computer and light scanner.

Baseline shift A command on the style menu that allows the raising or lowering of text in an anchored box.

Bastard setting Unstandardized setting.

Beard The area of a hot metal base supporting the character's descender.

Bell Gothic Typeface with an economic width used for Stock Market prices and racing cards developed by the Bell Telephone Corporation in the USA for telephone directories.

Biblical P A reversed in-line P (back to front) used as a focal point to emphasize a paragraph or intro. Also on a computer screen to indicate invisible instructions which can be revealed at the touch of a 'show invisibles' command.

Binary A computer programming system using two numbers opposed to the decimal which operates out of the base figure ten. The only binary figures used are 0 and 1. The Microsoft Windows system adopts a three figure format.

Bit Smallest unit of information a computer memory can hold.

Bit-blaster Laser printer producing highspeed 'outputs' (proofs) of computer generated material.

Bitmapped font Each character is represented as a dot pattern in contrast to a scaleable font where every character is mathematically described. When your computer is unable to locate a corresponding printer font it will create a bitmapped font.

Black letter The form of Gothic or Old English mediaeval letter type form based on script lettering from the monasteries.

Blanket Clamped around the offset litho cylinder, the rubber surfaced blanket transfers the printed image from printing plate to newsprint paper.

Blanket to blanket Blanketed cylinders act as opposing impression cylinders printing both sides of the web simultaneously

Bleach out Photographic effect by exposing a negative to a hard surface bromide paper. This will eliminate tone value giving a stark appearance. A computer command determining percentages of contrast will provide the same.

Bleed A page instruction that extends the print area out to the trimmed edge allowing pictures to run out of the page, usually by 3 mm.

Blend A Quark XPress instruction that will allow a colour or a black to commence at each side of the box and vignette together at the centre. This can also be achieved in reverse. Both colour and black are activated by using the colour palette under the menu View, then the Show Colours command.

Block Letterpress illustrative printing surface. The image etched after screening and mounting on a lead block to achieve type-high in order to print.

Blow up Enlargement of pictures, type or artwork.

Board Surfaces on which various types of illustrations are best produced. These are referred to as fashion, hard toothed and Bristol board.

Body copy Major part of the text in a publication.

Bold A thickened typographical version of a Roman typeface.

Borders Described in the Apple-Mac system as 'frame' – decorative lines of any weight surrounding set pieces of type, entire pages or pictures.

Brad *British Rate and Data*, an essential book listing all UK publications and their advertising specifications.

Broadsheet A large-size and more serious newspaper.

Bromide Light sensitive paper on which photographic images are chemically developed.

Burnt out Over-exposure of a picture due usually to a wrongly selected camera shutter speed.

Bus Information is transmitted electronically along a path within the computer mainframe. Buses connect computer devices, e.g. processors, input devices and RAM.

Button Illustration of a push-button like control appearing on screen in a dialogue box to be clicked on in order to perform, confirm or cancel an action.

Byline Authorship of a news story of feature displayed at the beginning of an article.

Byte Amount of computer memory space needed for storage. It represents 8 binary digits (bits).

C & LC Abbreviation for capital and lowercase letters.

C-series Range of sizes for envelopes that will fit the A-series International size paper.

Camera-ready art Elements of a finished page ready for reproduction.

Caps An abbreviation for capital letters (uppercase), e.g. ABC.

Caption Description with a picture.

Carry forward To continue the text forward into the next column.

Cartridge paper Drawing paper, sometimes used for high quality printing work. Has good weight and high opacity.

Cassette Light-proof container for unused and unexposed film, or in its larger form for laser output printers.

Casting off Assessment of copy length in terms of the amount of type taken to fill a given space.

Catchline Identifying heading set at the top of a story to be eliminated before it gets to publication.

CD-ROM Compact disk read-only memory, 120 mm in diameter capable of storing some 550 megabytes of information. This information is designated 'read-only memory' since a CD-ROM drive can access but not write information.

Centred type or centre alignment Multiple lines of headline of uneven length centralized upon each other. The Quark command is on the Style menu under the name 'Alignment'.

Centre spread The two facing pages in the middle of a newspaper or magazine.

Character A letter, number, punctuation mark or symbol.

Character count Number of letters and spaces in a line of type.

Chip The name for 'integrated circuit', an electronic circuit contained in a single piece of semiconducting silicon.

Choose A Quark XPress term for selecting a menu entry by clicking on to its menu title, holding down the mouse button and dragging the arrow pointer over the entry and then displaying the submenu.

Cicero European type measure that approximates a pica. It is equal to 4.55 mm.

Claris A trade name. Claris Corporation covers MacDraw, MacPaint, MacProject and MacWrite. All useful software tools for the printing industry.

Classified Small personalized advert, usually of column width, applying to jobs, births, deaths and selling.

Clean proof or output The finalized proof of copy requiring no further correction.

Click Pressing and then releasing the mouse button.

Clipboard The last item cut or copied is stored in the computer's memory. Information can be pasted from clipboard to active document. Display clipboard by choosing 'Show Clipboard' on the Edit menu.

Close-up Correction mark indicating a reduction in space between characters.

CMYK Represents the four process colours – cyan, magenta, yellow and black.

CMYK image A four-channel, computer command used to print colour separation.

Cold colour The blue-coloured tone values of the spectrum.

Collate Sections of books and magazines gathered together in correct order. A software program called Cavalcade can control large numbers of associated tasks.

Colour separation Used in the colour printing process to create four separate colour plates that will unify in the finished product.

Colour swatch Specified colours in a sample book.

Colour transparencies Can be abbreviated to 'tranny'. A photo-positive film in full colour.

Colour wheel An Apple-Mac dialogue box appears when the change control button in the control panel is activated. The wheel allows you to adjust hue, saturation and brightness.

Command A computer menu word describing an action for it to perform.

Command key A key which when held down while another key is pressed, causes a command to take effect.

Condensed type A description of established compressed fonts. A computer can create condensation at the requirement of the operator.

Contact print An entire negative film is exposed to a sheet of light-sensitive bromide paper in order to choose the appropriate pictures for blowing up.

Contrast The relationship between highlights, middle tones and shadows of a picture.

Control panel Computer term describing a desk accessory used to change operating commands.

Copy A manuscript, photograph or artwork considered for publication.

Copyfitting Establishing how much copy will fit in a given space using a particular font. A computer can artificially condense or kern in order to achieve this end.

Copy protect Software publishers protect their disks to prevent them from being illegally duplicated by pirates. They make them uncopyable.

Copyright The proprietorial rights in a creative work. A symbol of a circle containing a letter C will show in the finished publication that its contents are protected under the law.

Core memory A computer's main storage capacity.

Crop To select part of a picture or artwork and discard the unselected area. An editorial decision achievable on screen.

Crosshead The smallest of display headlines, used to interrupt the flow of body type. Set left or centred on a column measure, its value is to rest the reader's eye.

Cross reference An indication that a story continues on another page.

Current start-up disk This contains the system files the computer is currently using.

Cursor A flashing indicator on a visual display screen informing the operator of the position of the next input or correction.

Cut-off The length of a printed sheet as determined by the preset press, and the circumference of the printing cylinder. The term is also used to describe an item cut-off from the rest of the text by print rules.

Cut-off rule A fine black line used in newspapers to indicate the end of one story and the beginning of the next. Applied both horizontally and vertically.

Cut-out Illustration with the background masked out to leave the image against the page background. Software such as Adobe Photoshop have a command that will drop out the background at a keystroke.

Cyan The colour blue used in four-colour printing.

Dagger Second symbol after an asterisk to indicate a cross reference or footnote.

Daisy wheel Flat disk with letters attached to its stalks used in typewriters to change the font.

Deactivate A computer term for clicking outside the active item. This use of the mouse will deactivate the item.

Deep etch A hot metal/letterpress term which explains the process of etching away the non-printing areas of a litho or zinc plate, either with acid or by mechanical means.

Default The settings and values used by software programs to perform its functions. These can be changed when the revised settings and values become the new defaults. The identity of defaults will be revealed during the computer's start-up process.

Definition The detail and sharpness of reproduction.

Delete Proofreader's mark on a proof meaning to remove. A computer keyboard contains a delete key.

Descender Portion of lowercase letters falling below the baseline. The letters g, j, p, q and y have descenders.

Developer Chemical used to remove unexposed area of a litho-plate, also to develop a negative and positive photograph.

Dialogue box Screen displayed box in response to a command involving additional specifications.

Diazo Photographic chemical coating used in plate-making. A copying process that uses light sensitive compound.

Didot Typographical measure used throughout Europe and named after Francois Didot. The measure is 0.376 mm.

Direct input Keyboard stroking text into a computer by an originator of copy (stories) to be retrieved later for editing.

Disk Internal hard disk is the permanent occupier in the mainframe of a computer. Floppy disks can be slotted into the machine for independent data information purposes or for downloading material from the hard disk.

Dither Computer term for the creation of an additional colour or shade of grey on screen by altering the values of adjacent pixels. It can also create an appearance of grey shades on a mono monitor and the appearance of additional colours on a colour monitor.

Document window On-screen window displaying an open document. These windows include the identity of the document, its title bar, zoom and close boxes, scroll bars, etc.

DOS Disk operating system, a computer term for machines using hard and soft disks.

Dot leaders Three dots that indicate a suggested word or pause in the flow of a sentence. Also ellipses.

Dots per inch A method of describing the resolution of a picture's screen. It also indicates the ability of a printer or monitor to produce different depths of screen (dpi).

Double-click The button on a computer mouse pressed twice in rapid succession without moving the mouse will generally open a document or dialogue box.

Download A command to transfer data from screen to printer.

Downloadable font A font not resident in the printer's memory and therefore one that has to be sent to the printer from the mainframe of the computer.

Drag Computer term to hold down the mouse button and shift the mouse will activate and reposition an item on the screen.

Drop cap A capital letter larger than the surrounding text at the beginning of a paragraph.

Drop-out halftone Highlight area of a picture with no screen dots which emphasizes added brightness. For instance the whites of eyes.

Dropped out type A computer term for reversed type, i.e. white on black.

Dummy Mock-up of a projected publication or page-by-page record of a publication being compiled.

Duotone Two negatives with different screen angles used to produce a two-colour reproduction from a single colour original.

Edit To control the contents of a publication in the position of editor. A computer control under the Edit menu. To check for spelling, grammar and length.

Edition One printed version of a publication.

Egyptian A family of slab-seriffed typefaces.

Electrostatic printing Process of copying or printing where an electrically charged drum receives the reflected light bounced off the original copy. The charge is lost on the areas affected by the light but the powdered toner retained by the charged areas is fused to the paper thus creating the image.

Ellipses Three points (. . .) used in place of omitted words. See Dot leaders.

E-mail Electronic mail. A computer method of communicating with other users worldwide.

Em Another word for a pica (12 points), six to an inch. One em or pica represents 0.166044 inches.

Em-dash A dash that is the width of two zeros or a quad twice the width of an en-dash.

Em-quad The square of any font size. Traditionally referred to as a mutton.

Em-space In hot metal terms, it is an em square in any size of type. A 12 point em is a square measuring 12 x 12 points.

Emulsion Photosensitive coating adhering to the surface of film or paper.

Engraving Sensitized copper, zinc, steel or plastic surface etched by acid or worked with engravers' tool.

Enter key A computer key that confirms entry or command.

EPS Encapsulated postscript – a graphic format for the storage of graphics, bit-mapping, grey-scale, particular to the Mac system.

Erase disk To remove information from a floppy disk prior to its reformatting.

Ergonomics The designing of working environments.

Exposure The time taken for the light reflected from the photographed subject to impinge on the camera film. See F number.

Extended A term applied to a typeface that is designed to be much wider than its regular face with a computer keyboard.

Face The design style of a given typeface.

Facsimile (fax) The precise electronic transmission of a document, usually over the telephone.

Family A series of typefaces which are related, e.g. Egyptian, Antique, Venetian and Old Faces.

File Collection of information stored on a computer disk or document.

File server A mass storage device that allows computers to use a shared common file and other applications throughout a network.

Filler Small stories on newspaper pages to fill areas of the page where the more important stories have fallen short. They can add to the readability and design of the page.

Film processor A machine for processing exposed film, producing negative and positive results. It develops, fixes, washes and dries at high speed.

Filter Coloured glass, gelatin or plastic placed over the lens to prevent certain colours reaching the film.

Financial setting A description of the most economic way to set stock exchange prices and race cards, etc. The most favoured face for this is Bell Gothic.

Flash A split second illumination used in combination with a camera. It allows more reflected light into the camera lens.

Flatbed Printing press that prints from a flat surface rather than from a cylinder.

Flat plan The projected plan of a newspaper or magazine drawn up in rectangles.

Floppy disk A flexible plastic computer disk, usually three and a half inches in diameter. It stores information magnetically and allows the operator to copy from the hard disk.

Flyer Launching of an editorial idea or an advertising hand bill.

Focal length Distance inside a camera between the film and lens.

Focal plane Where a sharp image is formed by light passing through the lens on to film, the area is termed the focal plane.

Fog Unanticipated light impinging on the film surface causing over-exposure.

Folder A rectangular computer icon indicating its ability to hold documents and applications.

Folio The area at the top or bottom of a page reserved for the page number, date and name of publication.

Font A complete set of characters of the same typeface, together with all its related furniture.

Footer The bottom margin of each page or document which can include a page number or chapter title, etc.

Format The framework a computer understands. A blank disk needs to be initialized to establish how information is to be stored.

Four-colour process (full colour) Printing in colour with the three subtractive primary colours yellow, magenta and cyan plus black.

Free sheet Slang term for publications given away free.

Front end Another term for direct input of text into a computer.

Full out Setting text square with no indentations of any kind. Justified is the computer term.

Full point A full stop.

Furniture Hot-metal term describing the pieces of metal used for locking together slugs of type for proofing. Also used to describe the extras in a type font such as symbols, arrows, hands and stars.

Galley A three-sided tray used in hot-metal letterpress to carry columns of metal type.

Gate fold A magazine page that folds out of the body of the magazine to double its size.

Generation Production of typesetting from computer.

Gigabyte Unit of measurement equal to 1024 megabytes. Compare byte, kilobyte and megabyte.

Gigo Slang term for bad results from a computer caused by faulty input. Garbage in, garbage out!

Gloss print Finished photograph that has been glazed by heating its surface.

Grainy Film speed that creates an artistic grain on the finished photograph.

Gravure Abbreviation of the term roto-gravure. A high-speed press producing sixteen pages across each cylinder. These are engraved (etched) into the copper surface and finally chromed to give the surface a longer running capability.

Grey scale Computer command producing ranges of density using the intensity of the screen's pixels. It constitutes a single-channel image consisting of up to 256 levels of grey and in colour format 8 bits of colour information per pixel.

Gutter Vertical space between adjacent columns or the area between facing pages.

Hairline The finest thickness of column rule (0.5 point). Can be commanded with the Apple-Mac through the Style menu.

Halftone Continuous tone image reproduction created by using a screen that breaks the image into various sized dots, constituting highlight and shadow.

H&J Hyphenation and justification is a command under the Edit menu creating specifications used to control both the breaking of words at the edge of text in a predefined pattern and the square justification of the setting.

Hanging indent The first line of a paragraph left hanging over an indent from the rest of the paragraph. Also 0 and 1.

Hanging quote For display reasons a quote or quote marks of a larger size than the body type is allowed to hang at the front and back of the paragraph.

Hard disk Permanent metal computer disk magnetically sensitized and sealed into a drive or cartridge. It can store huge amounts of information compared to a floppy disk.

Hardware The generic name for typesetting and word processing equipment.

Header See Folio.

Headline or heading Display lines of type summarize a story on the page.

Highlight Piece of screen text which requires additional commands is highlighted in Quark XPress by using the command key and letter A on the keyboard. It produces a green screen over the appropriate area enabling the instruction to be applied.

Hot metal Expression referring to the melting and recasting into typographical shapes of the lead or metal (lead, antimony and tin) for the letterpress process.

Horizontal/vertical scale A command in the Quark XPress style menu that will condense and expand the width and/or height of type characters.

Hypercard An Apple Computer Inc. trademark name for its sophisticated software covering everything from screens to illustration . . . setting eccentricities to chart composition.

Icon A piece of graphics on a computer menu to represent an object, concept or a message.

Imagesetter Device used to output computer-generated pages or images at high resolution on to film or paper.

Imposition The arrangement of pages for the presses that will ensure them appearing in the correct order.

Imprint The name, address and details of the printer, designer and perhaps proprietor of the publication. Usually placed in a discreet position in the newspaper or magazine in small body type.

Indent Setting a paragraph, on the first line, at less than the full column measure.

Initial or illumination The first letter of a printed piece decorated by a large capital letter or one reversed white on black or in colour. The term derives from the ancient monastic art of colour-illuminated letters.

Initialize Dividing the disk surface into tracks and sectors which the system uses to memorize the disks' contents.

Input Copy to be processed by the computer typesetter.

Insert Term used for the addition of a piece of written copy inside the existing flow of text.

Italic All fonts of type which slope to the right following the original skills of scriptwriting.

Jobbing The work of a commercial printer running a jobbing shop. One that does not publish.

Justify To set lines of body type correctly spaced by the computer to create a straight edge on both sides of the column.

Kern Deleting or inserting units of space between letters and words in order to achieve visual balance. Letters that will always require this balance are the shaped ones, i.e. A, W, V and Y, plus all the rounded letters. Computers allow a house kerning style to be achieved with a once-only command inserted into the memory. Entire alphabets of every font in the house can be kerned to the taste of the editor or designer.

Key In-house copy codes distinguishing typefaces, size and column width at the top of the copy set by computer. It will appear at the top of the screen before the copy begins to run.

Keyline Outline laid down by the artist or designer to contain an area of tint, colour, photograph or type, usually as a guide and not to be reproduced.

Kicker Scene-setting intro to a story with independent display or a specially displayed story on a page. In the USA, a strap line, a small display line leading to the main heading.

Kill To delete a story or illustration. To erase type already set in the computer.

Kilobyte (K) Equal to 1024 bytes. Compare byte, gigabyte and megabyte.

Knockout Area of background or a particular spot of colour that is not to be printed from the original picture.

KPH An acronym for keystrokes per hour achieved by the computer operative.

Laser printer Containing a memory, the machine will communicate with Apple-Mac equipment via an internal programme called AppleTalk. All instructions of type, pictures and design will download to the printer and reproduce on paper or film at whatever size the printer is constructed to, i.e. A2, A3 or A4. Toner (powder) is the ink medium inserted in cassette form.

Layout A hand drawn page. It has the format of the page already printed on it in a light colour that does not interfere with the design, indicating gutters and white space. The designer draws the page, indicating type and written instructions.

/LC An instruction for lowercase setting only as opposed to C/LC which is caps and lowercase.

Leader (ellipses) A row of three dots interrupting the text and indicating a pause.

Lead-in The first few words of a story that achieve emphasis by being set in capitals or bold.

Leading (pronounced ledding) Traditional typesetting term referring to the space between lines of type. Originally thin strips of lead were inserted between lines of metal type to increase the space.

Left-aligned Set left in old terminology. The text has a straight left edge and a ragged right edge.

Left indent The distance between the left edge of a text box or column and the starting point of the text.

Letraset Corporate name for a waxed-back rub-down transfer typeface. Also the company produces computer software.

Ligature Font cases in the hot metal system contained particular letters that were joined together for the sake of traditional English grammar, as in the two lower case ffs. Designers have extended the use of ligaturing in the case, for example, of mastheads and company logos. The technique has become simpler with computer setting as the removal of spacing units can be made with a minus-kerning command.

Lightface A thinner version of any regular typeface.

Light pen Light sensitive stylus used to edit on a VDU screen, or to read bar codes.

Linecaster Any hot metal line caster setting machine that will set in lines such as the Linotype and Intertype. Also the single letter casting machines such as Monotype and Ludlow.

Line drawing Any piece of artwork made up of lines and solid masses with no tone area.

Line negative A high-contrast negative giving open spaces for the reproduction of black, created by allowing light to pass through on to sensitized paper.

Link Quark XPress command that links boxes or adjoining pages thus allowing text to flow from one to the other.

Linotronic The range of computer setting systems using the laser technique.

Linotype Mergenthaler's trade name for his line casting machine. A typesetting machine that sets slugs of type in hot metal.

Lithography A planographic printing process. The image areas are separated from the non-image parts by the repulsion of oil and water, oil being the etch.

Live matter Type already set, original copy to be used in the publication as opposed to dead matter.

Lock Computer command that will fix an item to a page so that is cannot be moved or resized with the item tool.

Logo Trade mark or masthead.

Lowercase See /LC.

Ludlow caster Trade name for a hot-metal casting machine where brass matrices are assembled by hand in a metal 'stick' one letter at a time and then cast as one type line. Used for display headings of above 18 point.

Macintosh Apple-Mac is the name of the most popular publishing computer hardware.

MacDraw, MacPaint, MacWrite Software applications produced by the Apple-Mac organization.

Make-up The designing for a printed page with elements gathered together to make an attractive and readable product. The process of making up a page on paste-up or screen. Sometimes used as a name for the layout.

Manuscript MS for short. Usually handwritten copy ready for setting.

Margin Space bordering the written or printed area on a page. It is defined by the fields in the margin guides area of the dialogue box file.

Mark-up Setting instructions at the side of a drawn layout.

Mask Description of any material used to block off portions of the printed page.

Masthead Design or logo for the title of a newspaper.

Matrix or matrice Mould from which type is cast in metal.

Matt finish Non-glossy surface to a photograph.

Mean line More commonly called the x-line or x-height. An imaginary line that runs along the top of lowercase type characters.

Measure The measured and set width of a column of text.

Mechanical Also referred to as a paste-up. The mechanical is the master document from which a printing plate is made. It includes all design elements – text, pictures, lines, etc. – in position and ready to be photographed to make negative.

Megabyte Equal to 1024 kilobytes.

Memory Data storage area of a computer. See RAM and ROM.

Menu Commands displayed when the mouse button is clicked and held down on a menu title in the menu bar.

Menu bar Horizontal strip displayed at the top of the screen containing menu titles.

Metric system Decimal measurements now applied to the printing industry. Metre: 39.37 inches; centimetre: 00.39.37 inches; millimetre: 00.0394 inches.

Microsoft Microsoft Corporation. A versatile software product designed primarily for PCs. In 1995 its president Bill Gates introduced Windows 95, an update which includes in the package an entry into Microsoft's network in competition with Internet and the World Wide Web.

Misprint Typographical error, also a literal.

Misregister One or more of the four colours printing out of alignment, resulting in distortion.

Modem A device that translates computer data into digital form for transmitting to screens by telephone.

Modern Description of typefaces developed in the late eighteenth century.

Modern figures Numerals of the same size type as the capital letters on the same line, as opposed to old style figures which dropped below the baseline.

Moiré Wavy screen pattern which appears both on colour and mono reproductions when the screens are wrongly angled.

Mono (monochrome) A reproduction using only the colour black. Also an abbreviation for Monotype Corporation's printing and computer equipment.

Montage Several pieces of artwork and pictures assembled into one artistic format.

Monitor Another name for a visual display unit (a computer screen keyboard).

Mouse Small mechanical device which when moved on a deskpad will correspond to a pointer moving around the screen.

Multi-finder Multi-tasking Macintosh operating system making it possible to have more than one application open at the same time. It will also allow the operator to perform one task while the computer performs another.

Mutton Ancient name for an em measure. See Em-quad.

Network Collection of interconnected individually controlled computers which allows operators to share data and devices such as printers, storage and exchange E-mail.

Newsprint Paper produced specifically for newspaper production, usually weighing between 45 and 58 grams but can be heavier.

Notepad Desk accessory that allows entry and editing of small amounts of text while another document is being compiled.

Numeric keypad Placed on the right of the keyboard, these keys allow you to enter numbers and perform calculations quickly.

Nut Antique name for an en-quad.

Oblique (italic) A character sloping to the right.

OCR (optical character recognition) Interpretation of typewritten material by a scanner which will then store text for subsequent typesetting.

Offset Planographic litho printing method. Image and non-image areas on the same plane are separated by the principle of repulsion of oil to water. The oil bound ink is transferred from the printing plate of a rubber blanket and from there to the paper. Thus the word 'offset'.

Old style Type character designs developed in the seventeenth century.

On-line Connected to a centralized computer mainframe and already communicating with it. An assurance from a laser printer that it is on-line and awaiting downloading instructions.

Opaque Painted out areas on a film not required in the reproduction. The chemical used is referred to as 'a tin of opaque', it is impervious to light.

Open matter Well-spaced lines of text type producing large white areas.

Optical centre The visual vertical middle of the page considered to be 10 per cent above the centre line. It is construed as the main focal point.

Option key AppleMac keyboards allow for a combination of keys such as Option alone or option/shift, where when used simultaneously produces optional version of characters or shortcuts in command.

Ornamented Embellished type characters based on monastic illuminated capital letters. Used decoratively in modern design for dropped or stand-up caps preceding body type.

Orphan First line of a paragraph falling at the bottom of a column.

Overmatter (overflow) Copy too long for the space for which it is designed. A computer will produce an overflow symbol when this occurs and the overflow will either reproduce on the computer clipboard or on an instantly generated document frame below the one being worked upon.

Overrun Copies of a publication printed in excess of requirements.

Ozalid Method of copying used for proofing film. Blueprint used mostly in architectural studios.

Page proofs A hot-metal page or galley was 'pulled'. Ancient proof pulling presses had a static roller that pressed down on the metal or wooden type and the paper covered type was pulled manually beneath the roller by hand.

Pagination Consecutive planning and numbering of pages. Software such as Cavalcade will perform this entire operation across many publications simultaneously and display the results on thumbnail screen reproduction.

Palette Software terminology indicating tool measurements, document layout, style sheet, colour, trap information and library, all activated by mouse and arrow.

Pantone colour A universal reference for primary colours and all the variations emanating from them. Every one has a number matching the printing machine's reference. First developed by Letraset in Britain, it is now accepted across the industry.

Paragraph mark A symbol used to draw attention to the beginning of a paragraph, i.e. a solid square, a blob or a biblical P. In each case it would be indented or set hanging as desired. Paragraph marker in written copy indicates that a par starts here and is a three-sided bracket drawn around the first letter of the word.

Paste-up Output bromides of set type and headlines waxed on a stout card printed with a non-photographic grid of the format of the publication. The process of making up a page by the paste-up method from a page design or layout.

Parenthesis Rounded brackets at each end of a grammatical aside.

Pasted dummy Now sidelined by computer technology, it was a representation of the publication filled up with pasted proofs throughout the compiling of an issue.

Perfect binding Magazine terminology for a squared spine to the publication rather than staple-stitched.

Pica A basic unit of typographic measurement. There are six picas to an inch, twelve points to a pica (see also Point).

Picture box Pictures are imported into an active picture box by choosing 'get picture' (on the Edit menu). Pictures can be imported in any of the following formats: PAINT, PICT, PICT2, EPS, TIFF, RIFF and Scitex CT.

Pi font A stored computer font containing characters that have no place among alphabets of type. Rules, corners, ligatured letters, arrows, logos, trade marks, column headings, etc.

Point The smallest measure in printing. There are 72 of them to the inch and each one measures 0.013837 of an inch or 0.351mm.

Polygon (picture box) An eccentric shape for a picture with more than three sides can be achieved with this software command.

Positive Opposed to negative. The finished photographic image, reproducible as such. The stage preceding is the negative where the whites are black and blacks are white.

Postscript Adobe Systems Inc., developed the postscript system as a page description language used to describe fonts, graphics, and the layout of pages.

Primary letters Lowercase letters with no descenders.

Printer driver File used to exchange information between computer and laser printer.

Proof A rough reproduction to test for corrections. See Page proofs.

Proofreading Checking proofs for errors and literals.

Quad left, right or centre The headline style. The squaring of a character of type.

Quark XPress A comprehensive software programme favoured wherever the computer is used in the printing industry.

Quire One-twentieth of a ream.

Quoins Expandable metal wedges for holding type in place on a metal galley (tray).

Quotes Double or single commas set around something said by somebody.

Qwerty Name for the standard typewriter keyboard. The title comes from the first six characters of the keyboard. Computer setting uses the qwerty method.

Ragged right (unjustified) Text type set as poetry with lines of varying lengths.

Raised cap Initial letter of the introductory paragraph of a typeset story. It should line up with the bottom of the first line of body type in order to stand above the setting in its own white space.

RAM A computer's random access memory.

Range To range a headline right or left is to square it with the right or left of the column setting.

Rate card The price stated on a card by the advertising department for the various advertisement positions available in a publication.

Recto Page on the right side of a publication spread carrying an odd number. The even side is the verso. The back and front of a two-sided printed sheet.

Reel Roll of newsprint.

Reference mark Any symbol that will direct the reader's eye to information at the bottom of the page, e.g. asterisks, daggers, etc.

Register marks Crossed hairline motif in the margin of a page that allows for automatic colour registration.

Resolution The degree of clarity of a monitor or printer. It is usually measured in dots per inch (dpi). The higher the resolution, the finer the detail of the screen or printed page.

Retouching Traditionally photographs and transparencies were corrected or changed by hand with airbrush and paint. Adobe Photoshop and others have produced systems which perform the operation on screen.

Reversed type (white on black) Sometimes referred to as dropped-out type, it is white type on a black background usually achieved by two commands on the computer keyboard.

RGB (red, green, blue) Colour model based on the additive colour theory. It is used for video output systems and computer monitors.

Right-aligned Computer terminology for set right or squared on the right, ragged on the left.

River An unsightly river of white space running downwards through a column of type caused by the over-even construction of the words thus distracting the reader's eye.

ROM (read-only memory) Chips containing information the computer uses throughout the system, even that which gets itself started. Information in ROM is permanent.

Roman type A regular typeface coming in density halfway between bold and light.

ROP Acronym for run of paper or run of press. How the press is dressed for printing.

Rules Printed lines measured in points of thickness and used to divide stories and add display to the page. Panels of varying thickness of rule or decoration will provide variety.

Runaround A command on the Item pop-up menu that will provide stories and add display to the page.

Run-on To eliminate a paragraph by allowing the sentence to continue without the pause.

Running headline A line which repeats itself as an object of continuity when the story turns to another page.

Save The application of the command key coupled with the 'A' key on a keyboard will save every item on the current document.

Scaling Technology now allows horizontal and vertical scaling bars to be applied to the picture on screen defining its size before it is saved for the computer. Tracing paper and pencil were the traditional tools for this operation.

Scanner A flatbed device for inputting a picture or transparency digitally with all its tonal values intact. Illuminated internally, the subject can then be scaled and stored in the computer memory.

Screen See Halftone.

Screen angles The correct positioning of the four screens on a colour picture avoids the moiré effect which gives the impression of a pronounced watermark on the finished reproduction.

Screen font Bitmapped representation of a font used to display the on-screen character.

Scroll bars On the right and bottom of the screen are shaded bars containing small boxes. With the mouse arrow on these boxes the document can be moved vertically to display more of the left and right of the page or horizontally to do the same or move on to the page beneath.

SCSI Acronym for small computer system interface. Industrial standard interface providing speedy access to ancillary devices.

SCSI port Plug-in panel at the back of the computer to which SCSI devices are connected.

Sequential access The process of scanning colour originals to achieve separately the four subtractive primary colours of cyan, magenta, yellow and black (CMYK – the blacking being known in acronym as K). The colours are scanned individually and registered for final etching on the colour printing plates.

Serial interface From which information is transmitted one bit at a time.

Series The complete range of sizes from the same typeface.

Serif Typeface designed with horizontal cross strokes that create 'feet' at the bottom of each square letter and 'shoulders' at their tops, together with curlicues and sometimes blobs at the extremities of the capital letters.

Set solid Lines of text are described as set solid when the size equals the depth of the leading, e.g. 10 point type with 10 point leading.

Sheet fed Printing on separate sheets as opposed to a roll.

Shift key When pressed it causes subsequent letters to appear in caps.

Shoulder When widely set body type runs into a single column and then turns into a second column under the original wide setting, a shoulder is created in the second leg. To prevent the reader's eye from being diverted into the wrong area, bylines or pieces of half rule can be used to create the division required.

Show through When the printed word and picture shows through on to the next page and distorts or interferes with the image on the backing page.

Side bar See Filler.

Sidehead A small crossheading to relieve the greyness of the body type. Either set flush left or right of the column.

Sign-off The writers' names at the end of a story.

Small caps Caps that will line up top and bottom with the x-height lowercase line alongside.

Software Computer programs on floppy disk or diskette.

Smoothing Typographical bitmap image created by the computer to smooth jagged edges.

Specimen Sample of page set to specify the display method suggested and the mood of the editorial.

Spectrometer Electronic device that measures the colour of paper from its reflected light.

Spot colour Colours can be specified as spot or process colours. Separations of each spot colour on a page is printed on its own plate. However, process colours are broken down into CMYK, each of which is printed on its own separation plate.

Spread Pair of facing pages with gutters in the centre. The middle pages of a publication without a gutter are referred to as a centre spread.

Square back See Perfect binding.

Square serif Common to the Egyptian faces such as Karnak, Beton or Rockwell – sometimes they are thicker than the verticals of their parent character.

SS Same size.

Start-up disk It contains the system files a computer needs to get started. It must have a finder and systems file. Many desktop machines also have start-up disks containing files such as printing resources and scrapbook.

Step and repeat Software terminology for highlighting and duplicating individual text and pictures.

Stet Proofreader's mark to indicate that copy deleted should be reinstated.

Stipple A random dot screen effect used in artwork for eccentric light and shade.

Strapline Small display headline that will set the scene above the main much larger line. On occasions these 14 or 18 point lines are set alongside the main banner line.

Sub An abbreviation for the jobtitle of subeditor. Controls the subeditorial desk and staff will rewrite, trim or grammatically improve stories from their original source and prepare instructions for setting on the computer screen.

Subhead A more discursive line of type that follows a main banner line but in a smaller typeface.

Swash letters Typefounders developed decorative caps with long descenders and flourishes, which follow the principles of illuminated drop letters. Garamond is an example.

Swatch Colour specimens such as the Pantone colour chart.

Swelled rules Decorative rules thicker at their centre than at their ends.

System file The computer uses this file to start itself up. It contains folders for practically every function of the machine.

Tab key Key that directs tabulation. It will establish pointers on a dialogue box defining in millimetres the distance between set columns or indent paragraphs.

Tails Bottom margin of the page.

Teach text A tools system that allows the operator to read plain text documents.

Terminal Keyboard and visual display screen for the generation of typesetting and design.

Terminator SCSI device that maintains the integrity of the signal passing along the SCSI chain. The chain should never have more than two terminators, one of each end.

Text Body type as opposed to display headings. Monastic quill writing was called Textus.

Text box Any dialogue box where information can be typed.

Thumbnail Small drawings sometimes pocket cartoons. A computer term for small representations of pages.

TIFF/RIFF Acronyms for tag image file format and raster image file format. These are formats storing scanned images. Pictures in TIFF can be line art, greyscale or colour. RIFF files are created by the colour studio digital image process programme.

Tiling When printing with a laser printer this technique allows the page to be broken up on screen into squares or tiles. It applies when the size of the printer is for instance an A4 and the design is A2. This command can be found in the print dialogue box.

Tint Colour or black reduced by dot screening on a percentage basis.

Title bar The strip of grey at the top of a document on the screen containing the title. The use of the mouse and arrow allows the document to be dragged around the screen.

Tone value Variations of a colour or halftone.

Tooth Description of the surface texture of paper usually by number.

Tool palette In show tools from the View menu will display in icon form the variety of tools available.

Tracking See Kerning.

Transducer Electronic device for converting signals into readable material and sending them down the cable.

Transfer type Transparent plastic A4 sheet covered with type of various sizes that will dry transfer with press on to artwork. Letraset and Mecoanorma are the best-known producers.

Transitional typefaces Designed with characteristics of old and modern styles, such as Baskerville.

Transparency Photographic film as a positive from 8 mm upwards. It can be projected or viewed on a lightbox with the appropriate magnifier. Shiny side upwards is the correct way to view.

Trash Icon on the desktop used to drag and discard documents, folders and applications. On AppleMac the word used is wastebasket.

Trichromatic Using three colours.

Trim The process of cutting tabloid or broadsheet newsprint publications to the page size.

TRS (transpose) One element - a word, page or picture must be interchanged with another as a correction or preference.

Turn A newspaper story turning from one page and continuing on another. It carries a small cross reference, bold line at the end of the first part indicating which page the story will be turning to.

Type area The parameters of the area of a printed page.

Type family Related typefaces - for example Folio Bold, Folio Extended, Folio Regular, Folio Condensed and Folio Extra Bold.

Type gauge A ruler designed for the industry calibrated with millimetres, points and picas.

Type style A command in the style menu. Enables the operator to choose 13 styles to highlighted text: plain, bold, italic, underline, work underline, strike thru, outline, shadow, all caps, small caps, superscript, subscript, and superior.

Typefaces Designed fonts of type produced over the centuries in wood and brass by type designers. The perfect reproduction of their work was never possible until the digitally reconstructed versions appeared with the advent of the computer.

Typesetter Company or individual whose primary job is to set type.

Typographer Designer to type material can be called a graphic designer.

Unit value 100 units to each type character width. Of use in the kerning process.

Unjustified type A column of type set with a ragged edge, left or right. Common practice is to use the ragged format on the right. The computer avoids hyphenation, thus creating uniformly tight individual lines throughout. It will provide more visual white than justified columns.

Untrimmed size Outside dimensions of a printed page before the trim to the finished size. Important to designers when they need to instruct on the extra space needed to bleed a picture off the page.

UPC (unique product code) Acronym for the bars which identify goods and publications and include the price for which they sell.

Update Incorporating into a computer file the newest material or program. Also to keep edition of a newspaper topical throughout its run.

Uppercase Capital letters.

Van Dyke American printing term for a dye-line proof produced in sepia.

Variable space Technique used by a computer to create a justified set line of type.

Verso A left-hand page with an even number. A right-hand page is a recto.

Vignette Illustration of any kind fading towards the outside edges.

Visual See Layout.

Visual display unit (VDU) Computer terminal consisting of a cathode ray tube screen, key-board from input and commands together with a hand-driven mouse directing an arrow on screen.

Volume A book. The thickness of paper indicated as a volume number equal in thickness to 100 sheets of paper at a 100 grams in weight.

Weight Term used to describe the varying visual impact of work. Also type volume.

Warm (hot) colour Spectrum shades of the red and yellow range.

Wash drawing A water colour illustration, sometimes combined with black ink.

Wash-up Cleaning in the ink ducts of a printing press prior to changing the ink colour.

Web offset A system of printing in which the inked page image is transferred from a smooth printing plate on to a rubber blanket (roller) and then offset on to the paper as opposed to being printed directly on to the paper by a roll bearing a relief impress.

WF Acronym for wrong font. A proof reader's mark indicating a mix of fonts.

White space (visual white) Carefully placed white areas on a printed page that give emphasis to chosen pieces of type and pictures.

Widow One word left on the last line of a paragraph.

Word break The natural or logical point of break in a word at the end of a line to take the insertion of a hyphen. Computers are formatted to the taste of the printing house and its designers for this continuous operation.

Word processors A sophisticated form of electronic typewriter with some computer characteristics.

Workings The number of passes through the same printing press required to create the finished product, e.g., four impressions on a single-colour machine to achieve a four-colour reproduction.

Work station See Terminal.

WOB Traditional acronym for white type on a black background (reverse).

WOT White type on a tinted background of various percentages.

WYSIWYG Acronym for what you see is what you get. Refers to a computer screen display that will accurately show what will appear on the paper at the end of the job.

X-height Height of any lowercase character (x) of a given font measured from type baseline to the top of the letter exclusive if descenders and ascenders.

Xenon Photographic flash of intense light.

Yankee dryer Steam-heated cylinder for drying photographic paper to a glazed finish. Now almost extinct.

Zip-a tone A trade name for a rub-down plastic film tint, tone or colour having a pre-waxed back for adhesion to art work or pictures.

Index